T0320965

GPS and GNSS for Land Surveyors

Based on the success of the previous four editions, this new fifth edition includes Global Navigation Satellite Systems (GNSS) in the title, which is part of Global Positioning System (GPS). The book provides an introduction to the concepts needed to understand and use GPS and GNSS. Neither simplistic nor overly technical, the new edition is thoroughly updated with the changes in GPS and GNSS hardware, software, and procedures. It describes why modern GNSS positions can be acquired with more certainty, increased stability, and improved tracking in obstructed areas. The book offers a rare combination of knowledge and skills that every land surveyor needs to master.

FEATURES

- Written by a well-known land surveyor with extensive knowledge in satellite navigation and the ability to explain difficult concepts to a broad audience
- Includes a useful set of self-assessment exercises and explanations at the end of each chapter
- Takes a practical approach to the rapid and continuous technological progress in GNSS
- Provides the latest information on GNSS and GPS
- Minimizes the reliance on mathematical explanations and maximizes the use of illustrations and examples that allow the reader to visualize and grasp the concepts

Intended for both novices and professionals in the field, this book explains broad concepts in an accessible way. It provides support to undergraduate students in Civil Engineering, Geomatic Engineering, and those taking introductory GPS and GIS Mapping Courses, as well as professionals in the field, a practical approach to GPS and GNSS technology.

GPS and GNSS for Land Surveyors

Fifth Edition

Jan Van Sickle

Illustrations by Bob Marquez

CRC Press
Taylor & Francis Group
Boca Raton London New York

CRC Press is an imprint of the
Taylor & Francis Group, an **informa** business

Designed cover image: © Aaron Witt

Fifth edition published 2024

by CRC Press

6000 Broken Sound Parkway NW, Suite 300, Boca Raton, FL 33487-2742

and by CRC Press

4 Park Square, Milton Park, Abingdon, Oxon, OX14 4RN

CRC Press is an imprint of Taylor & Francis Group, LLC

© 2024 Taylor & Francis Group, LLC

First edition published by Sleeping Bear Factory Press 1996
Second edition published by CRC Press 2001
Third edition published by CRC Press 2008
Fourth edition published by CRC Press 2015

Library of Congress Cataloging-in-Publication Data

Names: Van Sickle, Jan, author.
Title: GPS and GNSS for land surveyors / Jan Van Sickle.
Other titles: GPS for land surveyors
Description: Fifth edition. | Boca Raton, FL : CRC Press, 2024. | Revised edition of the author's GPS for land surveyors, 2015. | Includes bibliographical references and index. |
Summary: "Based on the success of the previous four editions, this new edition includes Global Navigation Satellite Systems (GNSS) in the title, of which GPS is part. Neither simplistic nor overly technical, the new edition is updated with the changes in GPS and GNSS hardware, software, and procedures. It describes why modern GNSS positions can be acquired with more certainty, with increased stability, and improved tracking in obstructed areas. The book offers a rare combination of knowledge and skills that every land surveyor needs to master. Intended for both ingenues and professionals in the field, broad concepts are explained in an accessible way"-- Provided by publisher.
Identifiers: LCCN 2022060735 (print) | LCCN 2022060736 (ebook) |
ISBN 9781032521022 (hardcover) | ISBN 9781003405238 (ebook)
Subjects: LCSH: Artificial satellites in surveying. | Global Positioning System.
Classification: LCC TA595.5. V36 2024 (print) | LCC TA595.5 (ebook) |
DDC 526.9/82--dc23/eng/20230111
LC record available at https://lccn.loc.gov/2022060735
LC ebook record available at https://lccn.loc.gov/2022060736

ISBN: 978-1-032-52102-2 (hbk)
ISBN: 978-1-003-40523-8 (ebk)

DOI: 10.1201/9781003405238

Typeset in Times
by KnowledgeWorks Global Ltd.

Printed and bound in Great Britain by Bell & Bain Ltd, Glasgow

Dedication

This book is dedicated to Heidi, Jake, and Maxwell Van Sickle.

Contents

Preface to the Fifth Edition

Global Positioning System (GPS) is now a part of a growing international context—the Global Navigation Satellite System, GNSS. One definition of GNSS embraces any constellation of satellites providing signals from space that facilitate positioning, navigation, and timing (PNT) on a global scale. Current use of the term typically includes global systems such as the Russian Federation's GLONASS, an acronym for Globalnaya Navigatsionnaya Sputnikovaya Sistema; the Galileo system administered by the European Union (EU); and the Chinese BeiDou Navigation Satellite System (BDS). It can also include regional systems such as the Indian NavIC System and the Japanese Quasi-Zenith Satellite System (QZSS).

In a sense, GNSS surveying is GPS surveying because the measurement principles, procedures, and applications originally developed for GPS surveying apply. That is not to say that they remain unchanged. Developments in GNSS hardware, software, and procedures are accelerating. Keeping up is a real challenge. The extraordinary range of GNSS equipment and software requires the user to be familiar with an ever-expanding body of knowledge. My hope in offering this fifth edition of GPS and GNSS for Land Surveyors is to contribute some small assistance in that process. Please note, in this book, where the information presented applies to both GPS and the larger GNSS context, the term GNSS will be used. Where the information applies to GPS alone, the term GPS will be used.

This book has been written to find a middle ground. It is intended to be neither simplistic nor overly technical but an introduction to the concepts needed to understand and use GNSS, not a presentation of the latest research in the area. An effort has been made to explain the progression of the ideas at the foundation of satellite positioning and delve into some of the particulars.

This is a practical book, a guide to some of the techniques used in GNSS, from their design through observation, processing, *Static Positioning*, *Real-Time Kinematic (RTK)*, *Precise Point Positioning (PPP)*, and *Real-Time Networks*.

About the Author

Jan Van Sickle has been a licensed professional land surveyor for 43 years and is currently licensed in Colorado, California, Oregon, Texas, North Dakota, West Virginia, and Pennsylvania. He supervised control work using the first commercial GPS receiver suitable for geodetic work, the Macrometer V-1000. He led the team that produced control positions for more than 120 cities around the world, assisted in the first GPS survey of the Grand Canyon, was involved in the BIM of the White House, gave technical assistance in the reconstruction of the geodetic network of Nigeria, managed the imagery and gravity analysis in the South Gobi, delivered the 2021 keynote at the NSPS URISA Summit, taught at Penn State and the University of Colorado, and was a member of the Royal Institute of Chartered Surveyors. Jan earned his PhD degree in geospatial engineering from the University of Colorado. He is an ASPRS Certified Photogrammetrist and Mapping Scientist, Lidar, and the Principal Geodetic Engineer at Maxar Technologies. Website: www.janvansickle.com

1 The Signal

THE GLOBAL NAVIGATION SATELLITE SYSTEM (GNSS) SIGNAL

GNSS AND TRILATERATION

Positioning with global navigation satellite system (GNSS) can be compared to a surveying method called *trilateration*. Both rely on the measurement of distances exclusively, but there are some differences between them. For example, the distances are called ranges in GNSS and they are not initially measured to control points on the Earth but rather to the satellites orbiting it.

A PASSIVE SYSTEM

The ranges are measured from the data in the signals that are broadcast from the satellite's antennas to the receiver's antennas in the microwave part of the electromagnetic spectrum. The original design was called a passive system because of the one-way communication. The satellites transmit signals and the users simply receive them. So, just as millions of radios may be tuned to the same channel without disrupting a station's broadcast, millions of GNSS receivers may monitor satellite's signals without danger of overburdening the system. There is virtually no limit to the number of receivers that may simultaneously monitor the GNSS signals. It's a distinct advantage, but it also means that the signals must carry enough information from the satellite to the receiver for the receiver to determine its own position on the planet.

TIME

Time measurement is essential to GNSS surveying in several ways. The determination of ranges, like distance measurement in a modern trilateration survey, is done electronically. In both cases, distance is a function of the speed of light, an electromagnetic signal of stable frequency, and elapsed time. In a trilateration survey, frequencies generated within an electronic distance measuring (EDM) device in a *Total Station* can be used to determine the signal's travel time because the signal bounces off a reflector and returns to where it started. However, the signals from a GNSS satellite do not return to where they started; they travel one way, from the satellite to the receiver. The satellite can mark the moment the signal

DOI: 10.1201/9781003405238-1

departs and the receiver can mark the moment it arrives, but the measure-ment of the ranges in GNSS depends on the elapsed time. That means the signal itself must provide the receiver with some way to determine how long it took to make the trip.

CONTROL

Both GNSS and trilateration surveys begin from control. In any survey-ing, measurement of a distance to a control point without knowledge of that control point's position is useless, and since the control in GNSS are, initially, the satellites themselves, knowledge of their positions are critical. So, it is not enough that the GNSS signals provide a receiver with informa-tion to measure the range between itself and the satellite. That same signal must also communicate the position of the satellite at the very instant it leaves the satellite. This task is complicated somewhat by the fact that the satellite is always moving several kilometers per second with respect to the receiver.

ATMOSPHERE

In a GNSS survey, as in a trilateration survey, the signals must travel through the *atmosphere*. Compensation for the atmospheric effects on an EDM's electromagnetic signal can be estimated from local observations and applied at the signal's source in a trilateration survey. This is not pos-sible in GNSS. The GNSS signals begin in the virtual vacuum of space. They hit the Earth's atmosphere and then travel through much more of it than EDM signals ever do. Therefore, the GNSS signals must give the receiver some information about needed atmospheric corrections.

AT LEAST FOUR

It takes more than one measured distance to determine a new position in a trilateration survey or in a GNSS survey. Each of the several distances used to define one new point must be measured from a different control station. For trilateration, three distances are adequate for each new point. Each new GNSS surveyed point requires ranges measured to at least four satellites.

SATELLITE ID

Just as it is vital that every one of the three distances in a trilateration is correctly paired with the correct control station, the GNSS receiver must be able to match each of the signals it tracks with the satellite of its origin. Therefore, the GNSS signals themselves must also carry a kind of satellite identification. To be on the safe side, the signal should also tell the receiver where to find all of the other satellites in its constellation too.

To sum up, a GNSS signal must somehow communicate to its receiver: (1) what time it is on the satellite at the instant the signal is broadcast, (2) the instantaneous position of the rapidly moving satellite, (3) some information about necessary atmospheric corrections to be made to the signal, (4) some sort of satellite identification system to tell the receiver where the signal came from, and (5) where the receiver may find the other satellites in that constellation. How does a GNSS satellite communicate all that information to a receiver? It uses codes.

CODES

GNSS constellations include a wide variety of codes, but they are all binary, zeroes, and ones, the language of computers. In GPS, there are three basic legacy codes that have been around since the beginning of the system. They are the precise code, now known as the *P(Y)* code, the Coarse/Acquisition (*C/A*) code, and the *Legacy Navigation* (*LNAV*) message code. There are also new codes, including the L2C code, the L1C code, and a new military code called the M-code. These new codes have yet to reach *full operational capability* (*FOC*) but are nevertheless being broadcast by some of the GPS satellites now.

All GNSS codes contain the information receivers need to function, but they must travel a considerable distance from the satellites to the receivers to deliver it. The GPS satellites are in nearly circular orbits at a nominal altitude of about 20,000 km above the Earth. Some GNSS satellites, also known as *Space Vehicles* (*SVs*), are lower, GLONASS satellite's orbit at about 19,100 km, and, some are higher, Galileo satellites at 23,222 km. To make the trip, the codes ride on conveyances generated by the satellites themselves, harmonic radio waves, called *carriers.*

For a carrier wave to be able to convey information, it must have at least one characteristic that can be changed or *modulated.* The characteristic can be the phase, amplitude, or frequency. For example, the information, music, or speech received from an AM radio station is encoded onto the carrier wave by amplitude modulation, and the information on the signal from an FM radio station is there because of frequency modulation. GNSS carriers are phase modulated. Looking at some of the characteristics of waves makes a good start toward understanding how that works.

WAVELENGTH AND FREQUENCY

A wavelength with duration of 1 second, known as 1 cycle per second, has a frequency of 1 *Hertz* (*Hz*) in the International System of Units (SI) (see Figure 1.1).

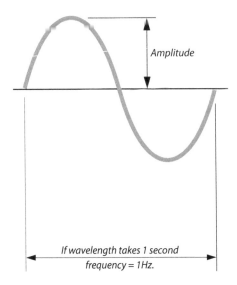

Amplitude

If wavelength takes 1 second
frequency = 1Hz.

FIGURE 1.1   Wavelength

A frequency of 1 Hz is rather low. The lowest sound human ears can detect has a frequency of about 20 Hz. The highest is about 20,000 Hertz, or 20 *kiloHertz (kHz)*. Most of the modulated carriers used in Total Station EDMs and all those in GNSS have frequencies that are measured in units of a million cycles per second, or *MegaHertz (MHz)*.

The fundamental frequencies assigned to GNSS carriers come from a part of the electromagnetic spectrum known as the *L-band*. Using the old (pre-1970) U.S. military classification, the L-band includes high frequencies from approximately 390 to 1550 MHz with wavelengths from 15 to 30 cm. The more current definition is from the Institute for Electrical & Electronics Engineers (IEEE), which classifies the range from 1 to 2 *GigaHertz (GHz)* (1000–2000 MHz) as the L-band. All the satellites in the global GNSS constellations, GPS, the European Union's *Galileo*, Russia's *GLONASS,* and China's *BeiDou*, and the regional GNSS constellations Japan's *QZSS* and India's NavIC, broadcast within that definition of the L-band. The carriers of the GPS signals supplying positioning, navigation, and timing (PNT) are L1 at 1575.42, L2 at 1227.60, and L5 at 1176.45 MHz.

The NAV Messages

Getting back to codes and the information they transport, one of the most important codes in GNSS is the *Navigation Message*, also known as the *Data Message* or NAV code. The oldest of them, the GPS Legacy Navigation (LNAV) message, includes some of the information the receivers need to determine positions.

The GPS LNAV code is broadcast on both the L1 and the L2 GPS carriers. It transmits information about the location of the GPS satellites called the *ephemeris* and data used in time conversions, *clock offsets* also known as clock corrections. It communicates the health of the satellites on orbit and information about the *ionosphere*, a layer of atmosphere through which the GPS signals must travel to get to the user. It also includes data called *almanacs* that provide a GPS receiver with enough snippets of ephemeris information of all the satellites in the constellation to approximate their positions within a few kilometers. Here are some of the parameters of the design of the LNAV message (see Figure 1.2).

Figure 1.2 shows one LNAV 1500-bit frame. A satellite sends it at 50 bits-per-second (bps), so a receiver gets one frame every 30 seconds. There are 25 frames in the full LNAV broadcast. Each one is made of 5 subframes. Subframes 1, 2, and 3 tell the receiver information about the satellite that sent it. The first subframe contains the satellite clock correction and *issue-of-data-clock* (*IODC*), which changes whenever the clock parameters are changed. It also includes the *GPS week* number, satellite health status, an estimate of the *user's range accuracy* (*URA*), etc.

Subframes 2 and 3 communicate the transmitting satellite's ephemeris, so the receiver can find it. These subframes also contain an *issue-of-data-ephemeris* (*IODE*) value, which will differ during an update, so the receiver

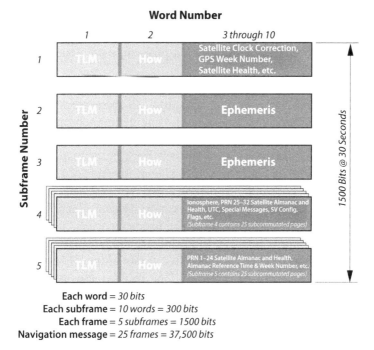

Word Number

Each word = 30 bits
Each subframe = 10 words = 300 bits
Each frame = 5 subframes = 1500 bits
Navigation message = 25 frames = 37,500 bits

FIGURE 1.2 One Frame of the Legacy NAV (LNAV) Message

only relies on the ephemeris data when both IODEs are the same. When subframes 2 and 3 IODE's are identical, the receiver can be sure the data is complete, timely, and not a mix across updates.

The data in subframes 1, 2, and 3 are repeated with every frame, but subframes 4 and 5 are different. Their data is not just about the satellite broadcasting the message. They concern all of the satellites in the constellation. Subframe 5 provides the almanac and health data for the first 24 satellites in the GPS constellation along with the almanac reference time and week numbers. Subframe 4 provides the almanac and health data for the GPS satellites 25–32 along with *Coordinate Universal Time (UTC)* data, special messages, various flags, etc. There are 25 pages of data in subframes 4 and 5. It takes a while to send all that.

Here is how it works. In the first 30 seconds, a receiver locked on to a satellite gets the data in subframes 1, 2, and 3, along with just the first page of subframes 4 and 5. In the next 30 seconds, it gets subframes 1, 2, and 3 again, but this time they are followed by the second page of subframes 4 and 5. In the subsequent 30-second increments, this continues with the third, fourth, fifth page, and so on for 25 repetitions. At 50 bps, it takes 750 seconds (12½ minutes) for the receiver to get all the information in all 37,500 bits of the LNAV. Once it's got the full LNAV, it stores it in its nonvolatile memory. That means that a receiver which has been in operation recently has the almanacs and position data within itself to do what is known as a *warm start* or *hot start,* so it's *time to first fix (TTFF)* can be short. A receiver is fixed when it has acquired and tracked at least four satellites and is capable of delivering a position, more about that later in this chapter.

Unfortunately, the accuracy of some aspects of the information included in the LNAV message degrades with time, so there are mechanisms in place to prevent the LNAV information from getting too old. The data in each satellite's message are updated by the government uploading facilities around the world, which, along with their tracking and computing counterparts, are known as the *Control Segment.* Subframes 1, 2, and 3 are updated every 2 hours. Subframes 4 and 5 are renewed, at minimum, every 6 days, though a daily update is more typical. Each satellite then transmits the updated information to the user's receivers.

The first two words in every subframe are the *telemetry (TLM)* word and the *handover word (HOW).* The first 8 bits of the TLM word are an asynchronization preamble, 10001011. The receiver uses that to find the beginning of the subframe from which it can start using the LNAV message. The next 14 bits (9–22) are reserved for those authorized to decode the encrypted P(Y) code. The 23rd bit is the integrity status flag that is currently set to 0. A setting of 1 would indicate an enhanced level of integrity assurance.

The first 17 bits of the HOW provide a truncated *time of week* (*TOW*) count, also known as the Z-count, one of the primary units for *GPS Time* (*GPST*). It's the number of seconds since the previous Saturday at midnight in units of 1.5 seconds, but its expressed 6-second steps in the HOW. That's exactly the time it takes to transmit one subframe, so it makes sense that it alerts the receiver of the moment the next data subframe will start. This word is known as the HOW because a receiver locked onto the C/A code can be handed over to the appropriate portion of the week-long P(Y) code and begin tracking it if it is authorized to do so. HOW also contains information valuable to the civilian user. When the 18th bit of the HOW is set to 1, it means that the use of the satellite is risky. It's the "use at your own risk" signal. When bit 19 is set to 1, it means *anti-spoofing* is on, in other words, the P code is encrypted as the Y code. Bits 20–22 are just IDs to identify which one of the five subframes is being broadcast.

There are new upgraded navigation messages for GPS. The military version is known as *MNAV,* and the civilian versions are known as *CNAVs*. The CNAVs were first broadcast in January of 2014 on L2 and L5. The L5-CNAV is modulated on a component of the L5 carrier called L5I. The L2-CNAV is modulated onto the L2 Civil Signal, L2C, part of the L2 carrier. They are considered pre-operational but are nevertheless available now for user familiarization and equipment development. The navigational content they provide is similar to that in LNAV, time, status, ephemeris, and almanac, but the new message formats are more flexible and robust. Unlike the just described LNAV's repetition of frames and subframes in a fixed pattern, these CNAVs send 12-second 300-bit messages in a pseudo-packetized format. Their messages need not follow a fixed order, and each of them can be one of several types. Not only does this design require less bandwidth, but since only a few of the available packet types have been defined the system can grow without jeopardizing compatibility. There is yet another CNAV format called CNAV-2. This one is on L1. It is modulated onto the new signal L1C broadcast by GPS Block III satellites, the first of which was launched in December 2018. It is also broadcast by Japanese Quasi-Zenith Satellite System (QZSS) satellites. In fact, the L1C signal was designed with international cooperation in mind. It is interoperable with Galileo's Open Service Signal, China's BeiDou, Japan's QZSS, and India's NavIC. There will be more about these improved GPS navigation messages in Chapter 8.

THE PRECISE P(Y) AND COARSE/ACQUISITION (C/A) CODES

Like the LNAV code, the P and C/A codes are designed to carry information from the GPS satellites to the receivers. Also, like the LNAV code, they are modulated onto carrier waves as are all the GNSS codes. However, the P and C/A codes are not vehicles for distributing information that has been uploaded by the Control Segment, they carry the raw data from which GPS receivers derive their position and time measurements. These codes are complicated, so complicated that they appear to be noise at first. In fact, like all GNSS positioning codes, they are known as *pseudorandom noise* (*PRN*) codes. Even though their zeros and ones appear to have no regularity, they are actually carefully designed. They have to be. They must be capable of repetition and replication.

P CODE

The P code is called the Precise code. It is a particular series of ones and zeroes generated at a rate of 10.23 million bits per second. It is carried on both L1 and L2, and it is very long, 2×10^{14} bits. In other words, it is 37 weeks long. So, each GPS satellite is assigned a week-long portion of the P code all its own and then repeats its portion every 7 days. This assignment of one particular week of the 37-week-long P code to each satellite helps a GPS receiver distinguish one satellite's transmission from another. For example, if a satellite is broadcasting the 14th week of the P code, it must be Space Vehicle 14 (SV 14).

The P code is reserved for the use of the U.S. military and has been so from the beginning of GPS. It is encrypted, which is accomplished by the modulation of a W-Code to generate the more secure Y code that replaces the P code. Therefore, the encrypted P code is formally known as the Y code but is most commonly called the P(Y) code and will be so referred to here. There is a flag in subframe 4 of the LNAV message that tells a receiver when the P code is encrypted into the P(Y) code. That flag has been activated by the Control Segment since January 1994. It was done to prevent *spoofing*. Spoofing is the generation and transmission of structured interference masquerading as an authentic GNSS signal. *Anti-spoofing* (*AS*) was first implemented on all GPS Block II satellites and it continues today throughout the GPS constellation. Commercial GPS receiver manufacturers are not authorized to use the P(Y) code, therefore, most developed proprietary *codeless, semi-codeless,* or *cross-correlation* techniques. These strategies do not de-encrypt the P(Y) code but nevertheless make it possible for the GPS receiver to have a dual-frequency, both L1 and L2, capability.

As the history of anti-spoofing shows, the concern used to be exclusively focused on the potential spoofing of military GNSS signals, but the threat to civil GNSS signals is just as worrying today. In short, all GNSS signals are vulnerable to spoofing, and the open civilian GNSS signals are actually more susceptible because they are not encrypted, their protocols are predictable and do not require authentication so, unfortunately, spoofing can succeed. One spoofing method is the repurposing of the simple recording and replaying of actual GNSS signals. It is a technique used in reputable receiver testing, but when used in spoofing, it is a malicious broadcast by a spoofer to confuse a receiver. Alternatively, a programmer with inexpensive off-the-shelf equipment can create and broadcast sufficiently realistic GNSS signals to degrade a receiver's position, velocity, and timing solutions. At its worst, spoofing can produce an entirely incorrect result. The potential for intentional harm is obvious, but it is also possible for spoofing to be accidental. A receiver can simply misconstrue an errant signal from a GNSS simulator or repeater as an authentic GNSS satellite transmission.

THE C/A CODE

The C/A code is a particular series of ones and zeroes as are the other codes. The rate of its generation is ten times slower than the P(Y) code. The C/A code rate is 1.023 million bits per second. Here satellite identification is accomplished by a strategy used by most GNSS constellations. In most cases, all the satellites in a GNSS constellation broadcast on the same frequency, on the same transmission channel and all at the same time. Under those conditions, it may seem impossible for receivers to know which signal comes from which satellite, but they can. They rely on the fact that each satellite broadcasts its own unique PRN sequence. For example, all GPS satellite's broadcast the C/A code on the same carrier frequency, L1 1575.42 MHz, but the receiver can still uniquely recognize each satellite from all the others because each of them sends its own particular C/A code and repeats it every *millisecond*.

STANDARD POSITIONING SERVICE (SPS) AND
PRECISE POSITIONING SERVICE (PPS)

The C/A code is the vehicle for the standard positioning service (SPS), which is used for nearly all civilian applications. The P(Y) code, on the other hand, provides the precise positioning service (PPS). The idea of SPS and PPS was developed by the Department of Defense many years ago. SPS was designed to provide a minimum level of positioning capability for nonmilitary applications considered consistent with national security. SPS accuracy was intended to be ±100 m horizontally, 95% of the time with a vertical accuracy of about ±175 m, but it turned out that the accuracy of the C/A point positioning was too good. The C/A-code point positioning was giving civilians access to accuracy of about ±20 m to ±40 m. That was not according to plan. So, in 1989 the Department of Defense instituted *Selective Availability* (*SA*), the intentional manipulation of the GPS broadcast ephemerides and the dithering of satellite clock stability to deny access to the full GPS accuracy for SPS users.

Fortunately, the SA is over now. It was discontinued on May 2, 2000, by presidential order. Actually, SA never did hinder the GPS surveying applications that relied on differential correction. In fact, it might be said that Selective Availability (SA) was an impetus for the development of differential GNSS, more about that in Chapter 3. However, satellite clock errors, the unintentional kind, still contribute errors to GPS positioning.

PPS via the P(Y) code was designed for higher positioning accuracy available only to users authorized by the Department of Defense. P(Y) code used to be the only military code. That is no longer the case. It has been joined by a new military signal called the M-code.

Time, UTC, and TAI

We'll return to codes and how they are modulated onto the carrier waves in a moment, but to help in understanding upcoming information on code generation, it is useful to first mention the foundation of that process, time. It is fundamental to all of GNSS. Let's start with Coordinated Universal Time (UTC), the civil time scale for the whole world.

The rate of UTC is exactly the same as *International Atomic Time (TAI)*, which is realized by the *International Bureau of Weight and Measurements (BIPM)* using a weighted average from 400 atomic clocks around the world. These clocks are so stable that they don't lose or gain a second in millions of years, and that presents a problem. They're more stable than the Earth's rotation rate. That means their time keeps getting out of sync with the planet's actual motion. That is ok for TAI, but it is definitely not ok for UTC. UTC simply cannot be allowed to get out of step with the Earth. Therefore, corrections of 1 second are periodically introduced into UTC to ensure that the offset between UTC and the planet is always less than 0.9 seconds. These corrections are called *leap seconds*. So, while the rate of UTC is identical with the rate of TAI, because of leap seconds, there is always a whole number of seconds between a particular instant of time in TAI and that same moment expressed in UTC. Since January 2017, that whole number of seconds has been 37, but that will change. Put another way, The *International Earth Rotation and Reference Systems Service (IERS)* monitors the Earth's rotation rate and advises the BIPM when a leap second should be inserted into UTC. When a positive leap second is added, the numbers denoting the moment of time in UTC is stepped up; when a second is omitted, the numbers denoting the moment of time in UTC is stepped down. Still, the rate of UTC stays consistent and stable all the time.

Here's the rub UTC is not available in real time. It is distributed monthly in a publication called *Circular T* by BIPM. That just won't do for GNSS constellations. They must have immediate precise timing information, not monthly updates. Remember those 400 atomic clocks around the world? Each GNSS constellation relies on predictions of exact time from one or more of the 70 timing laboratories whose clocks are part of that group. Russia's GLONASS depends on UTC(SU) at the Institute of Metrology for Time and Space in Moscow. BeiDou uses UTC (NTSC) National Time Service Center in China. Japan's QZSS UTC (NICT) timing is realized by the National Institute of Information and Communications Technology. The Indian NavIC system time is generated at the IRNSS Network Timing Facility at Byalalu. Galileo depends on an average of data from five European laboratories typically abbreviated UTC(k)'s. GPS relies on UTC(USNO), the Master Clock at the U.S. Naval Observatory to realize GPS Time.

GPS TIME

GPS Time, aka GPS System Time (GPST), began on UTC midnight (0 hr) on January 6, 1980. At that moment, UTC and GPS Time were exactly the same. However, as leap seconds were inserted in UTC, their moments of time diverged. Since 1980, 18 leap seconds have been inserted in UTC, but they have had no effect on GPS Time because GPS Time is continuous like TAI. There are no leap seconds involved in GPS Time which means it is off-set 18 seconds ahead of UTC as of this writing. Please note that while there are an integer number of seconds between a particular instant of time in UTC and the same moment expressed in GPS Time, their rates are virtually identical. Specifically, the rates of UTC(USNO) and GPS Time are always within 1 *microsecond*, 1 millionth of a second of each other, and usually less than 10–20 *nanoseconds* (*ns*), 10–20 billionths of a second of each other.

There is time information in subframes 1 and 4 of the GPS LNAV message. Two constants, A_0, time scale offset, and A_1, time scale rate of change on page 18 of subframe 4, provide a receiver with the offset and rate of GPS Time compared with UTC (USNO). Subframe 4 also notes future scheduled leap seconds. Subframe 1 contains coefficients that are used to estimate the difference between the satellite clock, *Space Vehicle* (*SV*) Time, and GPS Time.

SATELLITE CLOCKS

Each GNSS satellite carries its own very stable and accurate atomic clocks. They are also known as time or frequency standards. Most of them are regulated by the resonant frequencies from the electronic transition of cesium and rubidium atoms between two quantized energy levels. The first such clocks on orbit were those in the earliest GPS satellites, known as Block I. The first three GPS satellites had three *rubidium clocks*, but as the *cesium clocks* performed a bit better, subsequent Block 1 satellites had four: three rubidium and one cesium. As the GPS constellation grew and developed, more blocks of satellites followed. Each subsequent block of GPS satellites has been numbered consecutively with Roman numerals and sometimes with added alpha characters. The Block II and IIA satellites had four atomic clocks too, but there were two rubidium and two cesium, just like today's Block IIF satellites. There are three improved rubidium clocks on the Block IIR, Block IIR-M, and Block III satellites.

Russia's GLONASS satellite used cesium clocks exclusively for decades, but their new GLONASS-K series will have rubidium clocks as alternative *atomic frequency standards* (*AFS*). The regional Indian NavIC and the Japanese QZSS satellites also have rubidium clocks. Despite their higher mass and power consumption, Galileo and BeiDou are investigating the use of *hydrogen masers.*

The operation of the clocks in any one satellite is completely independent from the operation of the clock in any other satellite. Each satellite has its own SV time, so the government tracking stations of the Control Segment monitor all the satellite clocks. The Control Segment could continuously adjust them so all the satellite clocks would stay in near-perfect lockstep with the GPS Time standard. That is not done because constantly tweaking the satellite's onboard clocks would tend to reduce their useful life. Instead, each clock is allowed to drift up to 1 millisecond, 1 thousandth of a second, from GPS Time and their individual drifts are recorded and uploaded into subframe 1 of each satellite's LNAV message. These deviations from GPS Time available in the LNAV message are known as the satellite broadcast clock correction. Given the broadcast clock correction, a GPS receiver can relate the satellite's clock to GPS Time. This is part of the solution to the problem of directly relating the receiver's own clock to the satellite's clock. However, the receiver will need to rely on other aspects of the GPS signal for a complete time correlation. The drift of each satellite's clock is not constant, nor can the broadcast clock correction updates be frequent enough to completely define the drift. Therefore, the previously mentioned IODC, Issue of Data Clock, gets updated whenever the broadcast clock parameters change and thereby signifies broadcast clock correction currency.

GPS WEEK

As mentioned earlier, LNAV's subframe 1 contains information on the GPS week. GPS weeks are counted consecutively in integers from the first GPS week. GPS week 0 began at 00-hour UTC on January 6, 1980, and ended on January 12, 1980. It was followed by GPS week 1, GPS week 2, and so on.

However, at the end of GPS week 1023—August 15 through August 21, 1999—it was necessary to start the numbering again at 0 because the LNAV GPS week field has only 10 bits, so 1024 (2^{10}) is the largest week count it can accommodate. The following week, August 22 through August 28, 1999, would have been GPS week 1024, and that would have been beyond the capacity of the GPS week field in the LNAV message. When it reaches that limit, its reckoning of the GPS week starts again at 0. Such a rollover was required at 00 hours on August 22, 1999. It is worth noting that in UTC, the moment was 23:59:47 on August 21, 1999, considering leap seconds. In any case, a second rollover was required on the night of April 6 to 7, 2019, GPS week 2047. The next such rollover will be at midnight on November 20, 2038. To alleviate this problem, the modernized navigation messages L2-CNAV, CNAV-2, L5-CNAV, and MNAV have a 13-bit field for the GPS week count. This means that the GPS week, as counted in devices that use these messages, will not need to roll over until 157 years ($2^{13}/52$ years) after 1980 in 2137. In Table 1.1, a full GPS week and a part of another in May 2023 are shown.

Julian Date

Here is a little more about dates. It is usual in GPS practice to define particular dates in a sequential manner from the first of the year. Most practitioners of GNSS use the term *Julian date* to mean the day of the

---

TABLE 1.1
Seconds, Days, and Weeks

Gregorian Date	Day Name	GPS Week from 01/06/1980	Day of Week	Seconds from Beginning of Week at Midnight on That Day	Day of Year (DOY)	Modified Julian Day	Julian Day (CE)
7	Sunday	2261	0	0	127	60071	2460071.5
8	Monday	2261	1	86400	128	60072	2460072.5
9	Tuesday	2261	2	172800	129	60073	2460073.5
10	Wednesday	2261	3	259200	130	60074	2460074.5
11	Thursday	2261	4	345600	131	60075	2460075.5
12	Friday	2261	5	432000	132	60076	2460076.5
13	Saturday	2261	6	518400	133	60077	2460077.5
14	Sunday	2262	0	0	134	60078	2460078.5
15	Monday	2262	1	86400	135	60079	2460079.5
16	Tuesday	2262	2	172800	136	60080	2460080.5

year (DOY) counted consecutively from January 1 of that year. With this method, January 1 is day 1 and December 31 is day 365, except in leap years. For example, the 10 days of May 2023 from Sunday the 7th to Tuesday the 16th shown in Table 1.1 are 127 through 136 DOY. However, if the Julian dates were taken literally, they would be counted from noon on January 1, 4713 BC. Those are the dates on the far right in Table 1.1. under the heading Julian Day (CE).

The modified Julian date is sometimes referenced. It is in the column adjacent and is found by simply subtracting 2400000.5 from the Julian day (CE).

The days of the week numbers start at 0 each Sunday, continue consecutively to 6 on Saturday, and restart at 0 on Sunday. The column headed Seconds from the Beginning of Week at Midnight on That Day shows that GPS Time also restarts each Sunday at midnight (0:00 o'clock). The data in that column are the seconds elapsed since the last restart of GPS Time on Sunday 0:00 o'clock.

THE BROADCAST EPHEMERIS

Another example of time-sensitive information is found in subframes 2 and 3 of the LNAV message. They contain information about the position of the satellite with respect to time and is called the satellite's ephemeris. The ephemeris that each satellite broadcasts to the receivers provides information about its position relative to the Earth. Most particularly, it provides information about the position of the satellite *antenna's phase center* (*APC*). The ephemeris is given in the right ascension (RA) system of coordinates. There are six orbital elements; among them are the size of the orbit, that is, its semi-major axis, a, and its shape, that is, the eccentricity, e. Eccentricity is a measure of how far an orbit departs from circular. A perfectly circular orbit has an eccentricity of zero; higher numbers indicate more elliptical orbits. The *medium Earth orbits* (*MEOs*) of GPS satellites have an eccentricity under 0.02. In other words, they are nearly circular.

The orientation of the orbital plane in space is defined by other things. The RA of its *ascending node* is where the satellite crosses the Equatorial plane from south to north, Ω. The inclination of its orbital plane with respect to the equator is symbolized i. The argument of the *perigee*, ω, is the angle between the ascending node and perigee measured in the orbital plane. Perigee is the closest approach of the satellite to the Earth's center of mass; *apogee* is the most distant. The position of the satellite on its orbit, that is, the geocentric angle between perigee direction and satellite direction is known as the true anomaly, v. These parameters provide all the information the user's computer needs to calculate Earth-centered, Earth-fixed, X, Y, Z coordinates of the satellite at any moment in the World Geodetic System 1984, GPS Week 2139 [WGS84 (G2139)] *reference frame*.

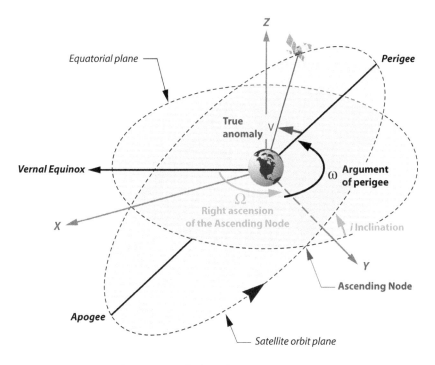

FIGURE 1.3 Inertial Frame Keplerian Elements of a Satellite Orbit

Figure 1.3 illustrates these Keplerian elements used to describe the motion of a satellite in an elliptical orbit around the Earth. However, it is important to note that the orbital path is not defined solely by the gravitational attraction between the satellite and the Earth, so this is not the classical two-body problem. There are other forces, i.e., solar flux, the gravity of the Sun and Moon, Earth's gravitational harmonics, etc., affecting the satellite's actual orbit. It follows that the satellites are unavoidably perturbed. They deviate from ideal smooth elliptical paths, so their actual paths through space are found by periodically changing the parameters shown to provide a least squares, best-fit of the satellite's orbits. As mentioned earlier, the values in the LNAV broadcast clock correction and the broadcast ephemeris vary with time. Remember that IODE is an acronym that stands for Issue of Data Ephemeris, an 8-bit unsigned integer broadcast in both subframes 2 and 3. It provides a means for receivers to detect the changes that provide corrections to the elliptical orbits provided by the Keplerian elements as they are made to the broadcast ephemeris.

ATMOSPHERIC CORRECTION

Subframe 4 addresses atmospheric correction. As with subframe 1, the data there offer a partial solution to a problem. The GPS Control Segment

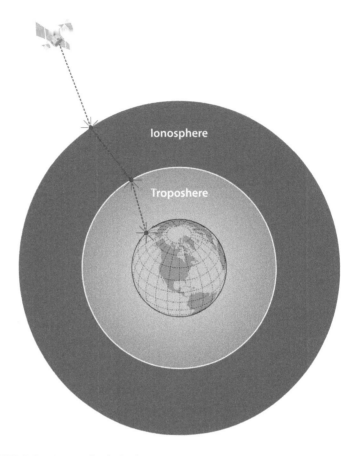

FIGURE 1.4 Atmospheric Delay

generates coefficients and, thereby, an estimate of the ionospheric delay based on a model created by Jack Klobuchar in the 1980s that was first used in GPS but has since been employed by other GNSS constellations. It will be discussed more in Chapter 2. For now, it is sufficient to say that a single-frequency receiver must rely on this ionospheric correction in subframe 4 of the LNAV message to remove part of the delay introduced by the atmosphere (see Figure 1.4). On the other hand, a receiver that can track more than one frequency can compare the differing propagation rates of say L1, L2, and L5 through the atmosphere and derive a better solution.

THE ALMANAC

As mentioned briefly earlier, the information in the almanac in subframes 4 and 5 tells the receiver where to find all the GPS satellites. Subframe 4 contains the almanac data for satellites with PRN numbers from 25 through 32, and subframe 5 contains almanac data for satellites with PRN numbers from 1 through 24. The GPS Control Segment uploads a new almanac every day to each satellite.

Mathematically, the GPS, Galileo, BeiDou, and QZSS almanac models are substantially the same, though their notation and arrangement of the equations are somewhat different. The GLONASS uses a more accurate but substantially more complex analytical almanac model. The almanacs are much smaller than the ephemerides because they contain coarse orbital parameters and incomplete ephemerides, but they are still accurate enough for a receiver to generate a list of visible satellites at power-up. They, along with a stored position and time, allow a receiver to find its first satellite. So even though a GNSS receiver must collect a complete ephemeris from each individual GNSS satellite to know its correct orbital position, it is convenient for a receiver to be able to have some information about where all the satellites in the constellation are. By reading the almanacs available, even just one of them, it can have that along with some auxiliary health and status information besides.

If a receiver has been in operation recently and has some left-over almanac and position data in its nonvolatile memory from its last observations, it can begin its search with what is known as a warm start. A warm start is also known as a normal start. In this condition, the receiver might begin by knowing the correct time within about 20 seconds, and its position within 100 km or so, and this approximate information helps the receiver estimate the range to satellites. For example, it will be able to restrict its search for satellites to those likely above its horizon rather than wasting time on those below it. Limiting the range of the search decreases the TTFF. It can be as short as 30 seconds with a warm start.

On the other hand, if a receiver has no previous almanac or ephemeris data in its memory, it will have to perform a cold start, also known as a factory start. Without previous data to guide it, the receiver in a cold start must search for all the satellites without knowledge of its own position, velocity, or the time. When it does finally manage to acquire the signal from one, it gets some help and can begin to download almanac data. That data will contain information about the approximate location of all the other satellites. As mentioned earlier, it takes 12½ minutes for a GPS receiver to acquire the full information from LNAV. The cold start TTFF is the longest. It takes less time for a warm start and least for a hot start. A receiver that has current almanacs, a current ephemeris, time, and position can have a hot start. A hot start can take from 1/2 to 20 seconds. Estimating how long each type of start will actually take is difficult, overhead obstructions interrupting the signal from the satellites, GNSS signals reflecting from nearby structures, etc., can delay the loading of the ephemeris necessary to lock onto the satellite's signals.

Satellite Health

Subframe 1 contains information about the health of the satellite the receiver is tracking when it receives the LNAV message and allows it to determine if the satellite is operating within normal parameters. Subframes 4 and 5 include health data for all of the satellites, data that is periodically uploaded by the Control Segment. These subframes inform users if a satellite is unhealthy before they rely on its signal. The codes in these bits may convey a variety of conditions. They may tell the receiver that all signals from the satellite are good and reliable or that the receiver should not currently use the satellite because there may be tracking problems or other difficulties. They may even tell the receiver that the satellite will be out of commission in the future, perhaps it will be undergoing a scheduled orbit correction. GPS satellites' health is affected by a wide variety of breakdowns, particularly clock trouble. That is one reason they carry multiple clocks.

Telemetry and Hand Over Words

As mentioned before, each of these five subframes begins with the same two words: the telemetry word (TLM) and the HOW. Unlike nearly everything else in the GPS LNAV message, these two words are generated by the satellite itself. The TLM is the first word in each subframe. It indicates the status of uploading from the Control Segment while it is in progress and contains information about the age of the ephemeris data. The HOW provides the receiver information on the TOW and the number of the subframe, among other things. For example, the HOW's Z count (the internally derived 1.5-second epoch) tells the receiver exactly where the satellite stands in the generation of positioning codes. In fact, the HOW helps the receiver go from tracking the C/A code to tracking the P(Y) code.

THE PRODUCTION OF A MODULATED CARRIER WAVE

All GNSS codes come to a receiver on a modulated carrier. The signal created by an electronic distance meter (EDM) in a Total Station is a good example.

EDM Ranging

An EDM in a Total Station only needs one frequency standard because its electromagnetic wave travels to a retroprism and is reflected back to where it started. The EDM is both the transmitter and the receiver of the signal,

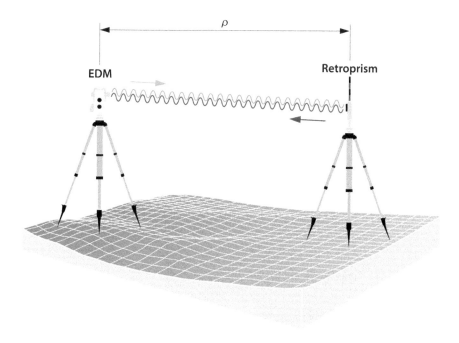

FIGURE 1.5   Two-Way Ranging

so it only needs one clock. In general terms, if the instrument has the time elapsed between the moment of transmission and the moment of reception, it can multiply by the speed of light, divide by 2, because the signal traveled the distance twice, out and back, and it will have the distance between itself and the retroprism. Illustrated in Figure 1.5, the fundamental elements of the calculation of the distance measured by an EDM, ρ, are the time elapsed between transmission and reception of the signal, Δt, and the speed of light, c.

$$\text{Distance} = \rho$$

$$\text{Elapsed Time} = \Delta t$$

$$\text{Rate} = c$$

GNSS RANGING

However, the one-way ranging used in GNSS requires two clocks. The signals broadcast from the satellites are collected by the receiver, not reflected back to the satellites. So, in general terms, the full time elapsed between the instant a GNSS signal leaves a satellite and arrives at a receiver, multiplied by the speed of light, is the distance, or the range, between them.

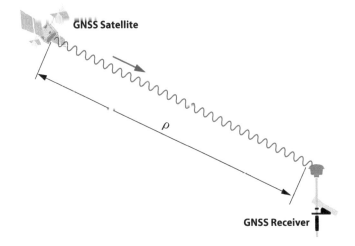

FIGURE 1.6   One-Way Ranging

It follows that a GNSS signal cannot be analyzed at its point of origin like the wave generated by an EDM. The measurement of the elapsed time between the signal's transmission by the satellite and its arrival at the receiver is complicated by the fact that the clock in the satellites marks the moment of transmission, and the clock in the receiver marks the moment of reception. To perfectly represent the distance between the satellite and the receiver, these two clocks would have to be perfectly synchronized with one another. The necessary level of perfection is not possible. As shown in Figure 1.6, the calculation of a range between a GNSS receiver and the satellite, ρ, relies on the multiplication of the time elapsed between a signal's transmission and reception, Δt, with the speed of light, c. Therefore, a discrepancy of 1 microsecond, or 1 millionth of a second, in the synchronization of the clock aboard the GNSS satellite and the clock in the receiver would cause a range error of about 300 m. Since that is far beyond the acceptable limits for surveying work, the problem is addressed mathematically.

OSCILLATORS

The time measurement devices used in both EDM and GPS measurements are clocks only in the most general sense. They are usually more correctly called *oscillators, cesium frequency standards,* or *rubidium frequency standards.* They don't produce a series of audible ticks. They keep time by varying a continuous beam of electromagnetic energy at extremely regular intervals and thereby producing steady sine waves, the foundation of *carriers.*

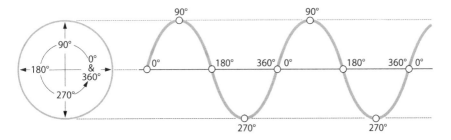

FIGURE 1.7 Generation of a Sine Wave with Phase Angles

Imagine a coherent signal produced by a cyclical rotation as it passes through a system (see Figure 1.7). The wavelength begins at the 0° phase angle. The first quarter rotation produces the 90° phase angle, second the 180° phase angle, third the 270°, and the 360° phase angle marks the end of one wavelength and the beginning of the 0° phase angle of the next. As the process repeats, it generates a continuous carrier.

For such the signal to carry data in GNSS, the time and distance between every phase angle from 0° to 360° must be incredibly regular, so it is essential that the rate of an oscillator's operation is very stable. The elapsed time and hence the length between the beginning and end of every wavelength is the same. Wavelength is often symbolized by the Greek letter lambda, λ.

A CHAIN OF ELECTROMAGNETIC ENERGY

GNSS oscillators are sometimes called clocks because the frequency of a modulated carrier, measured in Hertz, can indicate the elapsed time between the beginning and end of a wavelength, which is a useful bit of information for finding the distance covered by a wavelength. The length is approximately:

$$\lambda = \frac{c_a}{f}$$

where
λ = the length of each complete wavelength in meters
c_a = the speed of light corrected for atmospheric effects
f = the frequency in Hertz

For example, if an EDM transmits a modulated carrier with a frequency of 9.84 MHz and the speed of light is approximately 300,000,000 m per second (a more accurate value is 299,792,458 m per second, but the approximation 300,000,000 m per second will be used here for convenience), then:

$$\lambda = \frac{c_a}{f}$$

$$\lambda = \frac{300,000,000 \text{ mps}}{9,840,000 \text{ Hz}}$$

$$\lambda = 30.49 \text{ m}$$

the modulated wavelength would be about 30.49-m long, or approximately 100 feet.

The modulated carrier transmitted from an EDM can be compared to a Gunter's chain constructed of electromagnetic energy instead of wire links. Each full link of this electromagnetic chain is a wavelength of a specific frequency. The measurement between an EDM and a reflector is doubled with this electronic chain because, after it extends from the EDM to the reflector, it bounces back to where it started. The entire trip represents twice the distance and is simply divided by 2, but like the surveyors who used the old Gunter's chain, one cannot depend that a particular measurement will end conveniently at the end of a complete link or complete wavelength. A measurement is much more likely to end at some fractional part of a link or wavelength. The question is, where?

PHASE SHIFT

With the original Gunter's chain, the surveyor simply looked at the chain and estimated the fractional part of the last link that should be included in the measurement. Those links were tangible. Since the wavelengths of a modulated carrier are not known, the EDM must find the fractional part of its measurement electronically. Therefore, it does a comparison. It compares the phase angle of the returning signal to that of a replica of the transmitted signal to determine the *phase shift*. That phase shift represents the fractional part of the measurement. This principle is used in distance measurement by both EDM and GNSS systems.

First, it is important to remember that points on a modulated carrier are defined by phase angles, such as 0°, 90°, 180°, 270°, and 360° (see Figure 1.8). When two modulated carrier waves reach exactly the same phase angle at exactly the same time, they are said to be in phase, coherent,

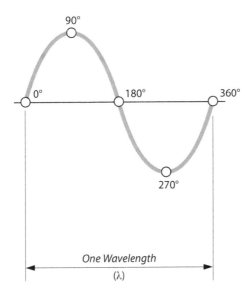

FIGURE 1.8   0° to 360° = One Wavelength

or *phase locked*. However, when two waves reach the same phase angle at different times, they are out of phase or phase shifted.

For example, in Figure 1.9, the sine wave shown by the dashed line has returned to an EDM from a reflector. Compared with the sine wave shown by the solid line, it is out of phase by one-quarter of a wavelength. The distance between the EDM and the reflector, ρ, is then:

$$p = \frac{(N\lambda + d)}{2}$$

where
   $N\lambda$ = the number of full wavelengths the modulated carrier has completed
   d = the fractional part of a wavelength at the end that completes the doubled distance

In this example, d is three-quarters of a wavelength because it lacks its last quarter, but how would the EDM know that? It knows because, at the same time, an external carrier wave is sent to the reflector, the EDM keeps an identical internal reference wave at home in its receiver circuits. In Figure 1.10, the external beam returned from the reflector is compared to the reference wave, and the difference in phase between the two can be measured.

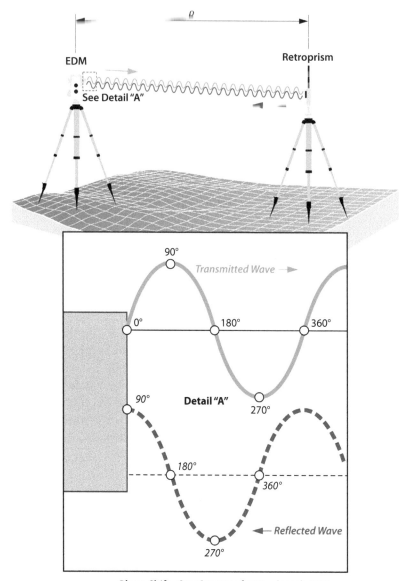

Phase Shift - One Quarter of a Wavelength (90°)

FIGURE 1.9   An EDM Measurement

Both EDM and GNSS ranging use the method represented in this illustration. In GNSS, the measurement of the difference in the phase of the incoming signal and the phase of the reference signal generated by the internal oscillator in the receiver reveals the distance at the end of a range. In GNSS, this is an aspect of *carrier-phase ranging*, and as the name implies, the *observable* is the carrier wave itself.

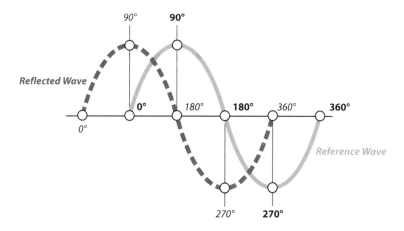

FIGURE 1.10   Reference and Reflected Waves

TWO OBSERVABLES

The word observable is used throughout GNSS literature to indicate the signals whose measurement contributes to solving the range or distance between the satellite and the receiver. The word is used to draw a distinction between the thing being measured, the observable, and the measurement, the observation. Two of the GNSS observables are the *pseudorange* and the *carrier phase*. The latter, also known as the *carrier beat phase*, is the basis of the techniques used for high-precision GNSS surveys. On the other hand, the pseudorange can serve applications when virtually instantaneous point positions are required or relatively low accuracy will suffice. These basic observables can also be combined in various ways to generate additional measurements that have certain advantages. It is in this latter context that pseudoranges are used in many GNSS receivers as a preliminary step toward the final determination of position by a carrier-phase measurement. The foundation of pseudoranges is the correlation of code carried on a modulated carrier wave received from a GNSS satellite with a replica of that same code generated in the receiver. Most of the GNSS receivers used for surveying applications are capable of code correlation. That is, they can determine pseudoranges. These same receivers can usually also do carrier-phase ranging, that is determine ranges using the unmodulated carrier. Let us look at the pseudorange first.

ENCODING BY PHASE MODULATION

A carrier wave can be modulated in various ways. As mentioned, radio stations use modulated carrier waves such as AM, amplitude modulated or FM, frequency modulated. When your radio is tuned to 105 FM, you are not actually listening to 105 MHz despite the announcer's assurances, it is well above the range of human hearing. 105 MHz is just the frequency of the carrier wave that is being modulated. Those modulations make the speech and music, the information being transmitted to you, intelligible. The information comes to you at a much slower frequency than does the carrier wave.

The GPS carriers L1, L2, and L5 could have been modulated in a variety of ways to carry the binary codes, the 0s and 1s, that are the codes, but neither amplitude nor frequency modulation is used. It is the alteration of the phases of the carrier waves that encodes them. The codes come to the receivers by phase modulation. One consequence of this method of modulation is that the signal can occupy a broader *bandwidth* than would otherwise be possible. The GPS signal is said to have a *spread spectrum* because of its intentionally increased bandwidth. In other words, the overall bandwidth of the GPS signal is much wider than the bandwidth of the information it is carrying. L1 is centered on 1575.42, L2 on 1227.60, and L5 on 1176.45 MHz, but they occupy more width on each side of these centers than do the signals they carry. Looking at L1, for example (see Figure 1.11), the C/A code is spread over of 2.046, L1C signal 4.092, and the P(Y) code 20.46 MHz, but the carrier itself occupies 24 MHz.

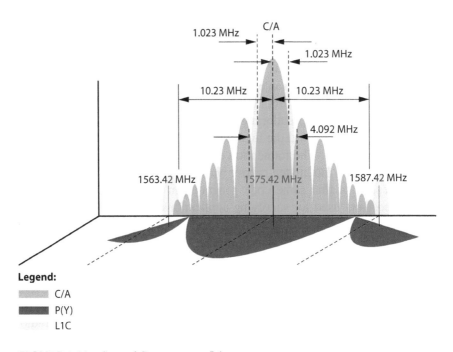

Legend:
C/A
P(Y)
L1C

FIGURE 1.11   Spread Spectrum on L1

This characteristic offers several advantages. It affords a better signal-to-noise ratio, more accurate ranging, less interference, and increased security. However, spreading the spectral density of the signal also reduces its power so that the GPS signal is sometimes described as a 25-watt light bulb in the sky. Clearly, the weakness of the signal makes it somewhat difficult to receive under obstructions.

A commonly used spread spectrum modulation technique is known as *binary phase shift keying (BPSK)*. This is the technique used to create the LNAV Message, the P(Y) code, and the C/A code. The *binary biphase* modulation is the switching from 0 to 1 and from 1 to 0 accomplished by phase changes of 180° in the carrier wave. Put another way, at the moments when the value of the code must change from 0 to 1 or from 1 to 0, the change is accomplished by an instantaneous reverse of the phase of the carrier wave. It is flipped 180° and each one of these flips occurs when the phase of the carrier is at the zero-crossing. Each 0 and 1 of the binary code is known as a code chip. 0 represents the normal state, and 1 represents the mirror image state.

The rate of all of the components of GPS signals are multiples of the standard rate of the oscillators. The standard rate is 10.23 MHz. It is known as the *fundamental clock rate* and is symbolized by F_0. For example, the GPS carriers are 154 times F_0, or 1575.42 MHz, 120 times F_0, or 1227.60 MHz, and 115 times F_0, or 1176.45 MHz. These represent L1, L2, and L5, respectively.

The codes are also based on F_0 10.23. The code chips of the P(Y) code, 0s or 1s, occur every microsecond. In other words, the *chipping rate* of the P(Y) code is 10.23 million bits per second (Mbps), the same as F_0, 10.23 MHz. The chipping rate of the C/A code is ten times slower than the P(Y) code. It is at 1.023 Mbps, one-tenth of F_0. Ten P(Y) code chips occur in the time it takes to generate one C/A code chip (see Figure 1.12). This is one of the reasons a P(Y) code derived pseudorange is more precise than a C/A code pseudorange and why the *coarse/acquisition code* is one of the names given the C/A code. Even though both codes are broadcast on L1, they are distinguishable from one another. They are transmitted in *quadrature,* meaning the C/A code modulation on the L1 carrier is phase shifted 90° from the P(Y) code modulation on the same carrier.

PSEUDORANGING

Strictly speaking, a pseudorange observable is based on a time shift. This time shift can be symbolized by $d\tau$, *d* tau. It is the time elapsed between the instant a GNSS signal leaves a satellite and the instant it arrives at a receiver. The concept can be illustrated by the process of setting a watch from a time signal heard over a phone.

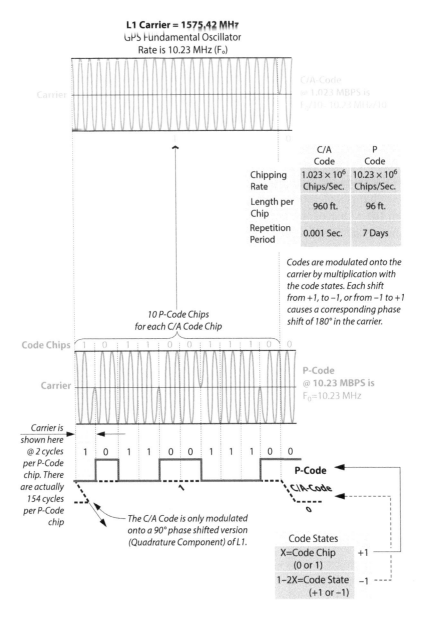

FIGURE 1.12   Code Modulation of the L1 Carrier

PROPAGATION DELAY

Imagine that a recorded voice said, "The time at the tone is 3 hours and 59 minutes." Supposing that the moment the tone was broadcast was indeed 3 hours and 59 minutes, a watch set at the instant the tone was heard, the watch would be wrong. The moment the tone is heard must be a bit later than the announced time. It is later because of the time it took the tone to travel through from the point of broadcast to the point of reception. This elapsed time would be approximately equal to the length traveled by the tone divided by the speed of electromagnetic energy, including light and radio signals. Imagine measuring that length by doing the division. In GNSS, that elapsed time is known as the propagation delay. It is used to measure length by a combination of codes. The idea is somewhat similar to the strategy used in EDMs. Please recall that an EDM generates an internal replica of its modulated carrier wave and compares it to the signal reflected back from the target. Similarly, a GNSS receiver uses a replica of a portion of a code as a *reference signal* to derive a pseudorange. In other words, the receiver generates this reference signal within itself and then compares that replica to the code that it receives from the satellite.

CODE CORRELATION

To conceptualize the process, one can imagine two codes generated at precisely the same time and identical in every regard: one in the satellite and one in the receiver. The satellite sends its code to the receiver, but on its arrival, the codes do not line up. They are identical but out of phase. However, time shifting the reference signal a little bit does make it fit the received satellite code. It is this time shift that reveals the propagation delay. That is the time it took the signal to make the trip from the satellite to the receiver, $d\tau$. It is the same idea described in the phone example, except the GNSS code is traveling much more space and atmosphere. Once the time shift of the replica code is accomplished, the two codes match fairly well and the time the satellite signal spends in transit has been measured, well, almost. It would be wonderful if that time shift could simply be divided by the speed of light and yield the true distance between the satellite and the receiver at that instant, and it is close, but there are limitations on the process that prevent such a perfect relationship.

AUTOCORRELATION

As mentioned earlier, the almanac information from the LNAV message of the first satellite a GPS receiver acquires tells it, among other things, which satellites can be expected to come into view. With this information, the receiver can start lining up its replica C/A codes with the signals it actually receives from the satellites. The process is called *autocorrelation*.

It uses code states. They can be explained that code chips (0 and 1) are transformed into *code states* (+1 and −1), with the formula:

$$\text{code state} = 1 - 2X$$

where X is the code chip value. For example, a normal code state is 11, which corresponds to a code chip value of 0. A mirror code state is −1. It corresponds to a code chip value of 1. The function of these code states can be illustrated by asking three questions: First, if a tracking loop of code states generated in a receiver does not match code states received from the satellite, how does the receiver know? In that case, the sum of the products of, say, for example, 10 of the receiver's code states with 10 code states from the satellite, divided by 10, does not equal 1. Second, what does the receiver do when the receiver's code states do not match the satellite's code states? It shifts the frequency of its search away from the center, i.e., 1575.42 MHz for L1, just a little bit. Since the satellite is always either moving toward or away from the receiver, there is always Doppler shift to accommodate, more about that in a moment. The receiver time-shifts its own generated replica code too. Both small shifts are iterative. They proceed in both time and frequency until the receiver's code states match those from the satellite. Third, how does the receiver know when a tracking loop of replica code states does match code states from the satellite?

Before the code from the satellite and the replica from the receiver are matched, the sum of the products of the code states is not 1. Following the correlation of the two codes, the sum of the products of each code state of the receiver's replica 10, with each of the 10 from the satellite, divided by 10, is exactly 1, and the receiver's replica code fits the code from the satellite like a key fits a lock (see Figure 1.13).

CORRELATION PEAK

As shown in the somewhat idealized triangular plot in Figure 1.14, the condition of maximum correlation between the receiver's replica and the C/A code received from the satellite has been achieved in the top center configuration, under which it says "correlation." In practice, the correlation peak is actually a bit rounded rather than being so emphatically triangular, but in any case, once the correlation peak has been reached, it is then maintained by the continual adjustment of the receiver generated code as described earlier.

It is usual to find three correlators used for tracking. One is at the prompt (P), punctual, or on-time position with the other two symmetrically located somewhat early (E) and late (L). The separation between the early, prompt, and late code phases is always symmetrical. The separation is often 1 chip, as illustrated here, but in a narrow *correlator*, they may be separated by as little as a 1/10th of a chip. Whatever the size of the *correlator spacing*, the symmetry is important to their purpose. That purpose is giving the receiver the information it needs to perform the shifts, the continual adjustments, alluded to above.

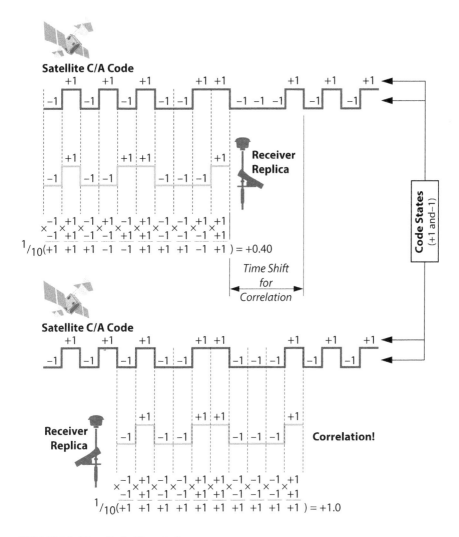

FIGURE 1.13   Code Correlation

Because there is an equal distance between them when the receiver's replica code is aligned with the code from the satellite, the early and late correlators will have the same amplitude on the triangle, as shown in the middle of Figure 1.14. On the other hand, when the codes are not aligned, the early and late correlator amplitudes will be different. The symmetry ensures that the difference in their amplitudes will be unequal in proportion to the amount that the codes are out of phase with one another. Using this information, the receiver can not only calculate the amount of the correlation error but also whether the receiver's replica code is ahead (early) or behind (late) the incoming satellite code.

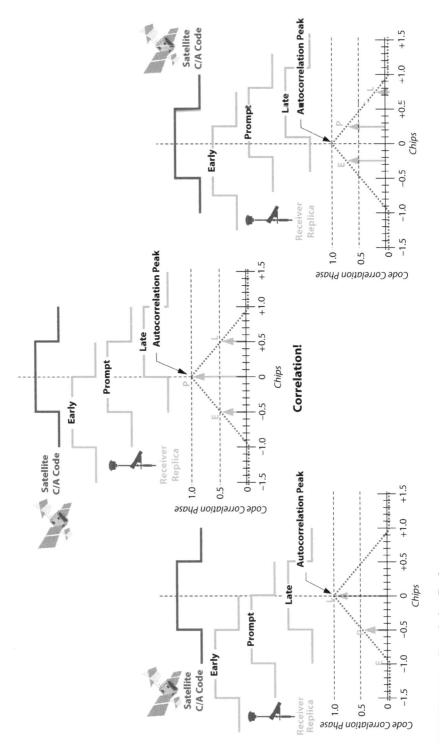

FIGURE 1.14   Correlation Peak

It can then correct the replica code generation to match the satellite's signal. This alignment is accomplished by a controller known as a *delay lock loop (DLL)*. In fact, the tracking of the code of most GNSS signals relies on some form of a DLL. The correlation typically takes about 1 to 5 milliseconds (ms) and does not exceed 20 ms. The same strategy, the early, prompt, and late tracking and correlation approach, is also utilized to align the receiver's replica carrier phase with the incoming satellite's carrier phase.

LOCK AND THE TIME SHIFT

Once correlation of the two codes is achieved, it is maintained by a *correlation channel* within the GNSS receiver, and the receiver is sometimes said to have achieved lock or to be locked on to the satellites. If the correlation is somehow interrupted later, the receiver is said to have lost lock. However, if the lock is present, the navigation message is available to the receiver. Remember that one of its elements is the broadcast clock correction that relates the satellite's onboard clock to system time, and this brings up a limitation of the pseudorange.

IMPERFECT OSCILLATORS

One reason the time shift, $d\tau$, found in autocorrelation cannot quite reveal the true range, ρ, of the satellite at a particular instant is the lack of perfect synchronization between the clock in the satellite and the clock in the receiver. Recall that the two compared codes are generated directly from the fundamental rate, F_o, of those clocks, and since these widely separated clocks, one on Earth and one in space, cannot be in perfect lockstep with one another, the codes they generate cannot be in perfect synch either. Therefore, a small part of the observed time shift, $d\tau$, must always be due to the disagreement between these two clocks. So, the time shift not only contains the signal's transit time from the satellite to the receiver it contains clock errors too. In fact, whenever GNSS satellite clocks and receiver clocks are checked against the carefully controlled systems time they are found to be drifting a bit. Their oscillators are imperfect. It is not surprising that they are not quite as stable as the Earth-based atomic clocks. Clocks on orbit are subject to the destabilizing effects of temperature, acceleration, radiation, and other inconsistencies. As a result, there are two clock offsets that bias every satellite to receiver pseudorange observable. That is one reason it is called a pseudorange (see Figure 1.15).

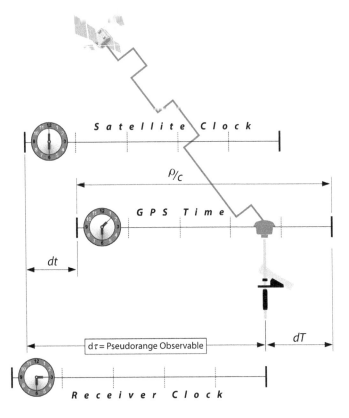

FIGURE 1.15   A GPS Pseudorange

A PSEUDORANGE EQUATION

However, clock offsets are only one of the errors in pseudoranges. Their relationship can be illustrated by the following equation (Fotopoulous 2000):

$$p = \rho + d_\rho + c\left(dt - dT\right) + d_{\text{ion}} + d_{\text{trop}} + \varepsilon_{\text{mp}} + \varepsilon_p$$

where
  p = the pseudorange measurement
  ρ = the true range
  d_ρ = satellite orbital errors
  c = the speed of light
  dt = the satellite clock offset from GPS Time
  dT = the receiver clock offset from GPS Time
  d_{ion} = ionospheric delay
  d_{trop} = tropospheric delay
  ε_{mp} = multipath
  ε_p = receiver noise

Please note that the pseudorange, p, and the true range, ρ, cannot be made equivalent without consideration of clock offsets, atmospheric effects, and other biases that are inevitably present.

This discussion of time can make it easy to lose sight of the real objective, which is the position of the receiver. Obviously, if the coordinates of the satellite and the coordinates of the receiver were known perfectly, then it would be a simple matter to determine time shift $d\tau$ or find the true range ρ between them. In fact, receivers placed at known coordinated positions can establish time so precisely that they are used to monitor atomic clocks around the world. Several receivers simultaneously tracking the same satellites can achieve resolutions of 10 nanoseconds or better. Also, receivers placed at known positions can be used as base stations to establish the relative position of receivers at unknown stations, a fundamental principle of most GNSS surveying.

It can be useful to imagine the true range term ρ, also known as the *geometric range*, actually includes the coordinates of both the satellite and the receiver. However, they are hidden within the measured value, the pseudorange, p, and all of the other terms on the right side of the equation. The objective then is to mathematically separate and quantify these biases so that the receiver coordinates can be revealed. Clearly, any deficiency in describing, or modeling, the biases will degrade the quality of the final determination of the receiver's position, but even if the biases were modeled completely, there would be a limit to how correctly a pseudorange could represent the range between a satellite and a receiver.

A 1-PERCENT RULE OF THUMB

Here is a convenient approximation. Suppose the maximum resolution available in a pseudorange is about 1 percent of the chipping rate of the code used. That assumption offers a basis to evaluate pseudoranging in general and compare the potentials of P(Y) code and C/A code pseudoranging.

A P(Y) code chip occurs every 0.0978 of a microsecond. In other words, there is a P(Y) code chip about every tenth of a microsecond. That's one code chip every 100 nanoseconds. Therefore, given the surmise, a P(Y) code-based measurement can have a maximum precision of about 1 percent of 100 nanoseconds, or 1 nanosecond. What is the length of 1 nanosecond? Well, multiplied by the speed of light its approximately 30 cm, or about a foot. So, under the 1-percent hypothesis, just about the best you can do with a P(Y) pseudorange is a foot or so. Because its chipping rate is ten times slower, the C/A code-based pseudorange is ten times less precise. Therefore, 1 percent of the length of a C/A code chip is 10 times 30 cm, or 3 m, around 10 feet.

However, the carrier-phase observable provides substantially better precision than does the pseudorange, and this can be illustrated with the same rule of thumb. First, the length of a single wavelength of each carrier is calculated using the formula:

$$\lambda = \frac{c_a}{f}$$

where
    λ = the length of each complete wavelength in meters
    c_a = the speed of light corrected for atmospheric effects
    f = the frequency in Hertz

The L1-1575.42 MHz carrier transmitted by GPS satellites has a wavelength of approximately 19 cm.

$$\lambda = \frac{c_a}{f}$$

$$\lambda = \frac{300 \times 10^6 \text{ mps}}{1575.42 \times 10^6 \text{ Hz}}$$

$$\lambda = 0.19 \text{ m}$$

The L2-1227.60 MHz frequency carrier transmitted by GPS satellites has a wavelength of approximately 24 cm.

$$\lambda = \frac{c_a}{f}$$

$$\lambda = \frac{300 \times 10^6 \text{ mps}}{1227.60 \times 10^6 \text{ Hz}}$$

$$\lambda = 0.24 \text{ m}$$

The L5-1176.45 MHz frequency carrier transmitted by GPS satellites has a wavelength of approximately 25 cm.

$$\lambda = \frac{c_a}{f}$$

$$\lambda = \frac{300 \times 10^6 \text{ mps}}{1176.45 \times 10^6 \text{ Hz}}$$

$$\lambda = 0.25 \text{ m}$$

Therefore, using the wavelength of any GPS carrier as the observable, the measurement resolved to 1% of the wavelength would be about 2 mm. The carrier-phase observable is preferred for the higher precision work most surveyors have come to expect from GPS.

CARRIER-PHASE RANGING

THE INTEGER AMBIGUITY PROBLEM

Even though the carrier-phase observable is at the center of high accuracy surveying applications, it introduces a difficulty that needs to be overcome. It is called the *integer ambiguity* problem, aka the integer-cycle ambiguity problem. It is like the hurdle encountered by the EDM. As you know, by comparing the phase of the signal returned from the reflector with the reference wave it kept at home, an EDM can measure how much the two are out of phase with one another. However, this measurement can only be used to calculate a small part of the overall distance. It only discloses the length of a fractional part of a wavelength used. This leaves a big unknown, namely the number of full wavelengths of the EDM's modulated carrier between the transmitter and the receiver at the instant of the measurement. This integer ambiguity is symbolized by N. Fortunately, the integer ambiguity can be solved in the EDM measurement process. The key is using carriers with progressively longer wavelengths. For example, the submeter portion of the overall distance can be resolved using a carrier with the wavelength of a meter. This can be followed by a carrier with a wavelength of 10 m which provides the basis for resolving the meter aspect of a measured distance. This procedure may be followed by the resolution of the tens of meters using a wavelength of 100 m. The hundreds of meters can then be resolved with a wavelength of 1000 m, and so on.

Such a method works in the EDM's two-way ranging system, but the GNSS one-way ranging makes the use of the same strategies impossible. GNSS ranging must use an entirely different strategy for solving the problem because the satellites broadcast only three carriers; L1, L2, and L5 the satellites. Those carriers have constant wavelengths, and they only propagate from the satellites to the receivers, one direction. Therefore, unlike an EDM measurement, the wavelengths of these carriers in GNSS cannot be periodically changed to resolve the integer ambiguity problem.

CARRIER-PHASE COMPARISONS

The unmodulated L1, L2, and L5 are the observables used in the GPS carrier-phase solution rather than the P(Y) and C/A codes. Understanding the carrier-phase is perhaps a bit more difficult than the pseudorange, but the basis of the measurements has some similarities. As you know, the foundation of a pseudorange measurement is the correlation of the codes received from a GPS satellite with replicas of those codes generated within the receiver. The foundation of the carrier-phase measurement is the combination of the unmodulated carrier itself received from a GPS satellite with a replica of that carrier generated within the receiver. As in the EDM example, it is the phase difference between the incoming signal and the internal reference that reveals the fractional part of the carrier-phase measurement. The incoming signal is from a satellite rather than a reflector, of course, but like an EDM measurement, the internal reference is derived from the receiver's oscillator, and at the moment of lock-on, the number of complete cycles between the receiver and the satellite is not immediately known.

BEAT

The carrier-phase observable is sometimes called the *reconstructed carrier-phase* or carrier beat phase observable. In this context, a *beat* is the pulsation resulting from the combination of two waves with different frequencies. An analogous situation occurs when two musical notes of nearly the same frequency are sounded at the same time. Their two frequencies interfere with one another and produce a pulsation called a beat. Musicians can tune their instruments by listening for the beat that occurs when two tones differ slightly. The pulsation is the result of constructive and destructive interference, and the beat phenomenon is by no means unique to musical notes; it can occur when any pair of oscillations with different frequencies are combined. In GNSS, a beat is created when a carrier generated in a GNSS receiver and a carrier received from a satellite are combined (see Figure 1.16).

At first, that might not seem sensible. How could a beat be created by combining two absolutely identical unmodulated carriers? There should be no difference in frequency between an L1 carrier generated in a satellite and an L1 carrier generated in a receiver. They both should have a frequency of 1575.42 MHz. If there is no difference in the frequencies, how can there be a beat? But there is a slight difference between the two carriers. Something happens to the frequency of the carrier on its trip from a GPS satellite to a receiver, its frequency changes. The phenomenon is described as the *Doppler effect*.

THE DOPPLER EFFECT

Sound provides a model for the explanation of the phenomenon. An increase in the frequency of a sound is indicated by a rising pitch; a lower

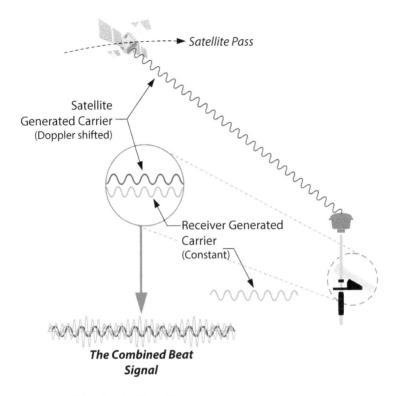

FIGURE 1.16   The Carrier Beat Phase

pitch is the result of a decrease in the frequency. A stationary observer listening to the blasting horn on a passing train will note that as the train gets closer, the pitch rises, and as the train travels away, the pitch falls. Furthermore, the change in the sound, clear to the observer standing beside the track, is not heard by the engineer driving the train. He hears only one constant, steady pitch. The relative motion of the train with respect to the observer causes the apparent variation in the frequency of the sound of the horn.

In 1842, Christian Doppler described this frequency shift now named for him. He used the analogy of a ship on an ocean with equally spaced waves. In his allegory, when the ship is stationary the waves strike it steadily, one each second, for example, but if the ship sails into the waves, they break across its bow more frequently. Now, if the ship then turns around and sails with the waves, they strike less frequently across its stern. The waves themselves have not changed; their frequency is constant, but to the observer on the ship, their frequency seems to depend on his motion.

GNSS AND THE DOPPLER EFFECT

From the observer's point of view, whether it is the source, the observer, or both that are moving, the frequency increases while they move together and decreases while they move apart. The Doppler effect is inversely proportional to the wavelength of the signal. Therefore, if a GNSS satellite is moving toward an observer, its carrier wave comes into the receiver with a higher frequency than it had at the satellite. If a GNSS satellite is moving away from the observer, its carrier wave comes into the receiver with a lower frequency than it had at the satellite. Since a GNSS satellite is virtually always moving with respect to the observer, any signal received from a GNSS satellite is Doppler shifted.

A CARRIER-PHASE APPROXIMATION

The carrier-phase observation in cycles is often symbolized by *phi*, φ, in GNSS literature. Other conventions include the use of superscripts to indicate satellite designations and the use of subscripts to define receivers. For example, in the following equation, φ_r^s is used to symbolize the carrier-phase observation between satellite s and receiver r. The difference that defines the carrier beat phase observation is (Wells 1986):

$$\varphi = \varphi_r^s = \varphi^s(t) - \varphi_r(T)$$

where $\varphi^s(t)$ is the phase of the carrier broadcast from the satellite s at time t. Please note that the frequency of this carrier is the same, nominally constant frequency, that is generated by the receiver's oscillator. $\varphi_r(T)$ is its phase when it reaches the receiver r at time T.

A description of the use of the carrier-phase observable to measure range can start with the same basis as the calculation of the pseudorange travel time. The time elapsed between the moment the signal is broadcast, t, and the moment it is received, T, multiplied by the speed of light, c, will yield the range between the satellite and receiver, ρ:

$$(T-t)c \approx \rho$$

Even using this simplified equation, it is possible to get an approximate idea of the relation between time and range for some assumed nominal values. These values could not be the basis of any actual carrier-phase observation because, among other reasons, it is not possible for the receiver to know when a particular carrier wave left the satellite. However, for the purpose of illustration, suppose a carrier left a satellite at 00 hours 00 minutes 00.000 seconds and arrived at the receiver 67 milliseconds later.

$$(T - t)c \approx \rho$$

$$(00\text{h}:00\text{m}:00.067\text{s} - 00\text{h}:00\text{m}:00.000\text{s})300,000 \text{ km/s} \approx \rho$$

$$(0.067)300,000 \text{ km/s} \approx \rho$$

$$20,100 \text{ km} \approx \rho$$

This estimate indicates that if the carrier broadcast from the satellite reaches the receiver 67 milliseconds later, the range between them is approximately 20,100 km. Carrying this example a bit farther, the wavelength, λ, of the L1 carrier can be calculated by dividing the speed of light, c, by the L1 frequency, f:

$$\lambda = \frac{c}{f}$$

$$\lambda = \frac{300,000,000 \text{ m/s}}{1,575,420,000 \text{ Hz}}$$

$$\lambda = 0.1904254104 \text{ m}$$

Dividing the approximated range, ρ, by the calculated L1 carrier wavelength, λ, yields a rough estimation of the carrier phase in cycles, φ:

$$\frac{\rho}{\lambda} \approx \varphi$$

$$\frac{20,100,000 \text{ m}}{0.1904154104 \text{ m}} \approx \phi$$

$$105,553,140 \text{ cycles} \approx \phi$$

The 20,100-km range implies that the L1 carrier would cycle through approximately 105,553,140 wavelengths on its trip from the satellite to the receiver. Of course, these relationships are much simplified. However, they can be made fundamentally correct by recognizing that ranging with the carrier-phase observable is subject to all of the same biases and errors as the pseudorange. For example, terms such as the receiver clock offset may be incorporated, again symbolized by dT as it was in the pseudorange equation. The imperfect satellite clock can be included; its error is symbolized by dt. The tropospheric delay d_{trop}, the ionospheric delay d_{ion} and multipath-receiver noise ε_ϕ are also added to the range measurement. The ionospheric delay will be negative here, more about that later. With these changes, the simplified travel time equation can be made a bit more realistic.

$$[(T+dT)-(t+dt)]c \approx \rho-d_{ion}+d_{trop}+\varepsilon_\phi$$

This more realistic equation can be rearranged to isolate the elapsed time (Tt) on one side, by dividing both sides by c, and then moving the clock errors to the right side (Wells 1986).

$$\frac{[(T+dT)-(t+dt)]c}{c} = \frac{\rho-d_{ion}+d_{trop}+\varepsilon_\phi}{c}$$

$$[(T+dT)-(t+dt)] = \frac{\rho-d_{ion}+d_{trop}+\varepsilon_\phi}{c}$$

$$(T-t+dT-dt) = \frac{\rho-d_{ion}+d_{trop}+\varepsilon_\phi}{c}$$

$$(T-t+dT-dt)+(dt-dT) = \frac{\rho-d_{ion}+d_{trop}+\varepsilon_\phi}{c}+(dt-dT)$$

$$T-t = dt-dT+\frac{\rho-d_{ion}+d_{trop}+\varepsilon_\phi}{c}$$

This expression now relates the travel time to the range. However, in fact, a carrier-phase observation cannot rely on the travel time for two reasons. First, in a carrier-phase observation, the receiver has no codes with which to tag any particular instant on the incoming continuous carrier wave. Second, since the receiver cannot distinguish one cycle of the carrier from any other, it has no way of knowing the initial phase of the signal when it left the satellite. In other words, the receiver cannot know the travel time and, therefore, it is hard to see how it can determine the number of complete cycles between the satellite and itself. This unknown quantity is called the integer ambiguity, aka the integer cycle ambiguity.

Remember, the approximation of 105,553,140 wavelengths in this example is for the purpose of comparison and illustration only. In actual practice, a carrier-phase observation must derive the range from a measurement of phase at the receiver, not from a known travel time of the signal.

The missing information is the number of complete phase cycles between the receiver and the satellite at the instant that the tracking began, the instant of lock-on. The critical unknown integer, symbolized by N, is the integer ambiguity, and it cannot be directly measured by the receiver. The receiver can count the complete phase cycles it receives from the moment it starts tracking until the moment it stops. It can also monitor the fractional phase cycles, but the integer ambiguity N is unknown.

An Illustration of the Integer Ambiguity Problem

The situation is somewhat analogous to an unofficial technique used by some nineteenth-century contract surveyors on the Great Plains. The procedure can be used as a rough illustration of the integer ambiguity problem in GNSS. It was known as the buggy wheel method of chaining. Some of the lines of the public land system that crossed open prairies were originally surveyed by loading a wagon with stones or stakes and tying a cloth to a spoke of the wheel. One man drove the team, another kept the wagon online with a compass, and a third counted the revolutions of the flagged wheel to measure the distance. When there had been enough turns of the improvised odometer to measure half a mile, they set a stone or stake to mark the corner and then rolled on, counting their way to the next corner.

A GNSS receiver is like the man assigned to count the turns of the wheel. He is supposed to begin his count from the moment the crew leaves the newly set corner, but instead, suppose he jumps into the wagon, gets comfortable, and takes an unscheduled nap. When he wakes up, the wagon is on the move. Trying to make up for his laxness, he immediately begins counting, but at that moment, the wheel is at a half turn, a fractional part of a cycle. He counts the subsequent half turn and then, back on the job, he intently counts each and every full revolution as they come around. His tally grows as the cycles accumulate, but he is in trouble and he knows it. He cannot tell how far the wagon has traveled; he was asleep for the first part of the trip. He has no way of knowing how far they had come before he woke up and started counting. He is like a GNSS receiver that cannot know how far it is from the satellite when it starts counting phase cycles. They can tell it nothing about how many cycles stood between itself and the satellite when the receiver was locked on and began tracking.

The 360° cycles in the carrier-phase observable are wavelengths λ, not revolutions of a wheel. Therefore, the integer ambiguity included in the complete carrier-phase equation is an integer number of wavelengths, symbolized by λN (Fotopoulos 2000). So, the complete carrier-phase observable equation can now be stated as:

$$\Phi = \rho + d_\rho + c(dt - dT) + \lambda N - d_{ion} + d_{trop} + \varepsilon_{m\Phi} + \varepsilon_\Phi$$

where
  Φ = the carrier-phase measurement
  ρ = the true range
  d_ρ = satellite orbital errors
  c = the speed of light
  dt = the satellite clock offset from GPS Time
  dT = the receiver clock offset from GPS Time
  λ = the carrier wavelength
  N = the integer ambiguity in cycles
  d_{ion} = ionospheric delay
  d_{trop} = tropospheric delay
  $\varepsilon_{m\Phi}$ = multipath
  ε_Φ = receiver noise

EXERCISES

1. What is the function of the information in subframe 5 of the GPS LNAV message?
   a. Once a receiver is locked onto a satellite, it helps the receiver to determine the position of the satellite that is transmitting the GPS LNAV message.
   b. Once a receiver is locked onto a satellite, it helps the receiver to correct the part of the delay of the signal caused by the ionosphere.
   c. Once a receiver is locked onto a satellite, it helps the receiver to correct the received satellite time to GPS Time.
   d. Once a receiver is locked onto a satellite, it helps the receiver acquire the signals of the other satellites.

2. Which height most correctly expresses a nominal altitude of GPS satellites above the Earth?
   a. 20,000 miles
   b. 35,420 km
   c. 20,183,000 m
   d. 108,000 nautical miles

3. Which of the following statements about the clocks in GPS satellites is not correct?
   a. The signal for each satellite is independent from the other satellites and is generated from its own onboard clock.
   b. The clocks in GPS satellites may also be called oscillators or frequency standards.
   c. Every GPS satellite is launched with very stable atomic clocks onboard.
   d. The clocks in any one satellite are allowed to drift up to one nanosecond from GPS Time before they are tweaked by the Control Segment.

4. The Global Positioning System was originally designed to be a passive system. What does a passive system mean in that context?
   a. The ranges are measured with signals in the microwave part of the electromagnetic spectrum.
   b. Only the satellites transmit signals; the users receive them.
   c. A GPS receiver must be able to gather all the information it needs to determine its own position from the signals it bounces off the satellites.
   d. The signals from a GPS receiver return to the satellite.

5. Which comparison of EDM and GPS processes is correct?
   a. EDM and GPS signals are both reflected to their sources.
   b. EDM measurements require atmospheric correction, GPS ranges do not.
   c. EDMs and GPS satellites both transmit modulated carriers.
   d. Phase differencing is used in EDM measurement but not in GPS.

6. What information below is critical to understanding the relationship between GPS Time and UTC?
   a. Multipath
   b. The broadcast ephemeris
   c. An anti-spoofing flag
   d. Leap seconds

7. The P(Y) code and the C/A code are created by shifts from 0 to 1 or 1 to 0, known as code chips. There is a corresponding change of 180° in the GPS carrier waves. What changes?
   a. The fundamental clock rate
   b. The frequency
   c. The phase
   d. The amplitude

8. Which of the following statements about the P(Y) code and the C/A code is not correct?
   a. The chipping rate for the C/A code is ten times slower than the chipping rate for the P(Y) code.
   b. They are broadcast in binary biphase modulation.
   c. Their chipping rates are equal to the fundamental clock rate.
   d. They are broadcast in quadrature.

9. Which of the following is the most correct description of the Doppler effect?
   a. The distortion of electromagnetic waves due to the density of charged particles in the Earth's upper atmosphere.
   b. The systematic changes that occur when a moving object or light beam passes through a gravitational field.
   c. The systematic changes that occur when a moving object approaches the speed of light.
   d. The shift in frequency of acoustic or electromagnetic radiation sensed by an observer when the source is moving relative to the observer.

10. Which of the following numbers describe the frequencies of the L1, L2, and L5 carrier signals?
    a. 1755.42, 1227.60, 1176.45
    b. 1575.42, 1227.60, 1176.45
    c. 1175.42, 1226.70, 1576.45
    d. 1542.77, 1260.52, 1167.85

ANSWERS AND EXPLANATIONS

1. Answer is (d)

   Explanation: Subframe 5 provides the almanac and health data for the first 24 satellites in the GPS constellation along with the almanac reference time and week numbers. The purpose of subframe 5 is to help the receiver acquire the signals of the other satellites. The receiver must first lock onto a satellite to have access to the LNAV message, of course, but once the LNAV message can be read the positions of all the other satellites can be computed.

2. Answer is (c)

   Explanation: The GPS orbital configuration is well settled. Even though there is some eccentricity, the GPS satellite's orbit is nearly circular and has a nominal altitude of 20,183 km or 20,183,000 m. The orbit is approximately 12,500 statute miles or 10,900 nautical miles above the surface of the planet. Their high altitude allows each GPS satellite to be viewed simultaneously from a large portion of the Earth at any given moment. However, GPS satellites are well below the height, 35,420 km, required for the sort of geosynchronous orbit used for communications satellites.

3. Answer is (d)

   Explanation: Each GPS satellite has its own SV time. Government tracking facilities monitor the drift of each satellite clock's deviation from GPS Time. The drift is allowed to reach a maximum of 1 millisecond. The satellite broadcast clock correction in subframe 1 of the LNAV message allows the receiver to relate each satellite's SV time to GPS Time.

4. Answer is (b)

   Explanation: Some of the forerunners of GPS were systems that involved transmissions from the users, but the original design of the GPS system excluded anything that would reveal the location of the GPS receiver such that the satellites transmit and the users receive.

5. Answer is (c)

   Explanation: While GPS receivers can and do make phase measurements on the unmodulated carrier waves, they also make use of the code information available on the modulated carrier transmitted by GPS satellites to determine pseudoranges. GPS uses a one-way system. The modulated carrier with code information travels from the satellite to the receiver, where it is correlated with a reference. This one-way method requires a frequency standard in both the satellite and the receiver. An EDM's measurement is also based on a modulated carrier. However, EDM uses a two-way system. The modulated carrier is transmitted to a retroprism. It is reflected and returns to the EDM. The EDM can then determine the phase delay by comparing the returned modulated wave with a reference in the instrument. This two-way method requires only one frequency standard in the EDM since the modulated carrier is reflected.

6. Answer is (d)

   Explanation: GPS Time is kept within 1 millisecond of UTC. However, UTC is adjusted by leap seconds from time to time, and GPS Time is not. Conversion of GPS Time to UTC requires knowledge of the leap seconds applied to UTC since January 1980.

7. Answer is (c)

Explanation: Phase modulations are used to mark the divisions between the code chips. Whenever the C/A code or the P(Y) code switches from a binary 1 to a binary 0 or vice versa, its carrier has a sharp mirror-image shift in phase. About every millionth of a second, the phase of the C/A code carrier can shift, but the P(Y) code carrier can have a phase shift about every 10-millionth of a second.

8. Answer is (c)

Explanation: The code chips of the P(Y) code, 0s or 1s, occur every microsecond. The P(Y) code chipping rate is 10.23 million bits per second (Mbps), the same as the fundamental clock rate, F_o, 10.23 MHz. The chipping rate of the C/A code is ten times slower than the P(Y) code. It is one-tenth of F_o, 1.023 Mbps. Ten P(Y) code chips occur in the time it takes to generate one C/A code chip.

9. Answer is (d)

Explanation: If the source of radiation is moving relative to an observer, there is a difference between the frequency perceived by the observer and the frequency of the radiation at its source. The shift is to higher frequencies when the source moves toward the observer and to lower frequencies when it moves away. This effect has been named for its discoverer, C.J. Doppler.

10. Answer is (b)

Explanation: The two legacy GPS carriers L1 at 1575.42 MHz and L2 at 1227.60 MHz are joined now by a third carrier. It is known as L5, and its frequency is 1176.45 MHz.

2 Biases and Solutions

BIASES

A LOOK AT THE ERROR BUDGET

The understanding and management of errors is indispensable for finding the true geometric range ρ from either a pseudorange or carrier-phase observation.

$$p = \rho + d_\rho + c(dt - dT) + d_{\text{ion}} + d_{\text{trop}} + \varepsilon_{m_p} + \varepsilon_p \ (\text{pseudorange})$$

$$\Phi = \rho + d_\rho + c(dt - dT) + \lambda N - d_{\text{ion}} + d_{\text{trop}} + \varepsilon_{m_\Phi} + \varepsilon_\Phi \left(\text{carrier phase}\right)$$

Both equations include environmental and physical limitations called *biases*. There are atmospheric errors among the biases; the ionospheric effect, d_{ion}, and the tropospheric effect, d_{trop}. The tropospheric effect may be somewhat familiar to Total Station and electronic distance measuring (EDM) device users, even if the ionospheric effect is not. Other biases, clock errors symbolized by $(dt - dT)$ and receiver noise, ε_p and ε_φ, multipath, ε_{m_p} and ε_{m_Φ}, and orbital errors, d_ρ, are unique to satellite surveying methods. As you can see, each of these biases comes from a different source. They are each independent of one another, but they combine to obscure the true geometric range. The objective here is to discuss each of them separately.

USER EQUIVALENT RANGE ERROR

The summary of the total error budget affecting a pseudorange expressed in units of distance is called the *user equivalent range error (UERE)*, also known as User Range Error (URE). It is usually divided into two categories. The first category is the *Signal-in-Space Range Error (SISRE)*, which includes the errors attributable to the space and control segment such as signal imperfections, satellite ephemeris errors, clock errors, etc. It does not include errors contributed by atmospheric delays, multipath, or errors specific to the user's receiver and environment. Those are in the second category, the *User Equipment Error (UEE)*. The total UERE can then be written: $UERE = \sqrt{SISRE^2 + UEE^2}$. Another way to consider it is to look at the effect of the errors. The atmosphere and orbital errors increase and

DOI: 10.1201/9781003405238-2

decrease with the length of the baselines between receivers, whereas those due to receiver noise and multipath do not. In other words, errors localized at the receiver itself stay the same regardless of the length of the baseline.

THE IONOSPHERIC EFFECT, d_{ion}

One of the errors in GNSS positioning is attributable to the atmosphere. The long relatively unhindered travel of the GNSS signal through the virtual vacuum of space changes as it passes through the Earth's atmosphere. Through both refraction and diffraction, the atmosphere alters the apparent speed and, to a lesser extent, the direction of the signal. This causes an apparent delay in the signals transit from the satellite to the receiver. The magnitude of these delays is determined by the state of the ionosphere at the moment the signal passes through, so it's important to note that its density and stratification vary.

IONIZED PLASMA

The ionosphere extends from about 50 km to about 1000 km above the Earth's surface and is the first part of the atmosphere that the signal encounters as it leaves the satellite. It is ionized plasma comprised of negatively charged electrons which remain free for long periods before being captured by positive ions. The sun plays a key role in the creation and variation of these aspects, and the daytime ionosphere is rather different from the ionosphere at night.

IONOSPHERE AND THE SUN

The sun's extreme ultraviolet radiation and x-rays ionize gas molecules in the ionosphere. Free electrons are released. As their number and dispersion vary, so does the electron density. This density is often described as *total electron content* (*TEC*), a measure of the number of free electrons in a column through the ionosphere centered on the signals path between the receiver and the satellite and having a cross-sectional area of 1 square meter. 10^{16} is one TEC unit. Typical values within that column vary from 10^{16} to 10^{19}. The higher the electron density, the larger the delay of the signal, but the delay is by no means constant. It varies with the season, magnetic activity, geographic location, local time, etc.

The ionospheric delay changes slowly through a daily cycle. It is usually least between midnight and early morning and most around local noon or a little after. During the daylight hours in the midlatitudes, the ionospheric delay may grow to be as much as five times greater than it was at night. It is also nearly four times greater in November, when the Earth nears its

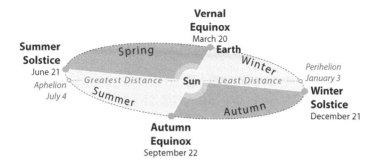

FIGURE 2.1 Earth's Orbit

perihelion, its closest approach to the sun than it is in July near the Earth's *aphelion*, its farthest point from the sun. The effect of the ionosphere on the GNSS signal usually reaches its peak in March, about the time of the vernal equinox (see Figure 2.1).

IONOSPHERIC STRATIFICATION

The ionosphere has layers, sometimes known as the mesosphere and thermosphere, that are themselves composed of D, E, and F regions (see Figure 2.2). Neither the boundaries between these regions nor the upper

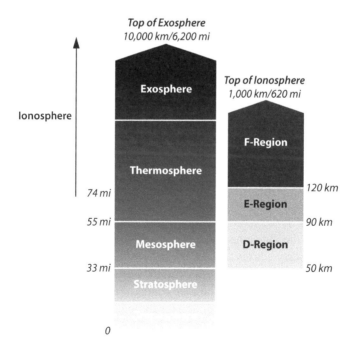

FIGURE 2.2 Atmospheric Model

layer of the ionosphere can be defined strictly. Here are some general ideas on the subject. The lowest detectable layer, the D region, extends from about 50 to 90 km. It has almost no effect on GNSS signals and virtually disappears at night. The E region, between 90 and 120 km, is also a day-time phenomenon. Its effect on the signal is slight, but it can cause the signal to scintillate. The layer that affects the propagation of electromagnetic signals the most is the F region. It extends from about 120 to 1000 km. The F region contains the most concentrated ionization in the atmosphere. In the daytime, the F layer can be further divided into F1 and F2. F2 is the most variable. F1, the lower of the two, is most apparent in the summer. These two layers combine at night. Above the F layer is fully ionized. It is sometimes known as the photosphere or the H region.

The ionosphere is not homogeneous. Its behavior in one region of the Earth is liable to be unlike its behavior in another. For example, iono-spheric disturbances can be particularly harsh in polar regions. However, the highest TEC values and the widest variations in the horizontal gradi-ents occur in the band of about 60° of *geomagnetic latitude*. That band lies 30° north and 30° south of the Earth's magnetic equator.

The Ionospheric Effect

The error introduced by the ionosphere can be very small, but it may be large when the vernal equinox is near and/or sunspot activity is severe. For example, the TEC is maximized during the peak of the 11-year solar cycle. It also varies with magnetic activity, location, time of day, and even the direction of observation. The severity of the ionosphere's effect on a GNSS signal also depends on the amount of time that signal spends travel-ing through it. For example, a signal originating from a satellite near the observer's horizon passes through a larger amount of the ionosphere and is more affected by it than is a signal from a satellite near the observer's zenith.

Group Delay and Phase Delay

The ionosphere is *dispersive*, which means that the apparent time delay contributed by the ionosphere depends on the frequency of the signal. This dispersive property causes the codes, the modulations on the carrier wave, to be affected differently than the carrier wave itself during the signal's trip through the ionosphere. The P(Y) code, the C/A code, the navigation message, and all the other codes appear to be delayed, or slowed, affected by what is known as the *group delay*. But the carrier wave itself appears to speed up in the ionosphere. It is affected by what is known as the *phase delay*. It may seem odd to call an increase in speed a delay. It is sometimes

called *phase advancement*. In any case, it is governed by the same properties of electron content as the group delay, phase delay just increases
negatively. So, while the magnitude is the same in both cases, the algebraic
sign of d_{ion} is negative in the carrier-phase equation and positive in the
pseudorange equation.

DIFFERENT FREQUENCIES ARE AFFECTED DIFFERENTLY

Another consequence of the dispersive nature of the ionosphere is that
its effect on a higher-frequency carrier wave is less than it is on a lower-
frequency wave. In the case of GPS, that means that L1, 1575.42 MHz,
is not affected as much as L2, 1227.60 MHz, and L2 is not affected as
much as L5, 1176.45 MHz. This fact provides one of the greatest advantages of a multifrequency receiver over the single-frequency receivers.
The separations between the L1 and L2 frequencies (347.82 MHz), the L1
and L5 frequencies (398.97 MHz), and even the L2 and L5 frequencies
(51.15 MHz) are large enough to facilitate the estimation of the ionospheric bias. Therefore, by tracking all the carriers, a multiple-frequency
receiver can model and remove nearly all of it. There are now several
possible combinations, L1/L2, L1/L5, and L2/L5. The frequency dependence of the ionospheric effect is described by the following expression
(Klobuchar, 1983 in Brunner and Welsch, 1993):

$$v = \frac{40.3}{cf^2} \cdot \text{TEC}$$

where:
      v = the ionospheric delay
      c = the speed of light in meters per second
      f = the frequency of the signal in Hz
  TEC = the quantity of free electrons per square meter

As the formula illustrates, the delay is inversely proportional to the
square of the frequency; in other words, the higher the frequency, the less
the bias. For example, at 1227.60 (L2), it is 65% larger than at 1575.42
MHz (L1), and at 1176.45 MHz (L5), it is 80% larger than 1575.42 (L1).

BROADCAST CORRECTION

The dispersive nature of the ionosphere enables a receiver tracking more
than one frequency to remove nearly all the ionospheric bias, but a single-
frequency GPS receiver cannot do that. It must rely on the broadcast ionospheric correction in subframe 4 of the navigation message. The basis of
which was created by Jack Klobuchar in the 1980s and so is known as the

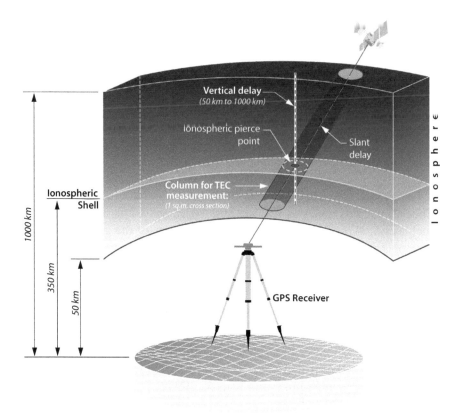

FIGURE 2.3 Klobuchar Model

Klobuchar Model (Figure 2.3). It was first used in GPS but has since been employed by other GNSS constellations. It models the ionosphere as if all the electron content were concentrated into an imaginary thin shell at the height of 350 km.

Given the approximate position of the receiver, day of the year, time of day, and using averaged solar condition and activity from the Control Segment, it generates eight time-varying coefficients and thereby estimates the ionospheric delay at the receiver's zenith. That value multiplied by an obliquity factor provides the delay along the path from the receiver to satellite, known as the slant delay, at the *Ionospheric Pierce Point* (*IPP*). The IPP is the place where that path intersects the imaginary shell. The Klobuchar model is predicated on averaged conditions which necessarily present an idealized calm ionosphere. Therefore, estimates derived from it do not reflect the actual everyday turbulence in the ionosphere, so it is not surprising that the broadcast correction based on this model does not remove more than about 50% to 60% of the total delay. So, where the baseline between GNSS receivers is short, the effect of the ionosphere can be small, but as the baseline grows, so does the significance of the ionospheric bias because the ionosphere at one end of a long baseline can be very different from the ionosphere at the other end.

THE SATELLITE CLOCK BIAS, *dt*

Another error can be attributed to the satellite clock bias. It can be large if the broadcast clock correction is not used by the receiver to bring the time signal acquired from a satellite's onboard clock in line with a GNSS constellation time standard. As time is a critical component in the functioning of GNSS, it is important to look closely at the principles behind this bias.

RELATIVISTIC EFFECTS ON THE SATELLITE CLOCK

Albert Einstein's special and general theories of relativity apply to the clocks involved here. The clocks in the GNSS satellites are traveling at great speed. GPS satellites are moving at 3.874 kilometers per second which means the clocks on those satellites appear to run slower than the clocks on Earth by about 7 microseconds a day. However, this apparent slowing of the clocks in orbit is counteracted by the weaker gravity around them. The weakness of gravity makes the clocks in the satellites appear to run faster than the clocks on Earth by about 45 microseconds a day. Therefore, on balance, the clocks in the GPS satellites in space appear to run faster by about 38 microseconds a day than the clocks in GPS receivers on Earth. So, to ensure the clocks in the satellites will actually produce the correct fundamental frequency of 10.23 MHz in space, their frequencies are set to 10.22999999543 MHz before they are launched into space.

There is yet another consideration, the eccentricity of the orbit of GNSS satellites. GPS orbits have an eccentricity of 0.02. The effect on the clocks can be as much as 45.8 nanoseconds. Fortunately, the offset is eliminated by a calculation in the GPS receiver itself; thereby avoiding what could be ranging errors of about 14 meters. The receiver is moving too; of course, so an account must be made for the motion of the receiver due to the rotation of the Earth during the time it takes the satellite's signal to reach it. This is known as the Sagnac effect, and it is 133 nanoseconds at its maximum. Luckily these relativistic effects can be accurately computed and removed from the system.

SATELLITE CLOCK DRIFT

Clock drift is another matter. As discussed in Chapter 1, the onboard GNSS satellite clocks are independent of one another. The rates of these rubidium and cesium oscillators are more stable if they are not disturbed

by frequent tweaking and adjustment is kept to a minimum. While GPS
Time itself is designed to be kept within one microsecond, 1 microsecond
or one-millionth of a second, of UTC (USNO), excepting leap seconds, the
satellite clocks can be allowed to drift up to a millisecond, 1 millisecond
or one-thousandth of a second, from GPS Time.

There are three kinds of time involved here. The first is UTC per the
U.S. Naval Observatory (USNO). The second is GPS Time. The third is
the time determined by each independent GPS satellite. The relationship
was described in Chapter One, but it bears repeating. It is as follows. The
Master Control Station (MCS) at Schriever (formerly Falcon) Air Force
Base near Colorado Springs, Colorado, gathers the GPS satellite's data
from monitoring stations around the world. After processing, this information is uploaded back to each satellite to become the broadcast ephemeris,
broadcast clock correction, and so forth. The actual specification for GPS
Time demands that its rate be within one microsecond of UTC as determined by USNO, without consideration of leap seconds. Leap seconds are
used to keep UTC correlated with the actual rotation of the Earth, but
they are ignored in GPS Time. In GPS Time, it is as if no leap seconds have
occurred at all in UTC since 00:00:00, January 6, 1980. In practice, the
rate of GPS Time is much closer than 1 microsecond of the rate of UTC.
It is usually within 10–20 nanoseconds of the rate of UTC. By constantly
monitoring the satellite's clock error, dt, the Control Segment gathers data
for its uploads of the broadcast clock corrections. You will recall that clock
corrections are part of the navigation message.

THE RECEIVER CLOCK BIAS, dT

Another error which can be caused by the GNSS receiver clock is its
oscillator. Both a receiver's measurement of phase differences and its generation of replica codes depend on the reliability of this internal frequency
standard.

TYPICAL RECEIVER CLOCKS

GNSS receivers are usually equipped with *quartz crystal-controlled oscillators*, aka clocks, which are relatively inexpensive and compact. They have
low power requirements and long-life spans. For these types of clocks, the
frequency is generated by the piezoelectric effect in an oven-controlled quartz
crystal disk, a device sometimes symbolized by *OCXO*. Their reliability
ranges from a minimum of about 1 part in 108 to a maximum of about 1 part
in 1010, a drift of about 0.1 nanoseconds in 1 second. Even at that, quartz
clocks are not as stable as the atomic frequency standards in the GNSS satellites and are more sensitive to temperature changes, shock, and vibration.

THE ORBITAL BIAS, d_ρ

Orbital bias contributes an error. It is addressed in the broadcast ephemeris. The orbital motion of GNSS satellites is not only a result of the Earth's gravitational attraction; there are several other forces that act on the satellite. The primary disturbing forces are the nonspherical nature of the Earth's gravity, the attractions of the sun and the moon and solar radiation pressure. A good model of these forces is the actual motion of the satellites themselves and the government facilities distributed around the world track them for that reason, among others (see Chapter 3). For GPS, those facilities are known collectively as the Control Segment, *Ground Segment*, or the *Operational Control System (OCS)*, continuously.

THE TROPOSPHERIC EFFECT, d_{trop}

The *troposphere* is that part of the atmosphere closest to the Earth. It extends from the surface to about 9 km over the poles and about 16 km over the equator, but here will be combined with the tropopause and the stratosphere, as it is in much of GNSS literature. Therefore, the following discussion of the tropospheric effect will include the layers of the Earth's atmosphere up to about 50 km above the surface.

Tropospheric effect is independent of frequency. The troposphere and the ionosphere are by no means alike in their effect on the satellite's signal. While the troposphere is refractive, its refraction of a GNSS satellite's signal is not related to its frequency as it is in the ionosphere. The refraction is tantamount to a delay in the arrival of a GNSS satellite's signal. It can also be conceptualized as a distance added to the range the receiver measures between itself and the satellite. The troposphere is part of the electrically neutral layer of the Earth's atmosphere, meaning it is not ionized. The troposphere is also nondispersive for frequencies below 30 GHz or so. Therefore, signals in the L-band are equally refracted. This means that the range between a receiver and a satellite will be shown to be a bit longer than it is. However, as it is in the ionosphere, density affects the severity of the delay of the GNSS signal as it travels through the troposphere. When a satellite is close to the horizon, the delay of the signal caused by the troposphere is maximized, whereas the tropospheric delay of the signal from a satellite at zenith, directly above the receiver, is minimized. The situation is analogous to atmospheric refraction in astronomic observations; the effect increases as the energy passes through more of the atmosphere. The difference is that in GNSS positioning the delay, not the angular deviation, caused by the changing density of the atmosphere is of primary interest. The GNSS signal that spends the least amount of time going through the troposphere will be the least delayed by it. In other words, the height of the

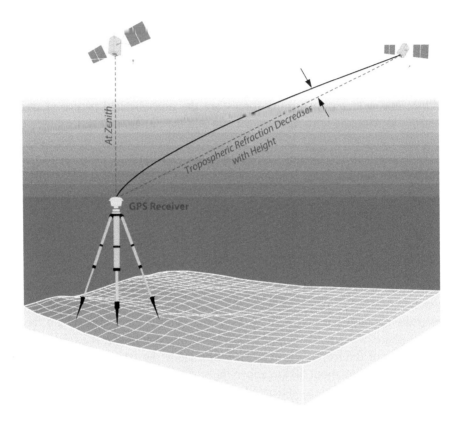

FIGURE 2.4 Tropospheric Effect

receiver affects the delay. The index of tropospheric refraction decreases as height increases (see Figure 2.4). The elevation angle and even the azimuth from which the signal arrives affect the size delay too. For example, a GNSS signal arriving from directly above the receiver will have less tropospheric delay than a signal that comes in at another path, say from an elevation angle of just 5°.

Modeling the troposphere is one technique used to reduce the bias in GPS data processing. Several have been developed. Along that line, it is important to note that refraction in the troposphere has a dry (hydrostatic) component and a wet component. The dry component contributes the larger portion of range error, perhaps 80% to 90%. The Saastamoinen model can calculate the dry component from latitude, height, and measured atmospheric pressure. The size of the delay attributable to the wet component depends on the highly variable water vapor distribution in the atmosphere. It is more difficult to quantify. Even though the wet component of the troposphere is nearer to the Earth's surface, measurements of temperature and humidity at the ground level are not strong indicators of conditions on the path between the receiver and the satellite. While instruments such as water vapor radiometers and radiosondes can be sent aloft to provide some idea of the conditions along the line between the satellite and the receiver, the high cost restricts their use to only the most high-precision GNSS work. In most cases, this aspect must remain in the purview of mathematical modeling. Therefore, some of the models can provide both the wet and the dry components and do so without any on-site measurements. They estimate temperature, pressure, and humidity from the latitude and time of any site. One was developed at the University of New Brunswick. It is called *UNB3m*. It can be used to present an example of the relative weight of the two tropospheric delay components. According to the UNB3 model, the dry component of the delay at the zenith of a receiver at the equator is about 2.3 m, whereas the wet component is about 0.27 m. The most prominent model of this type recommended by the International Earth Rotation and Reference Systems Service (IERS) is *global pressure and temperature 2 (GPT2)*. GPT2 relies on a global 5° grid of numerical mean weather data from which it provides estimates of the pressure, temperature, water vapor pressure, etc., at a receiver site.

These a priori models can be up to 95% effective when the satellites are at reasonably high elevation angles. However, the residual 5% can be quite difficult to remove. Therefore, it is advisable to limit GNSS observations to those signals above 15° or so elevation mask angle to ameliorate the effects of atmospheric delay.

There are other practical consequences of atmospheric biases. As mentioned earlier, the character of the atmosphere is never homogeneous; therefore, the importance of atmospheric modeling increases as the distance between GNSS receivers grows. Consider a signal traveling from one satellite to two receivers that are close together. That signal would be subjected to very similar atmospheric effects, and, therefore, atmospheric bias modeling would be less important to the accuracy of the measurement of the relative distance between them. On the other hand, a signal traveling from the same satellite to two receivers that are far apart may pass through levels of atmosphere quite different from one another. In that case, atmospheric bias modeling would be more important. In other words, the importance of atmospheric correction increases as the differences in the atmosphere through which the GNSS satellite signal must pass to reach the receivers increase. Such differences can generally be related to length.

MULTIPATH

Multipath is an uncorrelated error. It is a range delay symbolized by ε_{mp} in the pseudorange equation and $\varepsilon_{m\varphi}$ in the carrier-phase equation. As the name implies, it is the reception of the GNSS signal via multiple paths

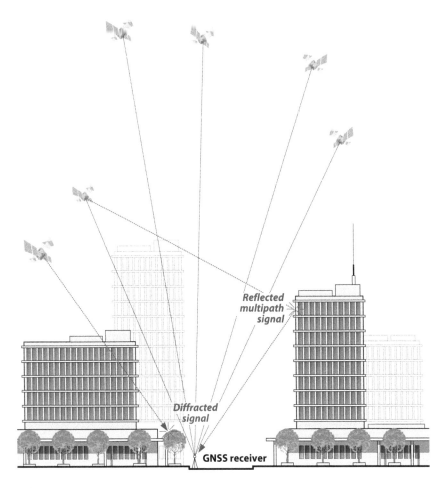

FIGURE 2.5 Multipath

rather than from a direct line of sight (see Figure 2.5). Multipath differs from both the apparent slowing of the signal through the ionosphere and troposphere and the discrepancies caused by clock offsets. The range bias from multipath is the result of the reflection and diffraction of the GNSS signal.

Multipath occurs when part of the signal from the satellite reaches the receiver after one or more reflections or scattering from the ground, a building, or another object. These reflected signals can interfere with the signal that reaches the receiver directly from the satellite. Since carrier-phase multipath is based on a fraction of the carrier wavelength and code multipath is relative to the chipping rate, the effect of multipath on pseudorange solutions is orders of magnitude larger than it is in carrier-phase solutions. The effect of multipath on a carrier-phase measurement can reach a quarter of a wavelength. The effect of multipath on a pseu-dorange measurement can reach 1.5 times the length of a chip, though it is more often a few meters. However, multipath in carrier phase is harder to mitigate than multipath in pseudoranges. Multipath can cause the cor-relation peak mentioned in Chapter 1 to become skewed. For example, the strategy of spacing the early, punctual, and late correlators at 1/10th of a chip rather than the standard 1 chip is not nearly as effective in mitigating carrier-phase multipath as it is in pseudorange solutions.

LIMITING THE EFFECT OF MULTIPATH

The high frequency of the GNSS codes tends to limit the field over which multipath can contaminate pseudorange observations. Once a receiver has achieved lock, that is, its replica code is correlated with the incoming signal from the satellite; signals outside the expected chip length can be rejected. Multipath delays of less than one chip, those that are the result of a single reflection, are the most troublesome. Fortunately, there are factors that distinguish reflected multipath signals from direct, line-of-sight, sig-nals. For example, reflected signals at the frequencies used for L1, L2, and L5 tend to be weaker and more diffuse than the directly received signals. Another difference involves the circular polarization of the GPS signal.

The signals received directly from the GPS satellites are *Right-Hand Circular Polarized, RHCP.*

RHCP means that it rotates clockwise when observed in the direc-tion of propagation. Polarized waves oscillate in more than one direction. The electrical field vectors of the GPS signal have a constant magnitude, but their direction rotates so that the electrical field vector of the wave describes a helix in the direction of propagation. Said another way, cir-cularly polarized waves are those where the angle of the electric vector rotates around an imaginary line traveling in the direction of the propaga-tion of the wave. The rotation may be either to the right or left. The GPS signal is a Right-Hand Circularly Polarized (RHCP) wave (see Figure 2.6).

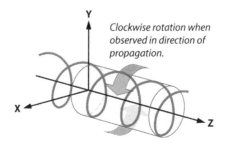

FIGURE 2.6   Right-Hand Circular Polarization (RHCP)

You can illustrate it this way. With your right hand give the thumbs up signal. Now, instead of pointing your thumb up, point it in the direction that the GPS signal is propagating. Your curling fingers show you the direction of the rotation of the field.

However, when the signal is reflected as it is in multipath, the polarization is reversed. Reflected, multipath signals become *Left-Hand Circular Polarized, LHCP*. While some multipath affected signals may be LHCP, it is possible for them to arrive at the receiver in-phase, usually through multiple reflections. Even then, however, the signal power is typically diminished substantially, and these characteristics allow some multipath signals to be identified and rejected at the antenna.

ANTENNA DESIGN AND MULTIPATH

GNSS antenna design can play a role in minimizing the effect of multipath in several ways. Ground planes, usually metal sheets, are used with many antennas to reduce multipath interference by eliminating signals from low elevation angles. Generally, larger ground planes, multiple wavelengths in size, have a more stabilizing influence than smaller ground planes. However, such ground planes do not provide much protection from the propagation of waves along the ground plane itself. When a GNSS signal's wave front arrives at the edge of an antenna's ground plane from below, it can induce a surface wave that can travel horizontally on the top of the plane, overlie the direct signal and reach the antenna (see Figure 2.7).

This problem can be mitigated using a *choke ring antenna*. Choke ring antennas, based on a design first introduced by the Jet Propulsion Laboratory (JPL), can reduce antenna gain at low elevations. The design contains a series of concentric circular troughs that are a bit more than a quarter of a wavelength deep and can prevent the formation of these surface waves. However, neither ground planes nor choke rings remove the effect of reflected signals from above the antenna very effectively. There are signal processing techniques that can reduce multipath. A widely used strategy is the 15° cutoff or *mask angle*. This technique calls for tracking satellites only after they are more than 15° above the receiver's horizon. Careful attention in placing the antenna away from reflective surfaces, such as nearby buildings, water, or vehicles, is another way to minimize the occurrence of multipath. Several antenna designs are possible in GNSS, but the satellite's signal has such a low power density, especially after propagating through the atmosphere, that antenna efficiency is critical.

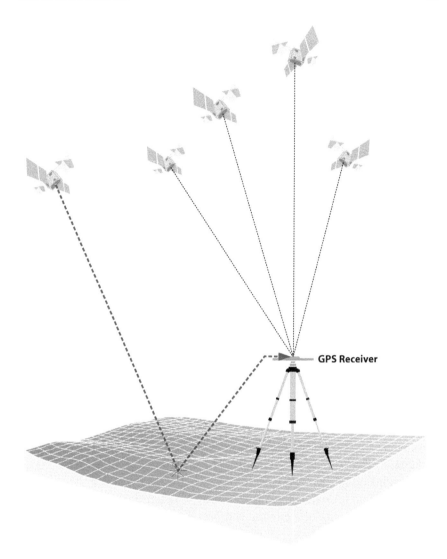

FIGURE 2.7 Multipath Propagation

RECEIVER NOISE

Receiver noise is directly related to thermal noise, dynamic stress, and so on in the GNSS receiver itself. Receiver noise is also an uncorrelated error source. The effects of receiver noise on carrier-phase measurements symbolized by ε_φ, like multipath, are small when compared to their effects on pseudorange measurements, ε_p. Generally speaking, the receiver noise error is about 1% of the wavelength of the signal involved. In other words, in code solutions, the size of the error is related to chip width. For example, the receiver noise error in a GPS C/A-code solution can be around 3 m which is about an order of magnitude more than it is in a P(Y) code solution, about 3 cm. In carrier-phase solutions, the receiver noise error contributes millimeters to the overall error.

SOME METHODS OF DATA COLLECTION

There are a variety of ways to limit the effects of the biases in GNSS work, but whether the techniques involve the methods of data collection or ways of processing the data, the objective is the management of errors.

STATIC AND KINEMATIC

GNSS work is sometimes divided into three categories; Positioning, Navigation, and Timing (PNT). Most often, GNSS surveying is concerned with the first of these, positioning. In general, there are two techniques used in surveying. They are kinematic and static. In static GNSS surveying sessions, the receivers are motionless with respect to the surface of the Earth during the observation. Because static work most often provides higher accuracy and more redundancy than kinematic work, it is often done to establish control. The results of static GNSS surveying are processed after the session is completed; they are *post-processed.*

In kinematic GNSS surveying, the receivers are either in periodic or continuous motion. Kinematic GNSS is done when real time, or near real time, results are needed. When the singular objective of kinematic work is positioning, the receivers move periodically using the start and stop methodology originated by Dr. Benjamin Remondi in the 1980s. When the receivers are in continuous motion, the objective may be determination of the location, altitude, and velocity of a moving platform (i.e., navigation) or positioning.

SINGLE POINT

Single point GNSS is the most familiar and ubiquitous application of the technology. It is the solution used by cell phones; GNSS-enabled cameras, car navigation systems, and many more devices. Single point positioning is also known as autonomous positioning, point positioning, the point solution, the navigation solution, or absolute positioning. Typically, the results of this solution are in real time or near real time. It is characterized by a single receiver measuring pseudoranges to a minimum of four satellites simultaneously (see Figure 2.8). In this solution, the receiver must rely on the information it receives from the satellite's navigation messages to learn the positions of the satellites, the satellite clock offset, the ionospheric correction, etc. Even if all the data in the navigation message contained no errors,

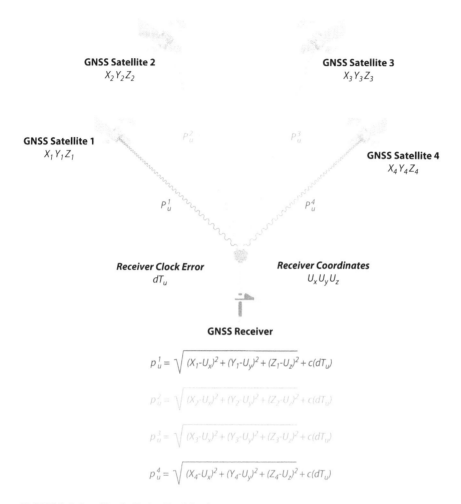

$$p_u^1 = \sqrt{(X_1-U_x)^2 + (Y_1-U_y)^2 + (Z_1-U_z)^2} + c(dT_u)$$

$$p_u^2 = \sqrt{(X_2-U_x)^2 + (Y_2-U_y)^2 + (Z_2-U_z)^2} + c(dT_u)$$

$$p_u^3 = \sqrt{(X_3-U_x)^2 + (Y_3-U_y)^2 + (Z_3-U_z)^2} + c(dT_u)$$

$$p_u^4 = \sqrt{(X_4-U_x)^2 + (Y_4-U_y)^2 + (Z_4-U_z)^2} + c(dT_u)$$

FIGURE 2.8   Single Point Positioning

four unknowns remain in the position of the receiver in three Cartesian coordinates, u_x, u_y, and u_z, and the receiver's clock error dT_u. Three pseudo-ranges provide enough data to solve for u_x, u_y, and u_z, and the fourth pseudo-range provides the information for the solution of the receiver's clock offset.

The ability to measure dT, the receiver's clock error, is one reason the moderate stability of quartz crystal clock technology is entirely adequate as a receiver oscillator. A unique solution is found here because the number of unknowns is not greater than the number of observations. The receiver tracks a minimum of four satellites simultaneously; therefore, these four equations can be solved simultaneously for every *epoch* of the observation. This meaning of epoch in GNSS is a very short period of observation time and is generally just a small part of a longer measurement. However, theoretically, there is enough information in any single epoch to solve these equations. This is the reason the trajectory of a receiver in a moving vehicle can be determined by this method. With four satellites available, the resolution of a receiver's position and velocity are both available through the simultaneous solution of these four equations. Single point positioning with its reliance on the navigation message is, in a sense, the fulfillment of the original idea of GPS.

Relative Positioning

One receiver is employed in single point positioning. A minimum of two receivers are involved in *relative* or *differential* positioning. The term differential GNSS, or DGNSS, sometimes indicates the application of this technique with coded pseudorange measurements, whereas relative GNSS sometimes indicates the application of this technique with carrier-phase measurements. However, these definitions are by no means universal and the use of the terms relative and differential GNSS are really interchangeable.

In relative positioning, one of the two receivers involved occupies a known position during the session. It is the base. The objective of the work is the determination of the position of the other, the rover, relative to the base. Both receivers observe the same constellation of satellites at the same time, and because, in typical applications, the vector between the base and the rover, known as a *baseline*, is so short compared with the orbital altitude of the GNSS satellites, there is an extensive correlation between the biases in the observations at the base and the rover so they can be significantly reduced. In other words, the two receivers record very similar errors, and since the base's position is known, corrections can be generated there that can be used to improve the solution at the rover.

DIFFERENCING

Please recall that the biases in both the pseudorange and the carrier-phase equations discussed at the top obscure the true geometric ranges between the receivers to the satellites, which then contaminate the measurement of the baseline between the receivers. In other words, to reveal the actual vectors between two or more receivers used in relative positioning, those errors must be diminished to the degree that is possible. Fortunately, some of those embedded biases can be virtually eliminated by combining the simultaneous observables from the receivers in processes known as differencing. Even though the noise is increased with each differencing operation, it is typically used in commercial data processing software for both pseudorange and carrier-phase measurements. There are three types of differencing: *single difference, double difference,* and *triple difference.*

The single difference category includes the *between-receivers single difference* and the *between-satellites single difference*.

BETWEEN-RECEIVERS SINGLE DIFFERENCE

A between-receivers single difference involves two receivers observing a single satellite (see Figure 2.9).

A between-receivers single difference reduces the effect of biases. Since the two receivers are both observing the same satellite at the same time, the difference between the satellite clock bias, dt at the first receiver and dt at the second receiver, Δdt, is obviously zero. Also, since the baseline is typically short compared with the orbital height of the GNSS satellites, the atmospheric biases, and the orbital errors, i.e., ephemeris errors, recorded by the two receivers at each end are similar. This correlation obviously decreases as the length of the baseline increases.

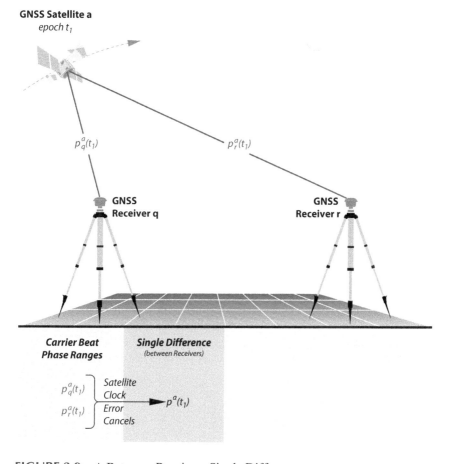

FIGURE 2.9 A Between-Receivers Single Difference

BETWEEN-SATELLITES SINGLE DIFFERENCE

The between-satellites single difference involves a single receiver observing two GNSS satellites simultaneously and the code and/or phase measurement of one satellite are differenced, subtracted, from the other (see Figure 2.10).

The data available from the between-satellites difference allows the elimination of the receiver clock error because there is only one involved. And the atmospheric effects on the two satellite signals are again nearly identical as they come into the lone receiver, so the effects of the ionospheric and tropospheric delays are reduced. However, unlike the between-receivers single difference the between-satellites single difference does not provide a better position estimate for the receiver involved. In fact, the resulting position of the receiver is not better than would be derived from single point positioning.

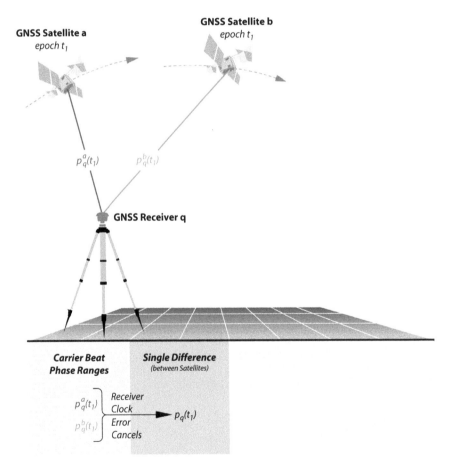

FIGURE 2.10   A Between-Satellites Single Difference

DOUBLE DIFFERENCE

When the two types of single differences are combined, the result is known as a double difference. A double difference can be said to be a between-satellite single difference of a between-receiver single difference. The improved positions from the between-receiver single difference step are not further enhanced by the combination with the between-satellite single difference. Still, including the between-satellite single difference is useful because the combination virtually eliminates clock errors; both the satellite and receiver clock errors (see Figure 2.11). The removal of the

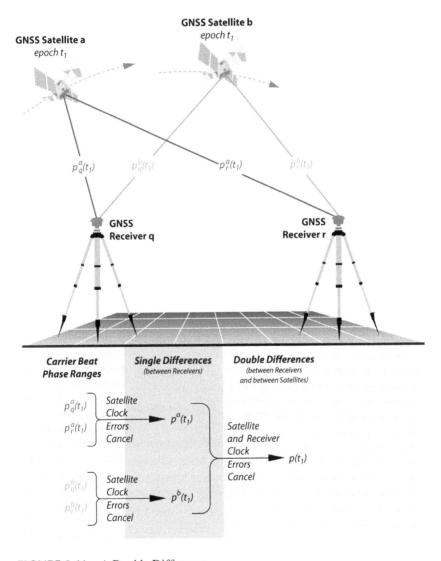

FIGURE 2.11   A Double Difference

receiver clock bias in the double difference makes it possible to segregate the errors attributable to the receiver clock biases from those from other sources. This segregation improves the efficiency of the estimation of the integer ambiguity in a carrier-phase observation, N. In other words, the reduction of all the non-integer biases makes the computation of the final accurate positions more efficient.

At this point, the double differenced carrier-phase integer ambiguities are often fixed by a method introduced by P. J. G. Teunissen and developed at the Delft University of Technology in the 1990s known as *LAMBDA* (*Least-squares AMBiguity Decorrelation Adjustment*). It is a two-step process. It is difficult to find the integer ambiguities when done in the search space supplied by the double difference. Therefore, the first step is to decorrelate them in a process also known as reduction. This produces new transformed ambiguities that cover a smaller spectrum than the original ambiguities. That way, the space within which the search for the solution is done can be scaled down. With the number of candidate coordinates reduced the estimation of the actual integer ambiguities are more easily solved. The second step is the search itself is thereby made faster and more efficient. The circumstances and redundancy of the observations themselves also affect the process. In other words, ambiguity resolution is most efficient when the baselines are short, there are lots of satellites, multiple frequencies, uninterrupted tracking, the dilution of precision is low, there is limited multipath, and the observation sessions are long. The initial ambiguity estimates yield floating-point numbers rather than integers, so the result is called a *float solution*. When some or all the ambiguities are ultimately resolved to their actual integer values, the superior *fixed solution* is achieved. Today's receivers typically accomplish ambiguity resolution quickly.

TRIPLE DIFFERENCE

A triple difference is the difference of two double differences over two different epochs. The triple difference has other names. It is also known as the *receiver satellite-time triple difference* and the *between-epochs* difference (see Figure 2.12). Triple differencing serves as a good preprocessing step because it can be used to detect and repair cycle slips.

A *cycle slip* is a discontinuity in a receiver's continuous phase lock-on a satellite's signal (see Figure 2.13). A power loss, a very low signal-to-noise ratio, a failure of the receiver software, or a malfunctioning satellite oscillator can cause a cycle slip. It can also be caused by severe ionospheric conditions. The most common cause, however, are obstructions that prevent the satellite signal from being tracked by the receiver; buildings, trees, etc. Under such circumstances, when the satellite reappears, the tracking

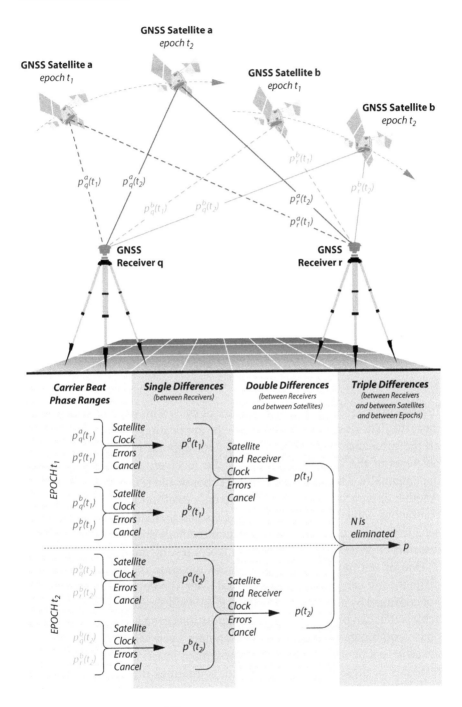

FIGURE 2.12 A Triple Difference

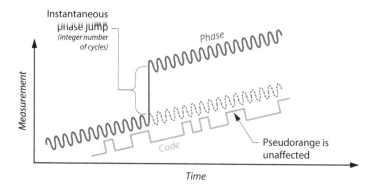

FIGURE 2.13   Cycle Slip in Double Difference

resumes. Coded pseudorange measurements are virtually immune from cycle slips, but carrier-phase positioning accuracy suffers if cycle slips are not detected and repaired. A cycle slip causes the critical component for successful carrier-phase positioning, a resolved integer ambiguity, N, to become instantly unknown again. In other words, lock is lost. When that happens, correct positioning requires that N be reestablished.

REPAIRING CYCLE SLIPS

In post-processing, the location and size of cycle slips must be determined; then, the data set can be repaired with the application of a fixed quantity to all the subsequent phase observations. One approach is to hold the initial positions of the stations occupied by the receivers as fixed and edit the data manually. This has proven to work, but would try the patience of Job. Another approach is to model the data on a satellite-dependent basis with continuous polynomials to find the breaks and then manually edit the data set a few cycles at a time. In fact, several methods are available to find the lost integer phase value, but they all involve testing quantities.

One of the most convenient of these methods is based on the triple difference. It can provide an automated cycle slip detection system that is not confused by clock drift, and once least-squares convergence has been achieved, it can provide initial station positions even using the unrepaired phase combinations. They may still contain cycle slips but the data can nevertheless be used to process approximate baseline vectors. Then the residuals of these solutions are tested, sometimes through several *iterations*. Proceeding from its own station solutions, the triple difference can predict how many cycles will occur over a particular time interval. Therefore, by evaluating triple difference residuals over that particular interval, it is not only possible to determine which satellites have integer jumps but also the number of cycles that have actually been lost. In a sound triple difference solution without cycle slips, the residuals are usually limited to fractions of a cycle. Only those containing cycle slips have residuals close to one cycle or larger. Once cycle slips are discovered, their correction can be systematic. For example, suppose the residuals of one component double difference of a triple-difference solution revealed that the residual of satellite PRN 16 minus the residual of satellite PRN 17 was 8.96 cycles. Further, suppose that the residuals from the second component double difference showed that the residual of satellite PRN 17 minus the residual of satellite PRN 20 was 14.04 cycles. Then one might remove 9 cycles from PRN 16 and 14 cycles from PRN 20 for all the subsequent epochs of the observation. However, the process might result in a common integer error for PRNS 16, 17, and 20. Still, small jumps of a couple of cycles can be detected and fixed in the double difference solutions.

Before attempting double difference solutions, the observations should be corrected for cycle slips identified from the triple difference solution. Even though small jumps undiscovered in the triple difference solution might remain in the data sets, the double difference residuals will reveal them at the epoch where they occurred. However, some conditions may prevent the resolution of cycle slips down to the one-cycle level. Inaccurate satellite ephemerides, noisy data, errors in the receiver's initial positions, or severe ionospheric effects all can limit the effectiveness of cycle-slip fixing. In difficult cases, a detailed inspection of the residuals might be the best way to locate the problem.

COMPONENTS OF THE CARRIER-PHASE OBSERVABLE

From the moment a receiver locks onto a satellite to the end of the observation, the carrier-phase observable can be divided into three parts. Two of them do not change during the session, and one of them does change (see Figure 2.14).

$$\phi = \alpha + \beta + N$$

where
    ϕ = total phase
    α = fractional initial cycle (phase measurement)
    β = observed cycle count
    N = carrier-phase ambiguity (cycle count at lock-on)

The *fractional initial phase* is established at the first instant of the lock-on. When the receiver starts tracking the satellite, it is highly unlikely to acquire the satellite's signal precisely at the beginning of a wavelength's

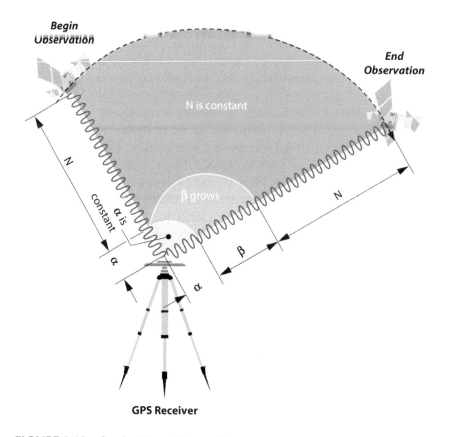

FIGURE 2.14   Carrier-Phase Observable

phase cycle. It will grab on at some fractional part of a phase, and this fractional phase will remain unchanged for the duration of the observation. It is called the fractional initial cycle, fractional instantaneous cycle, and is symbolized in the equation above by α.

The integer ambiguity N represents the number of full phase cycles between the receiver and the satellite at the first instant of the receiver's lock-on. It can also be labeled the carrier-phase ambiguity, integer-cycle ambiguity, or the cycle count at lock-on. It does not change from the moment of the lock to the end of the observation unless that lock is lost. But when there is a cycle slip, lock is lost, and by the time the receiver reacquires the signal, the normally constant integer ambiguity has changed. In that case, the receiver loses its place in its accumulated count of the integer number of cycles, symbolized as β.

The value β is the count of the number of full phase cycles coming in throughout the observation. Of course, the count changes from the moment of lock-on until the end of the observation. In other words, β is the receiver's record of the consecutive change in full phase cycles, 1, 2, 3, 4 ..., between the receiver and the satellite as the satellite flies over. If the observation proceeds without cycle slips the observed cycle count is the only one of three numbers that changes.

POST-PROCESSING

In many ways, post-processing is the heart of a GNSS control operation. On such projects some processing should be performed on a daily basis. Blunders from operators, noisy data, and unhealthy satellites can corrupt entire sessions and left undetected, such dissolution can jeopardize the whole survey. By processing daily, these weaknesses can be discovered when they can still be eliminated and a timely amendment of the observation schedule can be done if necessary. However, even after blunders and noisy data have been removed from the observation sets, GNSS measurements are still composed of fundamentally biased ranges. Therefore, GNSS data processing procedures are really a series of interconnected computerized operations designed to minimize the more difficult biases and extract the true ranges.

As has been described, the biases originate from a number of sources: imperfect clocks, atmospheric delays, cycle ambiguities in carrier-phase observations, and orbital errors. If a bias has a stable, well-understood structure, it can be estimated. In other cases, multifrequency observations can be used to measure the bias directly, as in the ionospheric delay, or a model may be used to predict an effect, as in tropospheric delay. Differencing is one of the most effective strategies in mitigating biases.

Correlation of Biases

As illustrated in the figures above, when two or more receivers observe the same satellite constellation simultaneously, a set of correlated vectors is created between the co-observing stations. Most GNSS networks consist of many sets of correlated vectors from each individual session. The longest baselines between stations on the Earth are usually relatively short when compared with the distances from the receivers to the GNSS satellites. Therefore, even when several receivers are set up on somewhat widely spaced stations, as long as the data is collected by the receivers simultaneously from the same constellation of satellites, they will record very similar errors. The resulting vectors will be correlated. It is the simultaneity of observation and the correlation of the carrier-phase observables that make the extraordinary GNSS accuracies possible because biases that are linearly correlated can be mitigated by differencing the data sets of a session.

ORGANIZATION IS ESSENTIAL

One of the difficulties of post-processing is the amount of data that must be managed, a structured approach must be implemented. One aspect of that structure is the naming of the GNSS receiver's observation files. Many manufacturers recommend a file naming format that can be symbolized by *pppp-ddd-s.yyf*. The first letters (*pppp*) of the file name indicate the point number of the station occupied. The day of the year, or Julian date, can be accommodated in the next three places (*ddd*), and the final place left of the period is the session number (*s*). The year (*yy*) and the file type (*f*) are sometimes added to the right of the period.

The first step in GNSS data processing is downloading the collected data from the internal memory of the receiver or data logger into a computer. When the observation sessions have been completed for the day, the data from each receiver is transferred. Nearly all GNSS systems used in surveying can accommodate post-processing in the field, but none can protect the user from a failure to back up this raw observational data onto some other form of semi-permanent storage.

Receiver and data logger memory capacity is sometimes limited, and it may be necessary to clear older data to make room for new sessions. Still, it is a good policy to create the necessary space with the minimum deletion and restrict it to only the oldest files in the memory. In this way, the recent data can be retained as long as possible, and the data can provide an auxiliary backup system. However, when a receiver's memory is finally wiped of a particular session, if redundant raw data are not available, re-observation may be the only remedy.

In most GNSS receivers, the navigation message, meteorological data, the observables, and really all the raw data are usually in a proprietary manufacturer-specific file format which is used to make it difficult to combine data from other vendors' receivers. This led some years ago to a proposal for an independent file format that allowed the easy exchange of the data collected by receivers from different manufacturers. It came to be called *RINEX* (*Receiver Independent Exchange Format*). The first version was developed by W. Gurtner in 1989 and published by W. Gurtner and G. Mader in 1990. Since then, GNSS data can be available in both a proprietary format and in RINEX.

The raw data are usually saved in several distinct files. For example, the computer operator will likely find that the phase measurements downloaded from the receiver will reside in one file and the satellite's ephemeris data in another, and so forth. Likewise, the measured pseudorange information may be found in its own dedicated file, the ionospheric information in another, and so on. The particular division of the raw data files is designed to accommodate the suite of processing software and the data management system that the manufacturer has provided its customers, so each will be somewhat unique.

CONTROL

All post-processing software suites require control. GPS static work sometimes satisfies that requirement by inclusion of passive NGS monuments in the network design. In this approach, the control stations are occupied by the surveyor building the network, but this will become less satisfactory as the North American Terrestrial Reference Frame 2022 (NATRF2022) and the other three TRFs are implemented by NGS. There is an alternative. Continuously Operating Reference Stations (CORS) will become the public's primary direct access to the new TRFs. They are known as *active control*. Their data can be used to support carrier-phase GPS static control surveys. More and more NGS CORS are tracking both GPS and GLONASS satellites. There are also other networks of CORS around the world operated by other organizations, i.e., the *International GNSS Service (IGS)*. The IGS is a collaboration between more than 350 organizations in more than 118 countries. It includes agencies, universities, and research institutions and operates a worldwide GNSS tracking network of over 500 CORS. For those doing their own post-processing rather than using an online processing service the most direct method of utilizing CORS is to download the data files posted on the Internet. The CORS data collected during the time of the survey can be combined with those collected in the field. They can be used to post-process the baselines and derive positions for the new points.

LEAST-SQUARES ADJUSTMENT

There are numerous adjustment techniques, but *least-squares* adjustment is the most precise and most used in GNSS. The foundation of least-squares adjustment is the idea that the sum of the squares of all the residuals applied to the GNSS vectors in their final adjustment should be held to the absolute minimum. Minimizing this sum requires first defining those residuals approximately. Therefore, in GNSS, the process is based on equations where the observations are expressed as a function of unknown parameters but parameters that are nonetheless given approximate initial values. Then by adding the squares of the terms thus formed and differentiating their sum, the derivatives can be set equal to zero. For complex work like GNSS adjustments, the least-squares method has the advantage that it allows for the smallest possible changes to the original estimated values.

Whereas random errors adhere to the principles of probability mistakes and blunders, systematic errors, do not. The latter should be eliminated from the dataset, such that only random errors remain—to the degree possible—before least-squares adjustment is attempted. A downside of least-squares adjustment is its tendency to spread the effects of even one mistake throughout the work. In other words, it can cause large residuals to show up for several measurements that are correct, and when that happens, it can be hard to know exactly what is wrong. The adjustment may fail the *chi-square* test. That tells you there is a problem, but unfortunately, it cannot tell you where the problem is. The chi-square test is based on probability, and it can fail when unmodeled biases such as multipath remain in the measurements.

Most programs look at the residuals in light of a limit at a specific probability, and when a particular measurement goes over the limit, it gets highlighted. Trouble is, you cannot always be sure that the one that got tagged is the one that is the problem. Fortunately, least-squares do offer a high degree of comfort once all the hurdles have been cleared. If the residuals are within reason and the chi-square test is passed, it is very likely that the observations have been adjusted properly.

NETWORK ADJUSTMENT

The solution strategies of GNSS adjustments themselves are best left to particular suites of software. Suffice it to say that the single baseline approach, that is, a baseline-by-baseline adjustment, has the disadvantage of ignoring the actual correlation of the observations of simultaneously occupied baselines. An alternative approach involves a network adjustment approach where the correlation between the baselines themselves can be more easily considered. For the most meaningful network adjustment, the endpoints of every possible baseline should be connected to at least two other stations. Thereby, the quality of the work itself can be more realistically evaluated. For example, the most common observational mistake, the mismeasured antenna height or HI, is very difficult to detect when adjusting baselines sequentially, one at a time, but a network solution spots such blunders more quickly. Most GNSS post-processing adjustments begin with a minimally constrained least-squares adjustment. That means that all the observations in a network are adjusted together with only the constraints necessary to achieve a meaningful solution; for example, the adjustment of a GNSS network with the coordinates of only one station is fixed. The purpose of the minimally constrained approach is to detect large mistakes, blunders, like a misidentifying one of the stations. The residuals from a minimally constrained work should come pretty close to the precision of the observations themselves. If the residuals are particularly large, there are probably mistakes; if they are small, the network itself may not be a strong as it should be. This minimally constrained solution is usually followed by an over-constrained solution. In other words, a least-squares adjustment where the coordinate values of more than one selected control station are held fixed.

USING AN ONLINE PROCESSING SERVICE

An alternative to the just described in-house processing is the use of an online processing service. There are several services available to GNSS surveyors. While they differ somewhat in their requirements, they are all based on the same idea. Static GNSS data collected in the field may be uploaded to the hosting organization's website from which the final positions are returned to the user often free of charge.

There can be advantages to the use of such a processing service. Aside from removing the necessity of having post-processing software in-house, there is the strength of the network solution available from them. In other words, rather than being processed against a single CORS in the vicinity of the work, the surveyor's data is often processed against a group, perhaps three, of the nearest CORS. Such a network-based solution improves the integrity of the final position markedly and may compensate for long baselines required by the sparseness of the CORS in the area of interest. Among the online resources available for processing services is the *National Geodetic Surveys Online Positioning User Service (OPUS)*. This service allows the user to submit RINEX 2 or RINEX 3 GPS files through an NGS OPUS upload Web page. Perhaps a bit of explanation is appropriate here.

The RINEX format has evolved since it was first rolled out in 1990. RINEX 2 became available in 1993. RINEX 3 was created in 2006/2007 to provide support for all GNSS constellations, signals, and observations. However, it is important to note that while there are plans for OPUS to incorporate GNSS constellations besides GPS, it will only process GPS data at this point. After the data file is uploaded, it is processed, and the computed *National Spatial Reference System (NSRS)* position is emailed back to the user, but the submitted data file must meet some requirements. Here are a few of them. It must be from a static observation with a length between 15 minutes and 48 hours. The observables must be carrier-phase multifrequency (L1/L2) full wavelength. The submitted epoch interval can be 1, 2, 3, 5, 10, 15, or 30 seconds but OPUS decimates all of them to 30 seconds. Only satellites at 10° above the horizon or higher are used. It is also very important that the *height of instrument (HI)* be right. In this case, that means entering the vertical height in meters from the mark to the *Antenna Reference Point (ARP)*. The ARP is that point on the antenna usually marked where the vertical axis of the antenna intersects the bottom mount, ground plane, or choke ring. It is designed to be easily accessible so its height can be measured conveniently, but it is important that the correct antenna type is also submitted to OPUS because there is a critical offset. That offset is a short vector from the ARP to the phase center of the antenna. Since the position measured by the receiver is at the phase center of the receiver's antenna, not the ARP, getting the *Phase Center Offset* (*PCO*) right is crucial. OPUS has models to do that, but only if it is given the correct antenna type. Additional OPUS options include choosing your own base stations. With the correct inputs in place, any GPS data file collected within the jurisdiction of the U.S. National Geodetic Survey will be processed automatically with NGS computers and software utilizing data from three CORS.

There are other similar services, such as Geoscience Australia's *Australian Online GPS Processing Service*, AUSPOS. It is also a free online data processing facility that utilizes the IGS Stations Network. Unlike OPUS, AUSPOS works with data collected anywhere on Earth.

SUMMARY

Relative positioning by carrier beat phase measurement is the primary vehicle for high-accuracy GNSS control surveying. Simultaneous static observations, often supported by CORS, double differencing in post-processing, and the subsequent construction of networks from GNSS baselines are the hallmarks of geodetic and control work in the field. The strengths of these methods generally outweigh their weaknesses, particularly where there can be an unobstructed sky and relatively short baselines and where the length of observation sessions is not severely restricted. Differencing is an ingenious approach to minimizing the effect of biases in GNSS measurements. It is a technique that largely overcomes the impossibility of perfect time synchronization. Double differencing is the most widely used formulation. Double differencing still contains the initial integer ambiguities, of course. The estimates of the ambiguities generated by the initial processing are usually not integers; in other words, some orbital errors and atmospheric errors remain; nevertheless it is possible to resolve the estimates for the ambiguities to the correct integers during subsequent processing. When the integers are fixed, the results are known as a fixed solution rather than a float solution. It is the fixed solution that makes the very high accuracy possible with GNSS. However, it is important to remember that multipath, cycle slips, incorrect instrument heights, and a score of other errors can be minimized or eliminated by good practice, which is fortunate because their resolution is not within the purview of differencing at all. The unavoidable biases that can be managed by differencing—including clock, atmospheric, and orbital errors—can have their effects drastically reduced by the proper selection of baselines, the optimal length of the observation sessions, and several other considerations included in the design of a GNSS survey. Such decisions require an understanding of the sources of these biases and the conditions that govern their magnitudes. The adage of "garbage in, garbage out" is as true of GNSS as any other surveying procedure. The management of errors cannot be relegated to mathematics alone.

EXERCISES

1. The Earth's atmosphere affects the signals from GNSS satellites as they pass through. Which of these statements about the phenomenon is true?
   a. The GNSS signals are refracted by the ionosphere. Considering pseudoranges, the apparent length traveled by a GNSS signal seems to be too short, and in the case of the carrier wave, it seems to be too long.
   b. The ionosphere is dispersive. Therefore, despite the number of charged particles the GNSS signal encounters on its way from the satellite to the receiver, the time delay is directly proportional to the square of the transmission frequency. Lower frequencies are less affected by the ionosphere.
   c. The ionosphere is a region of the atmosphere where free electrons do not affect radio wave propagation due mostly to the effects of solar ultraviolet and x-ray radiation.
   d. The TEC depends on the user's location, observing direction, the magnetic activity, the sunspot cycle, the season, and the time of day, among other things. In the midlatitudes, the ionospheric effect is usually least between midnight and early morning and most at local noon.

2. The clocks in GNSS satellites are rubidium and cesium oscillators, whereas quartz crystal oscillators provide the frequency standard in most GNSS receivers. Concerning the relationship between these standards, which of the following statements is not true?
   a. Quartz clocks are not as stable as the atomic frequency standards in the GNSS satellites and are more sensitive to temperature changes, shock, and vibration.
   b. Both GNSS receivers and satellites rely on their oscillators to provide a stable reference so that other frequencies of the system can be generated from or compared with them.
   c. The foundation of the oscillators in GNSS receivers and satellites is the piezoelectric effect.
   d. The oscillators in the GNSS satellites are also known as atomic clocks.

3. What is an advantage available to a user with a multifrequency GPS receiver that is not available to one using a single-frequency GPS receiver?

   a. A single-frequency GPS receiver cannot collect enough data to perform single, double, or triple difference solutions.

   b. A multifrequency receiver affords an opportunity to decrypt the P(Y) code, but a single-frequency receiver cannot.

   c. A multifrequency receiver has access to the Navigation code, but a single-frequency receiver does not.

   d. Over long baselines, a multifrequency receiver can rely on the dispersive nature of the ionosphere to model and reduce the ionospheric bias, whereas a single-frequency receiver cannot.

4. Which of the following statements is not correct concerning factors affecting the severity of the troposphere's effect on a GNSS signal?

   a. The troposphere is nondispersive for frequencies under 30 GHz.

   b. When a GNSS satellite is near the horizon, its signal is most affected by the atmosphere.

   c. The density of the troposphere governs the severity of its effect on a GNSS signal.

   d. The wet component of refraction in the troposphere contributes the larger portion of the range error.

5. In the between-receivers single difference across a short baseline, which of the following problems are not virtually eliminated?

   a. Satellite clock errors

   b. Atmospheric bias

   c. Integer cycle ambiguities

   d. Orbital errors

6. In the initial double difference across a short baseline, which of the following problems are not virtually eliminated?

   a. Atmospheric bias

   b. Satellite clock errors

   c. Receiver clock errors

   d. Integer ambiguity

7. Which of the following statements would not correctly complete the sentence beginning, "If cycle slips occurred in the observations …?"

   a. "… they could have been caused by intermittent power to the receiver."

b.   "… they might be detected by triple differencing."
c.   "… they are equally troublesome in pseudorange and carrier-phase measurements."
d.   "… it is best if they are repaired before double differencing is done."

8. What is the correct value for the maximum time interval allowed between GPS Time and a particular GPS satellite's onboard clock?
   a.   1 nanosecond
   b.   1 millisecond
   c.   1 microsecond
   d.   1 femtosecond

9. Which of the following biases are not mitigated by using relative positioning GNSS?
   a.   The ionospheric effect
   b.   The tropospheric effect
   c.   Multipath
   d.   Satellite clock errors

10. Which one of the following statements correctly defines TEC?
    a.   A total electron content in troposphere
    b.   A measure of the number of free electrons in a column through the ionosphere with a cross-sectional area of one square meter
    c.   A measurement unit for the density of the geoid
    d.   The total number of electrons in the F layer

11. Which statement concerning triple differences is most correct?
    a.   Triple differences are the difference between the carrier-phase observations of two receivers of the same satellite at the same epoch.
    b.   Triple differences are the differences of two single differences of the same epoch that refer to two different satellites.
    c.   Triple differences are the difference of two double differences of two different epochs.
    d.   Triple difference does depend on the integer ambiguity because this unknown is not constant in time.

12. What would cause the integer ambiguity, N, to change after a receiver has achieved lock-on?
    a.   An incorrect HI.
    b.   The setting of one of the satellites being observed.

    c. Inaccurate centering over the station.

    d. A cycle slip.

13. What is the difference between a float and a fixed solution?
    a. The integer ambiguity is resolved in a fixed solution but not in a float solution.
    b. The float solution is processed using a single difference, unlike a fixed solution.
    c. The float solution is more accurate than a fixed solution.
    d. The float solution may have cycle slips but not a fixed solution.

14. What is the idea underlying the use of least-squares adjustment of GNSS networks?
    a. The sum of the squares of all the residuals applied to the GNSS vectors in their final adjustment should be held to zero.
    b. The multiplication of the GNSS vectors by all the residuals in their final adjustment should be held to one.
    c. The sum of the squares of all the residuals applied to the GNSS vectors in their final adjustment should be held to the absolute minimum.
    d. The least of the squares of all the residuals applied to the GNSS vectors in their final adjustment should be held to zero.

15. What is usually the purpose of the initial minimally constrained least-squares adjustment in GNSS work?
    a. To repair cycle slips.
    b. To fix the integer ambiguity.
    c. To establish the correct coordinates of all the project points.
    d. To find any large mistakes in the work.

16. What may cause an adjustment to fail the chi-square test?
    a. Unmodeled biases in the measurements.
    b. Smaller than expected residuals.
    c. A measurement with too many epochs.
    d. The lack of cycle slips.

ANSWERS AND EXPLANATIONS

  1. Answer is (d)

    Explanation: The modulations on carrier waves encountering the free electrons in the ionosphere are retarded, but the carrier wave itself is advanced. The slowing is known as the ionospheric delay. In the case of pseudoranges, the apparent length of the path of the signal is stretched; it seems to be too long. However, when considering the carrier wave, the path seems to be too short. It is interesting that the absolute value of these apparent changes is almost exactly equivalent. The ionosphere is dispersive. Refraction in the ionosphere is a function of a wave's frequency, the higher the frequency, the less the refraction, and the shorter the delay in the case of modulations on the carrier wave. The ionospheric effect is proportional to the inverse of the frequency squared. The magnitude of the ionospheric effect increases with electron density. Also known as the total electron content or TEC, the electron density varies in both time and space. However, in the midlatitudes, the minimum most often occurs from midnight to early morning and the maximum near local noon.

2. Answer is (c)

   Explanation: Rubidium and cesium atomic clocks are aboard the current constellation of GNSS satellites. Their operation is based on the resonant transition frequency of the Rb-87 and CS-133 atoms, respectively. On the other hand, the quartz crystal oscillators in most GNSS receivers utilize the piezoelectric effect. Both GNSS receivers and satellites rely on their oscillators to provide a stable reference so that other frequencies of the system can be generated from or compared with them.

3. Answer is (d)

   Explanation: Sufficient data to calculate single, double, and triple differences can be available from both single and multifrequency receivers. The permission to decrypt the P(Y) code is not restricted by single or multifrequency capability but rather by Department of Defense authorization. GPS receivers have access to the navigation code whether they track L1, L2, L5, or all three. A single-frequency receiver's ionospheric correction can be computed from an ionospheric model. GPS satellites transmit coefficients for ionospheric corrections that most receivers' software use. The models assume a standard distribution of the total electron content; still, with this strategy, about a quarter of the ionosphere's actual variance will be missed. However, when receivers are close together, say 30 km or less, for all practical purposes, the ionospheric delay and carrier-phase advance is the same for both receivers. Therefore, phase difference observations over short baselines are little affected by ionospheric bias with single-frequency receivers. However, that is not the case over long baselines. Multifrequency receivers can utilize the dispersive nature of the ionosphere to overcome the effects. The resulting time delay is inversely proportional to the square of the transmission frequency. That means that L1, 1575.42 MHz, is not affected as much as L2, 1227.60 MHz, and L2 is not affected as much as L5, 1176.45 MHz. This fact provides one of the greatest advantages of a multifrequency receiver over single-frequency receivers. By tracking all the carriers, a multifrequency receiver has the facility of modeling and reducing much of the ionospheric bias.

4. Answer is (d)

Explanation: The troposphere is nondispersive for frequencies under 30 GHz. Refraction in the troposphere has a dry component and a wet component. The dry component is related to the atmospheric pressure and contributes about 90% of the effect. It is more easily modeled than the wet component. The GNSS signal that travels the shortest path through the atmosphere will be the least affected by it. Therefore, the tropospheric delay is least at the zenith and most near the horizon. GNSS receivers at the ends of short baselines collect signals that pass through substantially the same atmosphere and the tropospheric delay may not be troublesome. However, the atmosphere may be very different at the ends of long baselines.

5. Answer is (c)

Explanation: A between-receivers single difference is the difference in the simultaneous measurements from one GNSS satellite as measured by two different receivers during a single epoch. Single differences are virtually free of satellite clock errors. The atmospheric biases and the orbital errors recorded by the two receivers in this solution are nearly identical, so they, too, can be virtually eliminated. However, processing does not usually end at single differences in surveying applications because the difference between the integer cycle ambiguities at each receiver and the difference between the receiver clock errors remain in the solution.

6. Answer is (d)

Explanation: The differences of two single differences in the same epoch using two satellites are known as a double difference. The initial double difference combination is virtually free of receiver clock errors, satellite clock errors, and over short baselines, orbital and atmospheric biases. However, the integer cycle ambiguities remain.

7. Answer is (c)

   Explanation: A cycle slip is a discontinuity in a receiver's continuous phase lock-on a satellite's signal. When lock is lost, a cycle slip occurs. A power loss, an obstruction, a very low signal-to-noise ratio, or any other event that breaks the receiver's continuous reception of the satellite's signal can cause a cycle slip. The coded pseudorange measurement is virtually immune from this difficulty, but carrier-phase measurements are not. It is best if cycle slips are removed before the double difference solution. When a large residual appears in the component double differences of a triple difference, it is very likely caused by a cycle slip. This utility in finding and fixing cycle slips is a primary appeal of the triple difference. It can be used as a preprocessing step to weed out cycle slips and provide a first position for the receivers.

8. Answer is (b)

   Explanation: The GPS Time system is composed of the Master Control Clock and the GPS satellite clocks and is measured from 24:00:00, January 5, 1980. While GPS Time itself is kept within one microsecond of UTC, excepting leap seconds, the satellite clocks can be allowed to drift up to a millisecond from GPS Time. The rates of these onboard rubidium and cesium oscillators are more stable if they are not disturbed by frequent tweaking and adjustment is kept to a minimum.

9. Answer is (c)

   Explanation: Two or more GNSS receivers observing the same constellation of satellites simultaneously make relative positioning possible. These techniques can attain high accuracy in large part due to the extensive correlation between observations taken to the same satellites at the same time from separate stations. The distance between such stations on the Earth is short compared with the altitude of the GNSS satellites' two receivers operating simultaneously, collecting signals from the same satellites and through substantially the same atmosphere will record very similar errors of several categories. However, multipath is so dependent on the geometry of the location of a receiver it cannot be lessened in this way. Keeping the antenna from reflective surfaces, the use of a mask angle or the use of a ground plane or choke ring antenna are methods used to reduce multipath errors.

10. Answer is (b)

    Explanation: This density is often described as total electron content or TEC, a measure of the number of free electrons in a column through the ionosphere with a cross-sectional area of 1 square meter: 10^{16} is one TEC unit.

11. Answer is (c)

    Explanation: A triple difference is created by differencing two double differences at each end of the baseline. Each of the double differences involves two satellites and two receivers. A triple difference considers two double differences over two consecutive epochs. In other words, triple differences are formed by sequentially differencing double differences in time.

12. Answer is (d)

    Explanation: The integer ambiguity, usually symbolized by N, represents the number of full phase cycles between the receiver and the satellite at the first instant of the receiver's lock-on. N does not change from the moment of the lock is achieved unless there is a cycle slip.

13. Answer is (a)

    Explanation: The initial ambiguity estimates yield floating-point numbers rather than integers, so the result is called a float solution. When the process continues with some or all the ambiguities being resolved to their actual integer values, the superior fixed solution is achieved. Today's receivers typically accomplish ambiguity resolution quickly.

14. Answer is (c)

    Explanation: The idea at the foundation of least-squares adjustment is that the sum of the squares of all the residuals applied to the GNSS vectors in their final adjustment should be held to the absolute minimum.

15. Answer is (d)

    Explanation: Most GNSS post-processing adjustment begins with a minimally constrained least-squares adjustment. That means that all the observations in a network are adjusted together with only the constraints necessary to achieve a meaningful solution; for example, the adjustment of a GNSS network with the coordinates of only one station is fixed. The purpose of the minimally constrained approach is to detect large mistakes, like a misidentifying one of the stations. The residuals from a minimally constrained work should come close to the precision of the observations themselves. If the residuals are particularly large, there are probably mistakes; if they are small, the network itself may not be a strong as it should be.

16. Answer is (a)

   Explanation: The adjustment may fail the chi-square test. That tells you there is a problem; unfortunately, it cannot tell you where the problem is. The chi-square test is based on probability, and it can fail because there are still unmodeled biases in the measurements. Most programs look at the residuals considering a limit at a specific probability, and when a particular measurement goes over the limit, it gets highlighted. Trouble is you cannot always be sure that the one that got tagged is the one that is the problem. Fortunately, least-squares offer a high degree of comfort once all the hurdles have been cleared. If the residuals are within reason and the chi-square test is passed, it is very likely that the observations have been adjusted properly.

3 The Framework

TECHNOLOGICAL FORERUNNERS

CONSOLIDATION

In the early 1970s, the Department of Defense (DOD) commissioned a study to define its future positioning needs. That study found nearly 120 different types of positioning systems in place, all limited by their special and localized requirements. The study called for consolidation and NAVSTAR GPS (NAVigation System with Timing And Ranging, Global Positioning System) was proposed. Specifications for the new system were developed to build on the strengths and avoid the weaknesses of its forerunners. Here is a brief look at some of the earlier systems and their technological contributions toward the development of GPS.

TERRESTRIAL RADIO POSITIONING

Long before the satellite era, the developers of RADAR (RAdio Detecting And Ranging) were working out many of the concepts and terms still used in electronic positioning today. For example, the classification of the radio portion of the electromagnetic spectrum by letters, such as the L-band now used in naming the GPS carriers, was introduced during World War II to maintain military secrecy about the new technology. Actually, the 23-cm wavelength that was originally used for search radar was given the L designation because it was long compared to the shorter 10 cm wavelengths introduced later. The shorter wavelength was called *S-band*, the S for short. The Germans used even shorter wavelengths of 1.5 cm. They were called K-band, for kurtz, meaning short in German. Wavelengths that were neither long nor short were given the letter C, for compromise, and P-band, for previous, was used to refer to the very first meter-length wavelengths used in radar. There is also an X-band radar used in fire-control radars and other applications. In any case, the concept of measuring distance with electromagnetic signals (ranging in GPS) had one of its earliest practical applications in RADAR. Since then, there have been several incarnations of the idea.

SHORAN (SHOrt RAnge Navigation), a method of electronic ranging using pulsed 300 MHz very high frequency (VHF) signals, was designed for bomber navigation but was later adapted to more benign uses. The system

DOI: 10.1201/9781003405238-3

depended on a signal, sent by a mobile transmitter receiver-indicator unit, being returned to it by a fixed transponder. The elapsed time of the round trip was then converted into distances. It was not long before the method was adapted for use in surveying. Using SHORAN from 1949 to 1957, Canadian geodesists were able to achieve precisions as high as 1:56,000 on lines of several hundred kilometers. It was used in hydrographic surveys in 1945 by the United States Coast and Geodetic Survey. In 1951, SHORAN was used to locate islands off Alaska in the Bering Sea that were beyond positioning by visual means. Also, in the early 1950s, the U.S. Air Force created a SHORAN measured trilateration net between Florida and Puerto Rico. It was continued on to Trinidad and South America. That success led to the development of HIgh-precision ShoRAN (HIRAN). HIRAN's pulsed signal was more focused, its amplitude more precise, and its phase measurements more accurate. It was also applied to geodesy and was used to make the first connection between Africa, Crete, and Rhodes in 1943. Its most spectacular application was the arcs of triangulation joining the North American Datum (1927) with the European Datum (1950) in the early 1950s. By knitting together continental datums, HIRAN surveying might be considered to be the first practical step toward positioning on a truly global scale.

SATELLITE ADVANTAGES

These and other radio navigation systems proved that ranges derived from accurate timing of electromagnetic radiation were viable but useful as they were in geodesy and air navigation, they only whet the appetite for a higher platform. In 1957, the development and successful launch of *Sputnik 1*, the first Earth-orbiting satellite, led to a couple of realizations. First, if you were at a known location and measured the Doppler shift of its 20 MHz radio signals, you could work out the satellite's orbit. From that, it followed that if you knew the satellite's orbit then the shift could tell you your location. On top of that, it was clear that the potential for satellite coverage was virtually unlimited. The coverage of a terrestrial radio navigation system is limited by the propagation characteristics of electromagnetic radiation near the ground. To achieve long ranges, the ellipsoidal shape of the Earth favors low frequencies that stay close to the surface. One such terrestrial radio navigation system, LOng RAnge Navigation System-C (LORAN-C) was used to determine speeds and positions of receivers up to 3000 km from fixed transmitters. Unfortunately, its frequency had to be in the low-frequency range from 90 to 110 kHz. Many nations besides the United States used LORAN including Japan, Canada, and several European countries. Russia has a similar system called Chayka.

Omega, another low-frequency hyperbolic radio navigation system, was operated from 1968 to September 30, 1997, by the United States Coast Guard.

Omega was used by other countries too. It was capable of ranges of 9000 km. Its 10- to 14-kHz frequency was so low as to be audible (the range of human hearing is about 20 Hz to 15 kHz). Such low frequencies can be profoundly affected by unpredictable ionospheric disturbances and ground conductivity making modeling the reduced propagation velocity of a radio signal over land difficult. However, higher frequencies require line of sight. Line of sight is much less of a problem for Earth-orbiting satellites, of course. Signals from space overcome other low-frequency limitations too. It is possible to use a broader range of frequencies from space-based transmitters, such transmissions can be more reliable and can achieve almost limitless coverage. However, the development of the technology for launching transmitters with sophisticated frequency standards into orbit was not accomplished immediately.

OPTICAL SYSTEMS

Some of the earliest extraterrestrial positioning was done with optical systems. Optical tracking of satellites is a logical extension of astronomy. The astronomic determination of a position on the Earth's surface from star observations, certainly the oldest method, is actually very similar to extrapolating the position of a satellite from a photograph of it crossing the night sky. In fact, the astronomical coordinates, right ascension α and *declination* δ, of such a satellite image are calculated from the background of fixed stars. Photographic images that combine reflective satellites and fixed stars are taken with *ballistic cameras* whose chopping shutters open and close very fast. The technique causes both the satellites, illuminated by sunlight or Earth-based beacons, and the fixed stars to appear on the plate as a series of dots. Comparative analysis of photographs provides data to calculate the orbit of the satellite. Photographs of the same satellite made by cameras thousands of kilometers apart can thus be used to determine the camera's positions by triangulation. The accuracy of such networks has been estimated as high as ±5 m.

Other optical systems are much more accurate. One called *satellite laser ranging (SLR)* is similar to measuring the distance to a satellite using a sophisticated EDM. A laser aimed from ground-based tracking stations to satellites equipped with *laser retro-reflector arrays (LRA)* yields the range. In the early 1990s LRAs were installed on two GPS satellites, SVN 35 (PRN 05) and SVN 36 (PRN 06). The measurements made to the onboard corner cube reflectors allowed ground stations to separate the effect of errors attributable to satellite clocks from errors in the satellite's ephemerides. Both of those two satellites are long out of service, but the contribution of laser ranging to precise orbit determination was proven. Nearly all new GNSS satellites have LRAs that includes all the GLONASS, Galileo,

BeiDou, and NavIC satellites. The GPS Block III satellites, starting with SV-9, also carry retro-reflectors.

The same technique, called lunar laser ranging (*LLR*), is used to measure distances to the moon using corner cube reflector arrays left there during missions. There are five available arrays. Three of them set during Apollo missions, one during the Soviet Lunokhod 1 mission (rediscovered in 2010) and one from the Soviet Lunokhod 2 mission. More are planned. These techniques can achieve positions of millimeter precision when information is gathered from several stations. However, one drawback is that the observations must be spread over long periods, up to a month, and they, of course, depend on two-way measurement.

While some optical methods, like SLR and LLR, can achieve extraordinary accuracies, they can, at the same time, be subject to some chronic difficulties. Some methods require skies to be clear simultaneously at widely spaced sites. Even then, local atmospheric turbulence causes the images of the satellites to scintillate.

EXTRATERRESTRIAL RADIO POSITIONING

The earliest American extraterrestrial systems were designed to assist in satellite tracking and satellite orbit determination, not geodesy. Some of the methods used did not find their way into the GPS technology at all. Some early systems relied on the reflection of signals and transmissions from ground stations that would either bounce off the satellite or stimulate onboard transponders. Systems that required the user to broadcast a signal from the Earth to the satellite were not favorably considered in designing the then-new GPS system. Any requirement that the user reveals his position was not attractive to the military planners responsible for developing GPS. They favored a passive system that allowed the user to simply receive the satellite's signal. So, with their two-way measurements and utilization of several frequencies to resolve the integer ambiguity, many early extraterrestrial tracking systems were harbingers of modern EDM technology more than GPS.

MINITRACK

But elsewhere, there were ranging techniques useful to GPS. NASA's first satellite tracking system, *Minitrack*, relied on phase difference measurements of a single-frequency carrier broadcast by the satellites themselves and received by two separate ground-based antennas. This technique is called *interferometry*. Interferometry is the measurement of the interference between the phases of signals that originate at a common source but travel different paths to the receivers. The combination of such signals, collected by two separate receivers, invariably finds them out of step since one has traveled a longer distance than the other. Analysis of the signal's phase difference can yield very accurate ranges, and interferometry has become an indispensable measurement technique in several scientific fields.

VERY LONG BASELINE INTERFEROMETRY

For example, *very long baseline interferometry (VLBI)* did not originate in the field of satellite tracking or aircraft navigation but in radio astronomy. The technique was so successful it is still in use today. Radio telescopes, sometimes on different continents, record the microwave signals from quasars, star-like points of light many light-years from Earth (see Figure 3.1). These recordings are encoded with time tags controlled by hydrogen masers, the most stable of oscillators (clocks). The recordings are then brought together and played back at a central processor. Cross-correlation

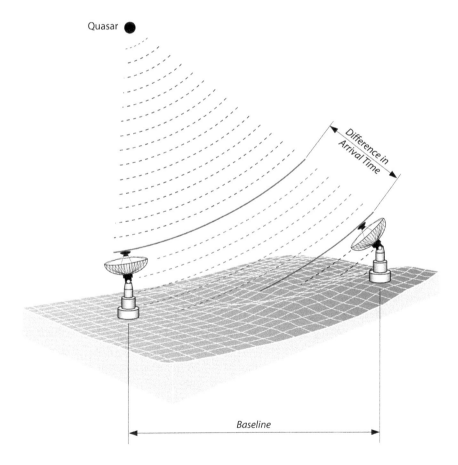

FIGURE 3.1   VLBI

of the time tags reveals the difference in the instants of wave front arrivals at the two telescopes. The discovery of the time offset that maximizes the correlation of the signals recorded by the two telescopes yields the distance and direction between them within a few centimeters over thousands of kilometers.

VLBI's potential for geodetic measurement was realized as early as 1967, but the concept of high-accuracy baseline determination using phase differencing was really proven in the late 1970s. A direct line of development leads from the VLBI work of that era by a group from the Massachusetts Institute of Technology (MIT) to today's most accurate GPS ranging technique, carrier-phase measurement. VLBI, along with other extraterrestrial systems like SLR, also provides valuable information on the Earth's gravitational field and rotational axis. The International Earth Rotation and Reference Systems Service (IERS) calculates the polar motion coordinates as part of the Earth Orientation Parameters (EOP), which are predicted on the basis of VLBI measurements and satellite ranging. VLBI is used in Universal Time (UT1) determination as well as the realizations of the International Terrestrial Reference Frame and the International Celestial Reference System. Without VLBI data, the high accuracy of the modern timing and coordinate systems that are critical to the success of GNSS would not be possible.

But the foundation for routine satellite-based geodesy actually came even earlier and from a completely different direction. The first prototype satellite of the immediate precursor of the GPS system that was successfully launched reached orbit on June 29, 1961. Its range measurements were based on the Doppler effect, not phase differencing, and the system came to be known as *TRANSIT*, or more formally, the *Navy Navigational Satellite System*.

TRANSIT

Satellite technology and the Doppler effect were combined in the first comprehensive Earth-orbiting satellite system dedicated to positioning. As mentioned earlier, tracking Sputnik 1 signal's Doppler shift provided valuable information. In 1957, physicists William Guier and George Weiffenbach at Johns Hopkins University's Applied Physics Laboratory (APL) found that it provided enough information to determine the exact moment of its closest approach to the Earth. Led by ARPA, later known as DARPA in 1972, and Dr. Richard Kershner of APL this discovery led to the creation of the Navy Navigational Satellite System (NNSS or NAVSAT) and the subsequent launch of six satellites specifically designed to be used for navigation of military aircraft and ships. This same system, eventually known as TRANSIT, was classified in 1964, declassified in 1967, and was widely

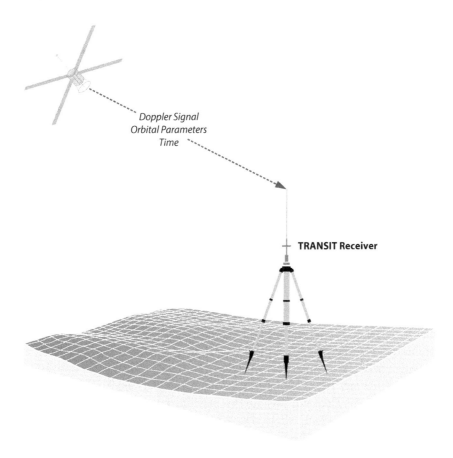

Doppler Signal
Orbital Parameters
Time

TRANSIT Receiver

FIGURE 3.2 TRANSIT

used in civilian hydrographic and geodetic surveying for many years until it was switched off on December 31, 1996 (see Figure 3.2).

The TRANSIT system had some nagging drawbacks. For example, its primary observable was based on the comparison of the nominally constant frequency generated in the receiver with the Doppler-shifted signal received from one satellite at a time. With a constellation of only six satellites, this strategy sometimes left the observer waiting up to 90 minutes between satellites and at least two passes were usually required for acceptable accuracy. With an orbit of only about 1100 km above the Earth, the TRANSIT satellite's orbits were quite low and, therefore, unusually susceptible to atmospheric drag and gravitational variations, making the calculation of accurate orbital parameters particularly difficult. Perhaps the most significant difference between the TRANSIT system and previous extraterrestrial systems was TRANSIT's capability of linking national and international datums with relative ease. Its facility at strengthening *geodetic coordinates* laid the groundwork for modern geocentric datums.

System 621B and Timation

In 1963, at about the same time the Navy was using TRANSIT, the Air Force funded the development of a three-dimensional radio navigation system for aircraft. It was called System 621B. The fact that it provided a determination of the third dimension, altitude, was an improvement of some previous navigation systems. It relied on the user measuring ranges to satellites based on the time of arrival of the transmitted radio signals, from that and knowing the instantaneous positions of the satellites, the user's position could be derived. The 621B program also utilized carefully design binary codes known as PRN codes or pseudorandom noise. Even though the PRN codes appeared to be noise at first, they were actually capable of repetition and replication. This approach also allowed all of the satellites to broadcast on the same frequency. Sounds rather familiar now, doesn't it? Unfortunately, System 621B required signals from the ground to operate.

A Naval project with a name that conflated time and navigation, Timation, began in 1964. The Timation 1 satellite was launched in 1967; it was followed by Timation 2 in 1969. Both of these satellites were equipped with high-performance quartz crystal oscillators, also known as XO. The daily error of these clocks was about 1 microsecond which translates to about 300 m of ranging error. The technique they used to transmit ranging signals was called side-tone ranging. The subsequent Timation 3 satellite, launched in 1974, had greatly improved clocks. It was the first satellite outfitted with two rubidium clocks. Though the clocks showed significant frequency drift, they proved to be space worthy atomic frequency standards, AFS, on orbit. It was a big step toward accurate satellite positioning, navigation, and timing (PNT). With this development, the Timation program demonstrated that a passive system using atomic clocks could facilitate worldwide navigation. For simplicity, the terms clock, oscillator, and frequency standard will be used interchangeably here. It was the combination of atomic clock technology, the ephemeris system from TRANSIT and the PRN signal design from the 621B program in 1973 that eventually became GPS. There was a Department of Defense directive to the United States Air Force in April 1973 that stipulated the consolidation of Timation and 621B into the navigation system called GPS. In fact, Timation 3 itself became part of the NAVSTAR–GPS program and was renamed Navigation Technology Satellite 1 (NTS-1). The next satellite in line, Timation 4, was known as NTS-2 (Figure 3.3). Its onboard cesium clock had frequency stability of 2 parts per 10^{13}. It was launched in 1977. Unfortunately, it only operated for 8 months.

NAVSTAR

Through the decades of the 1970s and 1980s, both the best and the worst aspects of the forerunner system informed GPS development. Many strategies

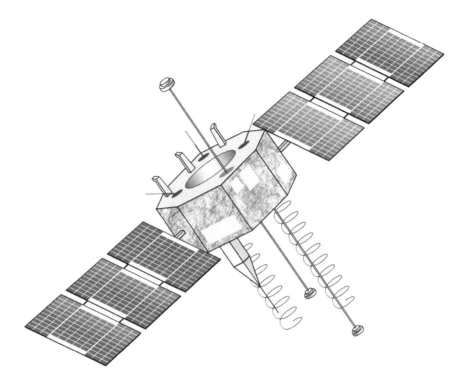

FIGURE 3.3   Timation 4 and NTS-2

used in TRANSIT were incorporated into GPS. For example, in the TRANSIT system, the satellites broadcast their own ephemerides to the receivers and the receivers had their own frequency standards. TRANSIT had three segments: the control segment, including the tracking and upload facilities; the space segment, meaning the satellites themselves; and the user segment, everyone with receivers. The TRANSIT system satellites broadcast two frequencies of 400 and 150 MHz to allow compensation for the ionospheric dispersion. TRANSIT's primary observable was based on the Doppler effect. All of these were used in GPS.

On the other hand, GPS improved on some of the shortcomings of the previous systems. For example, the GPS satellites were placed in nearly circular orbits over 20,000 km above the Earth, where the consequences of gravity and atmospheric drag are much less severe than the lower orbits assigned to TRANSIT, Timation, and some of the other earlier systems used. GPS satellites broadcast higher-frequency signals, L1 at 1575.42, L2 at 1227.60, and L5 at 1176.45 MHz which can be used to mitigate the ionospheric delay. The rubidium and cesium clocks pioneered in the Timation program and built into GPS satellites were a marked improvement over the quartz oscillators that were used in the TRANSIT satellites and other early satellite navigation systems. System 621B could achieve positional accuracies of approximately 16 m. TRANSIT's shortcomings restricted the practical accuracy of the system too. It could achieve submeter work, but only after long occupations on a station, at least a day, augmented by the use of a precise ephemerides for the satellites in post-processing. GPS provides much more accurate positions in a much shorter time than any of its predecessors, but these improvements were only accomplished by standing on the shoulders of the technologies that have gone before.

REQUIREMENTS

The genesis of GPS was military. It grew out of the congressional mandate issued to the Departments of Defense and Transportation to consolidate the myriad of navigation systems.

Its application to civilian surveying was not part of the original design. In 1973, the DOD directed the *Joint Program Office* (*JPO*) in Los Angeles to establish the GPS system. Specifically, JPO was asked to create a system with high accuracy and continuous availability in real time that could provide virtually instantaneous positions to both stationary and moving receivers. The forerunners could not supply all these features. The challenge was to bring them all together in one system.

SECURE, PASSIVE, AND GLOBAL

Worldwide coverage and positioning on a common coordinate grid were required of the new system—a combination that had been difficult, if not impossible, with terrestrial-based systems. It was to be a passive system, which ruled out any transmissions from the users, as had been tried with some previous satellite systems, but the signal was to be secure and resistant to jamming, so codes in the satellite's broadcasts would need to be complex.

EXPENSE AND FREQUENCY ALLOCATION

The U.S. Department of Defense (DOD) also wanted the new system to be free from the sort of ambiguity problems that had plagued Omega and other RADAR systems. DOD did not want the new system to require large expensive equipment like the optical systems did. Finally, frequency allocation was a consideration. The replacement of existing systems would take time, and with so many demands on the available radio spectrum, it was going to be hard to find frequencies for GPS.

LARGE CAPACITY SIGNAL

Not only did the specifications for GPS evolve from the experience with earlier positioning systems, but so did much of the knowledge needed to satisfy them. Providing 24-hour real time, high-accuracy navigation for moving vehicles in three dimensions was a tall order. Experience showed part of the answer was a signal that was capable of carrying a very large amount of information efficiently and that required a large bandwidth. So, the GPS signal was given a double-sided 10-MHz bandwidth. That was still not enough, so the idea of simultaneous observation of several satellites was also incorporated into the GPS system to accommodate the requirement. That decision had far-reaching implications.

THE SATELLITE CONSTELLATION

Unlike some of its predecessors, GPS needed to have not one but at least four satellites above an observer's horizon for adequate positioning, even more if possible. The achievement of full-time worldwide GPS coverage would require this condition to be always satisfied anywhere on or near the Earth. The specification for the GPS system required all-weather performance and correction for ionospheric propagation delay. TRANSIT had shown what could be accomplished with a multifrequency transmission from the satellites, but it had also proved that a higher frequency was needed. The GPS signal needed to be secure and resistant to both jamming and multipath. A spread spectrum, meaning spreading the frequency over a band wider than the minimum required for the information it carries, helped on all counts. This wider band also provided ample space for pseudorandom noise encoding, a fairly new development at the time. The PRN codes, like those used in System 621B, allowed the GPS receiver to acquire signals from several satellites simultaneously and still distinguish them from one another.

THE PERFECT SYSTEM?

The absolute ideal navigational system, from the military's point of view, was described in the Army POS/NAV Master Plan in 1990. It should have worldwide and continuous coverage. The users should be passive. In other words, they should not be required to emit detectable electronic signals to use the system. The ideal system should be capable of being denied to an enemy and it should be able to support an unlimited number of users. It should be resistant to countermeasures and work in real time. It should be applicable to joint and combined operations. There should be no frequency allocation problems. It should be capable of working on common grids or map datums appropriate for all users. The positional accuracy should not be degraded by changes in the user's altitude or by the time of day or the time of year. Operating personnel should be capable of maintaining the system. It should not be dependent on externally generated signals, and it should not have decreasing accuracy over time or the distance traveled. Finally, it should not be dependent on the identification of precise locations to initiate or update the system. A pretty tall order, GPS lives up to most, though not all, of the specifications.

GPS IN CIVILIAN SURVEYING

As mentioned earlier, application to civilian surveying was not part of the original concept of GPS. The civilian use of GPS grew up through partnerships between various public, private, and academic organizations. Nevertheless, while the military was still testing its first receivers, GPS was already in use by civilians. Geodetic surveys were actually underway with commercially available receivers early in the 1980s.

FEDERAL SPECIFICATIONS

The Federal Radionavigation Plan of 1984, a biennial document including official policies regarding radio navigation, set the predictable and repeatable accuracy of civil and commercial use of GPS at 100 m horizontally and 156 m vertically. This specification meant that a horizontal position derived from C/A code ranging for Standard Positioning Service could be defined by a circle with a radius of 100 m, 95% of the time. However, that same year civilian users were already achieving results better than that limit.

INTERFEROMETRY

By using interferometry, the technique that had worked so well with Prime Minitrack and VLBI, civilian users were showing that GPS surveying was capable of extraordinary results. In the summer of 1982, a research group at the MIT tested an early GPS receiver and achieved accuracies of 1 and 2 ppm of the station separation. Over a period of several years, extensive testing was conducted around the world that confirmed and improved on these results. In 1984, a GPS network was produced to control the construction of the Stanford Linear Accelerator. This GPS network provided accuracy at the millimeter level. In other words, by using the differentially corrected carrier-phase observable instead of code ranging, private firms and researchers were going far beyond the accuracies the U.S. government expected to be available to civilian users of GPS.

The interferometric solutions made possible by computerized processing developed with earlier extraterrestrial systems were applied to GPS by the first commercial users. The combination made the accuracy of GPS its most impressive characteristic, but it hardly solved every problem. For many years, the system was restricted by the shortage of satellites. The necessity of having four satellites above the horizon restricted the available observation sessions to a few, sometimes inconvenient, windows of time. Another drawback of GPS for the civilian user was the cost and the limited application of both the hardware and the software. In the beginning, GPS was too expensive and too inconvenient for everyday use, but the accuracy of GPS surveying was already extraordinary.

GPS SEGMENTS

THE SPACE SEGMENT

Though there has been some evolution in the arrangement, the current GPS constellation under full operational capability consists of 24 satellites. However, in recent times there are more than that, up to 31, on orbit and broadcasting. The constellation includes orbital spares. These measures ensure that each of the four slots in each of the six orbital planes is populated by a satellite and at a minimum maintaining the 24-satellite constellation without interruption. The orbital planes are labeled from A to F, and the slots are numbered so, for example, the first satellite slot in the A plane is A1. Each of the planes is inclined to the equator by 55° (see Figure 3.4) in a symmetrical, uniform arrangement that covers the globe completely, even though the coverage is not quite as robust at high latitudes as it is at midlatitudes.

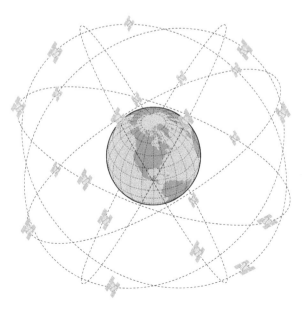

FIGURE 3.4   The GPS Constellation

The uniform design also means that multiple satellite coverage is available even when satellites need to be replaced. The satellites routinely outlast their anticipated design lives, but they do eventually require replacement, so spares are usually placed in slots adjacent to those most expected to fail. When the satellites reach the end of their life, they are moved up another 500 km to a disposal orbit, and their navigation signals are turned off.

GPS CONSTELLATION

ORBITAL PERIOD

NAVSTAR satellites are more than 20,000 km above the Earth in a *posigrade orbit*. A posigrade orbit is one that moves in the same direction as the Earth's rotation. Since each satellite is nearly three times the Earth's radius above the surface, its orbital period is 12 sidereal hours.

When an observer performs a GPS survey project, one of the most noticeable aspects of a satellite's motion is that it returns to the same position in the sky about 4 minutes (3 minutes and 56 seconds) earlier each day. This apparent regression is attributable to the difference between 24 solar hours and 24 sidereal hours. *Sidereal time* is also known as star time. GPS satellites retrace the same orbital path twice each *sidereal day*, but since their observers measure time in solar units, the orbits do not look quite so regular to them. This rather esoteric fact has practical applications; for example, if the satellites are in a particularly favorable configuration for measurement, the observer may wish to take advantage of the same arrangement the following day. However, he or she would be well advised to remember the same configuration will occur about 4 minutes earlier on the solar time scale. Both Universal Time (UT) and GPS Time are measured in solar, not sidereal units.

DILUTION OF PRECISION

As mentioned earlier, the GPS constellation was designed to satisfy several critical concerns. Among them were the best possible coverage of the Earth with the fewest number of satellites, the reduction of the effects of gravitational and atmospheric drag, sufficient upload, monitoring capability, and the achievement of maximum accuracy and reliability. The distribution of the satellites above an observer's horizon has a direct bearing on the reliability of the position derived from them. Like some of its forerunners, the quality of a GPS position is subject to a geometric phenomenon called *dilution of precision* (*DOP*). This number is somewhat similar to the strength of figure consideration in the design of a triangulation network. DOP concerns the geometric strength of the figure described by the positions of the satellites with respect to one another and the user's GNSS receivers. A low DOP

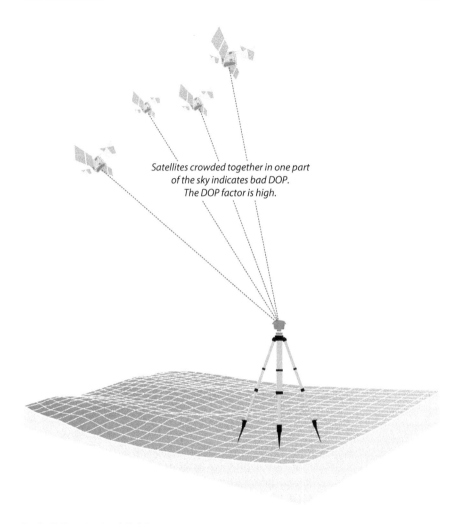

*Satellites crowded together in one part
of the sky indicates bad DOP.
The DOP factor is high.*

FIGURE 3.5 Bad DOP

factor is good; a high DOP factor is bad. In other words, when the satellites are in the optimal configuration for a reliable GNSS position the DOP is low, when they are not, the DOP is high (see Figure 3.5).

Four or more satellites must be above the observer's mask angle for the simultaneous solution of the clock offset and three dimensions of the receiver's position. However, if all those satellites are crowded together in one part of the sky, the position would be likely to have an unacceptable uncertainty, and the DOP, would be high. In other words, a high DOP is like a warning that the errors in a GNSS position are liable to be larger than you might expect.

But remember, it is not the errors themselves that are directly increased by the DOP factor; it is the statistical uncertainty of the GNSS positions

that is increased by the DOP factor. It is limited to random error propagation but, nevertheless, here is an approximation of the effect:

$$\sigma = DOP\,\sigma_o$$

where

σ = the uncertainty of the position

DOP = the dilution of precision factor

σ_o = the uncertainty of the measurements (User Equivalent Range Error, UERE)

Since a GPS position is derived from a three-dimensional solution, there are several DOP factors used to evaluate the uncertainties in the components of a GNSS position. For example, there is horizontal dilution of precision (HDOP) and vertical dilution of precision (VDOP), where the uncertainty of a solution for positioning has been isolated into its horizontal and vertical components, respectively. When both horizontal and vertical components are combined, the uncertainty of three-dimensional positions is called posi-tion dilution of precision (PDOP). There is also time dilution of precision (TDOP), which indicates the uncertainty of the clock. There is geometric dilution of precision (GDOP), which is the combination of all the above. Finally, there is relative dilution of precision (RDOP), which includes the number of receivers, the number of satellites they can handle, the length of the observing session, as well as the geometry of the satellites' configuration.

The user equivalent range error (UERE) is the total error budget affect-ing a pseudorange. It is the square root of the sum of the squares of the individual biases discussed in Chapter 2. Using a calculation like that men-tioned, the PDOP factor can be used to find the statistical error that will result from a particular UERE at the one sigma level (68.27%). For example, supposing that the PDOP factor is 1.5 and the UERE is 2 m, the statistical positional error would be 3 m at the 1 sigma level (68.27%). In other words, the standard deviation of the GPS position is the DOP factor multiplied by the square root of the sum of the squares of the individual biases (UERE). Multiplying the 1 sigma value times 1.96 would provide that 95% level of the statistical positional error estimate which would be 5.9 m.

As you can see in Figure 3.6, the size of the DOP factor is inversely proportional to the volume of body described by the satellite's positions and the position of the receiver. The larger the volume, the better the satel-lite geometry and the lower the DOP (see Figure 3.6). Obviously, having a receiver that tracks more satellites can lower the DOP too. It might be said that an ideal arrangement would be to have many satellites evenly spaced around the receiver, but at 15° above the horizon and a few near the receiver's zenith with that distribution, the DOP might be near 1, the lowest possible value. In practice, the low DOPs are generally around 2.

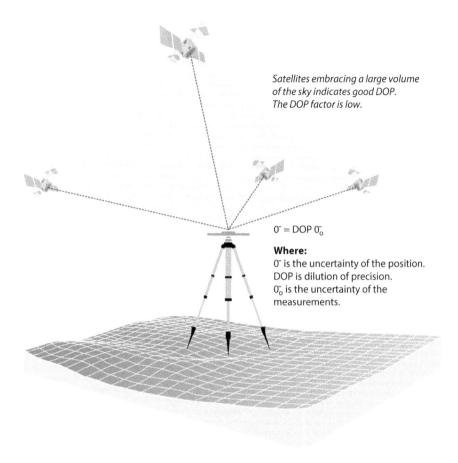

Satellites embracing a large volume
of the sky indicates good DOP.
The DOP factor is low.

$\sigma = DOP\ \sigma_0$

Where:
σ is the uncertainty of the position.
DOP is dilution of precision.
σ_0 is the uncertainty of the
measurements.

FIGURE 3.6 Good DOP

Obstructions can cause a high DOP. It will increase in forested areas, urban canyons, and places where the receiver's view of the sky is restricted and accuracy will decrease. Therefore, it's good that users of most GNSS receivers can set a PDOP mask to guarantee that data will not be logged if the PDOP goes above the set value. When a DOP factor exceeds a maximum limit in a particular location, indicating an unacceptable level of uncertainty exists over a period of time, that period is known as an *outage*. This expression of uncertainty is useful both in interpreting measured baselines and planning a GNSS survey.

Satellite Positions in Mission Planning

The position of the satellites above an observer's horizon is a critical consideration in planning a GNSS survey. So, most software packages provide various methods of illustrating the satellite configuration for a particular location over a specified period of time. For example, the configuration of the satellites over the entire span of the observation is important; as the satellites move, the DOP changes. Fortunately, the DOP can be worked out in advance. DOP can be predicted. It depends on the orientation of the satellites relative to the receivers. Since most GNSS software allow calculation of the satellite constellation from any given position and time, they can also provide the accompanying DOP factors. Another commonly used plot of the satellite's tracks is constructed on a graphical representation of half of the celestial sphere. The observer's zenith is shown in the center and the horizon is on the perimeter. The program usually draws arcs by connecting the points of the instantaneous azimuths and elevations of the satellites above a specified mask angle.

These arcs then represent the paths of the available satellites over the period and the place specified by the user. In Figure 3.7, the plot of the polar coordinates of the available satellites with respect to time and

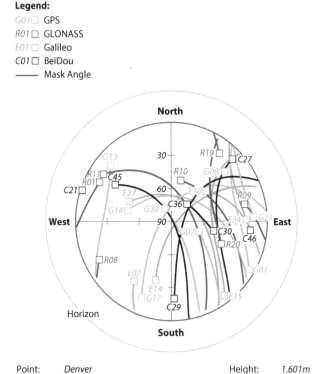

FIGURE 3.7   Polar Plot

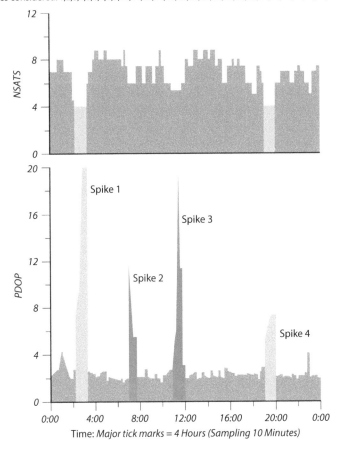

Number SVs and PDOP

Polnt: *Denver* Lat: *39:47:0 NT* Lon: *104:53:0 W* Almanace: *CURRENT.FPH 4/4/00*

Date: *Tuesday, November 14, 2000* Threshold Elevation: *15 (deg)* Time Zone: *'Mountain Std USA' - 7:00*

28 Satellites considered: *1,2,3,4,5,6,7,8,9,10,11,13,14,15,16,17,18,19,21,22,23,24,25,26,27,29,30,31*

FIGURE 3.8 Number of Space Vehicles (SVs) and PDOP

position is just one of several tables and graphs available to help the GNSS user visualize the constellation. Figure 3.8 is another useful graph that is available from many software packages. It shows the correlation between the number of satellites above a specified mask angle and the associated PDOP for a particular location during a particular span of time. There are four spikes of unacceptable PDOP, labeled here for convenience. It might appear at first glance that these spikes are directly attributable to the drop in the number of available satellites. However, please note that while spikes 1 and 4 do indeed occur during periods of 4 satellite data, spikes 2 and 3 are during periods when there are 7 and 5 satellites available, respectively. It is not the number of satellites above the horizon that determines the quality of GNSS positions, one must also look at their position relative to the observer, the DOP, among other things. The variety of tools to help the observer predict satellite visibility underlines the importance of their configuration to successful positioning.

SATELLITE BLOCKS

The 11 GPS satellites launched from Vandenberg Air Force Base between 1978 and 1985 were known as Block I satellites. The last Block I satellite was retired in late 1995. They were followed by the Block II satellites. The first of them was launched in 1989, and the last was decommissioned in 2007. During the time that the earliest of these satellites were in service, improved versions of the Block II satellites were built and launched. The first were called Block IIA satellites and their launches began in 1990. The last of them was decommissioned in 2019. The Block IIR satellites were next and the first successful launch was in July 1997. Several Block IIR satellites are still operational. In September 2005, the first of the next improved block, called Block IIR-M, was launched. The first Block IIF GPS satellites reached orbit in May of 2010, the last in 2016. Block III is the most recent. Block III satellites are on orbit and in service. There has been steady and continuous improvement in the GPS satellite constellation from the beginning.

SATELLITE NAMES

There has always been a bit of complication in the naming of the individual GPS satellites. The first GPS satellite was launched in 1978 and was known as NAVSTAR 1. It was also known as PRN 4 just as NAVSTAR 2 was known as PRN 7. The NAVSTAR number, or Mission number, includes the Block name and the order of launch, for example, I-1, meaning the first satellite of Block I, and the PRN number refers to the weekly segment of the P(Y) code that has been assigned to the satellite, and there are still more identifiers. Each GPS satellite has a Space Vehicle number (SVN), an Interrange Operation Number (IRON), a NASA catalog number, also known as the U.S. Space Command number, and an orbital position number. For example, the GPS satellite IIF-7 is PRN09. Its SVN is 68. Its NAVSTAR number is 71. Its orbital position is F3 and its U.S. Space Command Satellite Catalog number is 40105. In most literature, and to the GPS receivers themselves, the PRN number is the most important.

GPS SATELLITES

All GPS satellites have some common characteristics (Figure 3.9). They weigh about a ton and with solar panels extended are about 27 feet long.

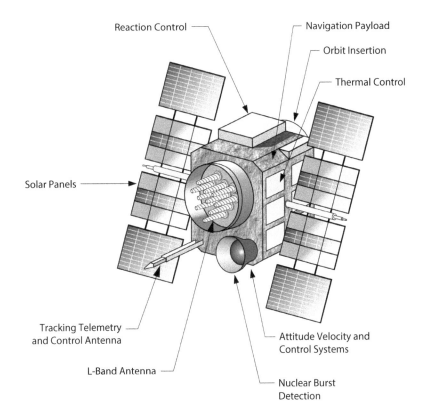

Reaction Control

Navigation Payload

Orbit Insertion

Thermal Control

Solar Panels

Tracking Telemetry
and Control Antenna

Attitude Velocity and
Control Systems

L-Band Antenna

Nuclear Burst
Detection

FIGURE 3.9   GPS Block II Satellite

They generate about 700 watts of power. They all have three-dimensional stabilization to ensure that their solar arrays point toward the Sun and their 12 helical antennae to the Earth. GPS satellites move at a speed of about 8700 miles per hour. Even so, the satellites must pass through the shadow of the Earth from time to time and onboard batteries provide power. All satellites are equipped with thermostatically controlled heaters and reflective insulation to maintain the optimum temperature for the oscillator's operation.

THE GPS OPERATIONAL CONTROL SYSTEM (OCS)

There are U.S. government tracking and uploading facilities distributed around the world. These facilities not only monitor the L-band signals from the GPS satellites and update their Navigation Messages but also track the satellite's health, maneuvers, and many other things, even battery recharging. Taken together, these facilities are known as the *GPS Operational Control System* (*OCS*) (Figure 3.10).

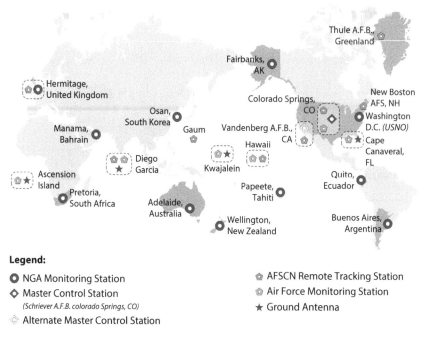

Legend:

○ NGA Monitoring Station

◇ Master Control Station
(Schriever A.F.B. colorado Springs, CO)

◈ Alternate Master Control Station

⬠ AFSCN Remote Tracking Station

⬡ Air Force Monitoring Station

★ Ground Antenna

FIGURE 3.10   Control Segment

The Master Control Station (MCS), once located at Vandenberg Air Force Base in California, now resides at the Consolidated Space Operations Center (CSOC) at Schriever (formerly Falcon) Air Force Base near Colorado Springs, Colorado, and has been manned by the 2nd Space Operations Squadron, 2SOPS, since 1992. There is an alternate MCS at Vandenberg Tracking Station in California.

The 2SOPS squadron has command and control of the GPS satellites. For example, they monitor the state of each satellite's onboard battery, solar and propellant systems. They resolve satellite anomalies, activate spare satellites and control anti-spoofing. They dump the excess momentum from the *wheels*, four devices that stabilize each satellite. In GNSS satellites attitude and rotational rates are controlled with these momentum wheels. Two rotations are necessary during the satellite's orbit to ensure the solar panels are pointing at the Sun. The angular momentum builds up and it accumulates in the wheels. It must be unloaded with magnetorquers during station-keeping to avoid orbital perturbations. Magnetorquers are magnetic coils that introduce the needed torque to do so through interaction with the Earth's magnetic field.

The 2SOPS squadron has continuous constellation tracking data available and aided by *Kalman filter* estimation to manage the noise in the data, they calculate and update the parameters in the navigation message (ephemeris, almanac, and clock corrections). They keep the information within limits because the older it gets, the more its veracity deteriorates. This process is made possible by a persistent two-way communication with the constellation managed by the Control Segment that includes both monitoring and uploading accomplished through a network of ground antennas and monitoring stations.

The data that feeds the MCS comes from monitoring stations. These stations track the entire GPS constellation. In the past there were limitations. When there were only six tracking stations it was possible for a satellite to go unmonitored for up to two hours each day. It was clear that the calculation of the ephemerides and the precise orbits of the constellation could be improved with more monitoring stations in a wider geographical distribution. It was also clear that if one of the six stations went down the effectiveness of the Control Segment could be considerably hampered. These ideas, and others, led to a program of improvements known as the *Legacy Accuracy Improvement Initiative, L-AII,* a *National Geospatial-Intelligence Agency* (*NGA*)-led project. It increased the data available to the GPS MCS. Completed in 2008, it added ten NGA GPS monitoring stations. The additional stations augmented the existing 6, more than doubling the number to 16 and tripling the amount of data on GPS satellite orbits which increased the accuracy of the broadcast ephemeris. L-AII also produced improvement in the MCS Kalman filtering and its ability to monitor the performance of the GPS space vehicles, benefitting all users of GPS.

Today, there are 6 Air Force and 11 NGA monitoring stations. The monitoring stations track all the satellites, in fact, every GPS satellite is tracked by at least 3 of these stations all the time and by an average of almost 5 over a 24-hour period. The monitoring stations collect range measurements, atmospheric information, satellite's orbital information, clock errors, velocity, right ascension, and declination and send them to the MCS. They also provide pseudorange and carrier-phase data to the MCS. The MCS needs this constant flow of information. It provides the basis for the computation of the almanacs, clock corrections, ephemerides, and other components that make up the navigation messages. The new stations also improve the geographical diversity of the Control Segment and that helps with the MCS isolation of errors, for example, making the distinction between the effects of the clock error from ephemeris errors. In other words, the diagnosis and solution of problems in the system are more reliable now because the MCS has redundant observations of satellite anomalies with which to work. Testing has shown that the augmented Control Segment and the better modeling have improved the accuracy of clock corrections and ephemerides in the navigation messages substantially. These upgrades may contribute to an increase in the accuracy of real time GPS by 10 to 15%.

When a message is calculated and needs to be sent back up to the satellites, some of the stations have ground antennas for uploading. Four monitoring stations are collocated with such antennas. The stations at Ascension Island, Cape Canaveral, Diego Garcia, and Kwajalein upload navigation and program information to the satellites via S-band transmissions. The station at Cape Canaveral also has the capability to check satellites before launch.

The modernization of the Control Segment has been underway for some time and it continues. In 2007, the Launch/Early Orbit Anomaly Resolution and Disposal Operations (LADO) mission PC-based ground system replaced the mainframe-based Command-and-Control System (CCS). Since then, LADO has been upgraded several times. It uses Air Force Satellite Control Network (AFSCN) remote tracking stations only, not the dedicated GPS ground antennas to support the satellites from spacecraft separation through checkout, anomaly resolution, and all the way to end of life disposal. It also helps in the performance of satellite movements and the presentation of telemetry simulations to GPS payloads and subsystems. Air Force Space Command (AFSPC) accepted the LADO capability to handle the most modern GPS satellites at the time, Block IIF, in October 2010. Another modernization program is known as the Next Generation Operational Control System or OCX is underway. It is being delivered in phases expressed as numbered blocks. In November 2017, the U.S. Air Force accepted the delivery of OCX Block 0 LADO support for GPS III. Seventeen OCX monitoring stations were completed in July 2021. The delivery of Blocks 1, 2, and 3F is expected in the near future. OCX will facilitate the full command and control of the new GPS signals like L5, as well as L2C and L1C, and the GPS III program.

KALMAN FILTERING

Kalman filtering is named for Rudolf Emil Kalman's 1960 linear recursive solution used to smooth the effects of system and sensor noise in large datasets. More specifically, it is a set of mathematical equations that provides an estimation technique based on least squares that teases an estimate of the actual signal, the signal with the minimum mean square error, from noisy sensor measurements. In GNSS, it derives an instantaneous position estimate from multiple statistical measurements on a time-varying signal.

Kalman filtering can be illustrated by the example of an automobile speedometer. Imagine the needle of an automobile's speedometer that is fluctuating between 64 and 72 mph as the car moves down the road (Figure 3.11). The fluctuation is analogous to signal noise. Despite the noise, the driver might estimate the actual speed at 68 mph. Although not accepting each of the instantaneous speedometer's readings literally, the number

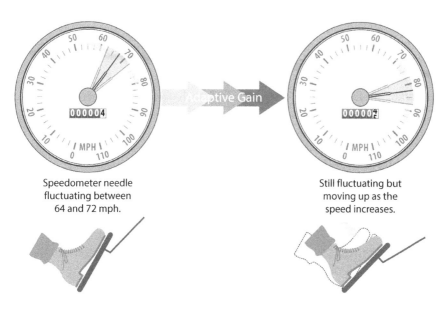

FIGURE 3.11 Kalman Filter Gain Pattern

of them is too large, he has nevertheless taken them into consideration and constructed an internal model of his velocity. If the driver further depresses the accelerator and the needle responds by moving up, his reliance on his model of the speedometer's behavior increases. Despite its vacillation, the needle has reacted as the driver thought it should. It went higher as the car accelerated. This behavior illustrates a predictable correlation between one variable, acceleration, and another, speed. Now he is more confident in his ability to predict the behavior of the speedometer. The driver is illustrating adaptive gain, meaning that he is fine tuning his model as he receives new information about the measurements. As he does, a truer picture of the relationship between the readings from the speedometer and his actual speed emerges without recording every single number as the needle jumps around. The driver's reasoning in this analogy is something like the action of a Kalman filter, but the Kalman can go further and take data inputs from multiple sources. This characteristic means it can predict unknown values more accurately than if individual predictions are made from one measurement method like the speedometer example. Suppose the driver wants to know where he is on the road at a given moment. So, let us add a GNSS receiver that supplies the positions of the vehicle as it moves, but suppose it is in an obstructed environment, so the positions from the receiver include a good deal of uncertainty too. When asked to determine where the vehicle is on the road at a given moment, the Kalman Filter goes through two steps repeatedly. First, there is an estimate of the state of the variation in the speedometer and the GNSS positions along with their uncertainties. Considering there, the algorithm makes predictions. Using the speed and position information, it adjusts its idea of the rate of change in the car's position over time and predicts where it will be. Then it repeats that process of state prediction followed by measurement update followed by state prediction again and again with each new measurement improving the next prediction as the vehicle travels down the road. Kalman Filtering has the capability to condense huge amounts of data into a manageable number of components. Without it, GNSS constellation control segments would be overwhelmed.

THE USER SEGMENT

In the early years of GPS, the military concentrated on testing navigation receivers, but civilians got involved much sooner than expected and took a different direction: receivers with geodetic accuracy. Some of the first GPS receivers in commercial use were single-frequency, six-channel, and codeless instruments. Their measurements were based on interferometry. As early as the 1980s, those receivers could measure short baselines to millimeter accuracy and long baselines to 1 ppm. It is true the equipment was cumbersome, expensive, and, without access to the LNAV, dependent on external sources for clock and ephemeris information. Nevertheless, they were the first available for civilian work in the field and their accuracy was impressive. During the same era, a parallel trend was underway. The idea was to develop a more portable, multifrequency, four-channel receiver that could use the navigation message. Such an instrument did not need external sources for clock and ephemeris information and could be more self-contained. Unlike the original codeless receivers that required all units on a survey brought together and their clocks synchronized twice a day, these receivers could operate independently. While the codeless receivers needed to have satellite ephemeris information downloaded before their observations could begin, this receiver could derive its ephemeris directly from the satellite's signal. Despite these advantages, the instruments developed on this model still weighed more than 40 pounds, were very expensive, and were dependent on P(Y) code tracking.

A few years later, a different kind of multichannel receiver appeared. Instead of the P(Y) code, it tracked the C/A code. Instead of using the L1 and L2 frequencies, it depended on L1 alone. On that single-frequency, it tracked the C/A code and measured the carrier-phase observable. This type of receiver established the basic design for many of the GPS receivers that surveyors use today. They are multichannel receivers, and they can recover all the components of the L1 signal. The C/A code is used to establish the signal lock and initialize the tracking loop. Then the receiver not only reconstructs the carrier wave but also extracts the satellite clock correction, ephemeris, and other information from the navigation message. Such receivers are capable of measuring pseudoranges, along with the carrier-phase and integrated Doppler observables.

Still, as some of the earlier instruments illustrated, the multifrequency approach does offer significant advantages. It can be used to limit the effects of ionospheric delay, it can increase the reliability of the results over long baselines, and it certainly increases the scope of GPS procedures available to a surveyor. For these reasons, a substantial number of receivers today utilize all frequencies using several configurations. In order to get the carriers without knowledge of the P(Y) code, some use a combination of C/A code correlation and codeless methods. By adding codeless technology such receivers can avail themselves of the advantages of a multifrequency capability while avoiding the difficulties of anti-spoofing, or AS, that is, the encryption of the P(Y) code on both L1 and L2. But even though the P(Y) code has been encrypted, the carrier-phase and pseudorange P(Y) code observables have been recovered successfully by many receiver manufacturers without decrypting the code itself. Still, others track all the available codes on all three GPS carriers' frequencies, L1, L2, and L5. However, please note that L5 nor the new civilian code on L2, L2C, are fully operational yet. When that happens cross-correlation with the P(Y) code will need to end because the P(Y) code will be replaced by the M code. In other words, receivers that exploit characteristics of the encrypted military P(Y) signal on the L2 frequency to achieve dual-frequency capability today will no longer have that option. As to when that will happen, there was an important announcement in the 2019 Federal Radionavigation Plan. In it, the U.S. government committed to maintaining the existing GPS L1 C/A, L1 P(Y), L2C, and L2 P(Y) signal characteristics, in other words, those that enable codeless and semi-codeless GPS access, until at least two years after there are at least 24 operational satellites broadcasting L5 with fully functional navigation messages. GPS users should be prepared for that and expect to utilize L2C and L5.

The military planned, built, and continues to maintain GPS. It is, therefore, no surprise that large component of the user segment of the system is military. While surveyors and geodesists have the distinction of being among the first civilians to incorporate GPS into their practice and are sophisticated users, their number is limited. In fact, it is quite small compared with the explosion of applications the technology has now among the general public. The use of GNSS in PNT applications in precision agriculture, machine control, aviation, railroads, and the marine industry are large and growing. With GNSS incorporated into smartphones, car navigation, financial systems, etc., it is no exaggeration that GNSS technology has transformed civilian businesses and lifestyles around the world. The noncommercial uses the general public finds for GNSS will undoubtedly continue to grow as the cost and size of the receivers continue to shrink.

EXERCISES

1. Which GNSS satellites are equipped with Laser Retro-reflector Arrays (LRA) and what is their purpose?
   a. Some GNSS satellites carry LRA to allow photographic tracking. The main purpose of the reflectors is to allow ground stations to distinguish the satellites, illuminated by Earth-based beacons, from the background of fixed stars.
   b. All GNSS satellites now on orbit carry LRA. The purpose of the reflectors is to allow the users to broadcast laser signals to the satellites that activate the onboard transponders they all carry.
   c. Nearly all new GNSS satellites have LRAs. A laser aimed from ground-based tracking stations to satellites equipped with them yields the range that allows ground stations to separate the effect of errors attributable to satellite clocks from errors in the satellite's ephemerides among other things.
   d. No GNSS satellites carry LRA. Such an arrangement would require the user to broadcast a signal from the Earth to the satellite meaning the system would no longer be passive.

2. Which of the following statements concerning the L-band designation is not true?
   a. The frequency bands used in radar were given letters to preserve military secrecy.
   b. The GPS carriers, L1, L2, and L5, are named for the L-band radar designation.
   c. The frequencies broadcast by the TRANSIT satellites were within the L-band.
   d. The L in L-band stands for long.

3. Which of the following is an aspect of the NAVSTAR GPS system that is an improvement not found in the retired TRANSIT system?
   a. Atomic frequency standards
   b. Satellites that broadcast more than one frequency
   c. A passive system, one that does not require transmissions from the users
   d. Satellites that broadcast their own ephemerides

4. Which of the following requirements for the ideal navigational system, from the military point of view, described in the Army POS/NAV Master Plan in 1990 does GPS not currently satisfy?
   a. Worldwide 24-hour coverage
   b. Resistant to countermeasures
   c. Real time capability
   d. Not dependent on externally generated signals

5. Which of the following statements best explains the fact that for a stationary receiver a GPS satellite appears to return to the same position in the sky about 4 minutes earlier each day?
   a. Over the same period of time vertical dilution of precision (VDOP) is frequently larger than horizontal dilution of precision (HDOP) for a stationary receiver.
   b. The apparent regression is due to the difference between the star time (sidereal time) and solar time over a 24-hour period.
   c. The loss of time is attributable to the satellite's pass through the shadow of the Earth.
   d. The approximately 4 minutes difference is due to the slowing of the clocks in the satellites explained in Einstein's special and general theories of relativity.

6. All of the following concepts described or developed in other contexts are now utilized in GPS. Which of them has been around the longest?
   a. Orbiting transmitters with accurate frequency standards
   b. Kalman filtering
   c. Measuring distances with electromagnetic signals
   d. The Doppler shift

7. Practically speaking, which of the following was the most attractive aspect of the first civilian GPS surveying in the early 1980s?
   a. Satellite availability
   b. GPS hardware
   c. Accuracy
   d. GPS software

8. Which of the following satellite identifiers is most widely used?
   a. Space Vehicle number
   b. Orbital slot
   c. PRN number
   d. NAVSTAR number

9. Satellites in which of the following categories are currently providing signals for positioning and navigation?
    a. Block III
    b. TRANSIT
    c. Block IIA
    d. Block I

10. How many stations of the Control Segment are tracking each GPS satellite at all times?
    a. Each GPS satellite is not tracked at all times. They can go unmonitored for up to 12 hours each day.
    b. Each GPS satellite is not tracked at all times. They can go unmonitored for up to 2 hours each day.
    c. Each GPS satellite is tracked at all times by three stations of the Control Segment.
    d. Each GPS satellite is tracked at all times by two stations of the Control Segment.

11. GDOP is a combination of which of the following?
    a. PDOP, HDOP, VDOP, SDOP
    b. PDOP, TDOP
    c. PDOP, VDOP, TDOP, XDOP
    d. SDOP, LDOP, MDOP, NDOP

12. Which of the following statements is correct?
    a. The larger the volume of the body defined by the lines from the receiver to the satellites, the better the satellite geometry and the lower the DOP.
    b. The volume of the body defined by the lines from the receivers to the satellites has no effect on DOP.
    c. The mask angle of 30° decreases the PDOP.
    d. The typical DOP is 0.

ANSWERS AND EXPLANATIONS

1. Answer is (c)
    Explanation: Two early GPS satellites carried onboard corner cube reflectors, SVN 36 (PRN 06) and SVN 37 (PRN 07). They were launched in 1994 and 1993, respectively. They are long out of service. However, the purpose of the corner cube reflectors, the LRA, was and is to allow Satellite Laser Ranging (SLR). They were not to distinguish the satellites from the background of fixed stars or the activation of transponders. A laser aimed from

ground-based tracking stations to satellites so equipped yields a range. That range contributes to precise orbit determination, which allows for the separation of the errors attributable to satellite clocks from errors in the satellite's ephemerides among other things. The satellite's ephemeris can be likened to a constantly updated coordinate of the satellite's position and when SLR nails down the difference between the satellite's broadcast ephemeris and the satellite's actual position this allows more proper attribution of the range error attributable to the satellite's clock. Today, nearly all new GNSS satellites have LRAs that includes all the GLONASS, Galileo, BeiDou, NavIC, and GPS Block III, starting with SV-9, carry retro-reflectors, LRA.

2. Answer is (c)

   Explanation: The original letter designations were assigned to frequency bands in radar to maintain military secrecy. The L-band was given the letter L to indicate that its wavelength was long compared to the shorter 10 cm wavelengths introduced later. The old (pre-1970) U.S. military classification of the L-band includes high frequencies from approximately 390 to 1550 MHz with wavelengths from 15 to 30 cm. The more current definition is from the Institute for Electrical & Electronics Engineers (IEEE), which classifies the range from 1 to 2 *GigaHertz* (*GHz*) (1000–2000 MHz) as the L-band. All the satellites in the global GNSS constellations; GPS, Galileo, GLONASS, and BeiDou, and the regional GNSS constellations QZSS and NavIC broadcast within that definition of the L-band. The carriers of the GPS signals supplying positioning, navigation and timing (PNT) are L1 at 1575.42, L2 at 1227.60, and L5 at 1176.45 MHz. While one might say that the L1 frequency is not exactly within the original military L-band designation, the frequencies broadcast by the TRANSIT satellites certainly were not. The frequencies used in the TRANSIT system were 400 and 150 MHz. These frequencies fall much below the L-band and into the VHF range.

3. Answer is (a)

   Explanation: Many of the innovations used in the TRANSIT system informed decisions made during the creation of the NAVSTAR GPS system. For example, the idea that satellites that broadcast more than one frequency allow compensation for the ionospheric dispersion. TRANSIT satellites used the frequencies of 400 and 150 MHz, while GPS uses 1575.42, 1227.60, and 1176.45 MHz. While the low frequencies used in the TRANSIT system were not as effective at eliminating the ionospheric delay, the idea is the same. No transmission from the user was required, the idea of a passive system, satellites that broadcast their own ephemerides to the receivers, were ideas that were in a sense inherited from TRANSIT. There are more similarities; division of the system into three segments: the control segment, the space segment, and the user segment. Another common attribute, each satellite and receiver contains its own frequency standard. However, the standards used in NAVSTAR GPS satellites are much more sophisticated than those that were used in TRANSIT. The stability of the rubidium and cesium atomic frequency standards used in GPS satellites and subsequent GNSS satellites far surpass the temperature-stabilized, crystal-controlled transistor oscillators that were used in the TRANSIT satellites. The TRANSIT navigational broadcasts were switched off on December 31, 1996.

4. Answer is (d)

Explanation: Not only does GPS have worldwide 24-hour coverage. It is resistant to countermeasures, i.e., anti-spoofing and the complexity of pseudorandom noise codes. There is more than one strategy that the Department of Defense can use to deny GPS to an enemy. At the same time, there can be a virtually unlimited number of users without over-taxing the system. It does have the real time capability as demonstrated by the method originated by Dr. Benjamin Remondi in the 1980s. However, The Control Segment's network of ground antennas and monitoring stations track the navigation signals of all the satellites. The MCS uses that information to generate updates for the satellites, which it uploads through ground antennas, and GPS is dependent on those externally generated signals for the continued health of the system. The satellites can operate for periods without uploads and orbital adjustments from the Control Segment. Those periods are growing longer and longer as the technology is improved, but they certainly cannot do without them entirely.

5. Answer is (b)

Explanation: The difference between 24 solar hours and 24 sidereal hours, otherwise known as star time, is 3 minutes and 56 seconds, or about 4 minutes. GPS satellites retrace the same orbital path twice each sidereal day, but since their observers on Earth measure time in solar units, the orbits do not look quite so regular to them and both Universal Time (UT) and GPS Time are measured in solar, not sidereal units. The apparent regression is not attributable to VDOP, HDOP, or the satellite's pass through the shadow of the Earth. The time dilation effects described in Einstein's special and general theories of relativity amount to microseconds and nanoseconds per day.

6. Answer is (d)

Explanation: The concept of measuring distance with electro-magnetic signals had its earliest practical applications in radar in the 1940s and during World War II. Development of the technology for launching transmitters with onboard frequency standards into orbit was available soon after the launch of Sputnik in 1957. TRANSIT 1B launched on April 13, 1960, was the first successfully launched navigation satellite.

Kalman filtering is a statistical method of smoothing and condensing large amounts of data and has been used in radio navigation ever since. It is an integral part of GPS. However, the Doppler shift was discovered in 1842, and certainly has the longest history of any of the ideas listed. The Doppler shift describes the apparent change in frequency when an observer and a source are in relative motion with respect to one another. If they are moving together, the frequency of the signal from the source appears to rise, and if they are moving apart, the frequency appears to fall. The Doppler effect came to satellite technology during the tracking of Sputnik in 1957. It occurred to observers at Johns Hopkins University's Applied Physics Laboratory (APL) that the Doppler shift of its signal could be used to find the exact moment of its closest approach to the Earth. However, the phenomenon had been described 115 years earlier by Christian Doppler using the analogy of a ship on the ocean.

7. Answer is (c)

Explanation: The interferometric solutions made possible by computerized processing developed with earlier extraterrestrial systems were applied to GPS by the first commercial users, but the software was cumbersome by today's standards. There were few satellites up in the beginning and the necessity of having at least four satellites above the horizon restricted the available observation sessions to difficult periods of time. GPS receivers were large, unwieldy, and very expensive. Nevertheless, in the summer of 1982, a research group at the Massachusetts Institute of Technology (MIT) tested an early GPS receiver and achieved accuracies of 1 and 2 ppm of the station separation. In 1984, a GPS network was produced to control the construction of the Stanford

Linear Accelerator This GPS network provided accuracy at the millimeter level. GPS was inconvenient and expensive, but the accuracy was remarkable from the outset.

8. Answer is (c)

   Explanation: The satellite known as Sacagawea has been given the Space Vehicle Number 77, or SVN77 is also known as PRN 14. This satellite, which was launched on November 5, 2020, currently occupies orbital slot B-6. It has the NAVSTAR number 80 or Mission number III-4. This designation includes the Block name and the order of launch of that mission. However, PRN are the most often used GPS satellite identifiers.

9. Answer is (a)

   Explanation: The first of 6 TRANSIT satellites to reach orbit was launched on June 29, 1961. The constellation of satellites known as the Navy Navigational Satellite System functioned until it was switched off on December 31, 1996, replaced by the GPS system. The first of 11 Block I GPS satellites was launched on February 22, 1978, and the last on October 9, 1985, however, no Block I satellites are operating today. The launches of the GPS satellites in Block IIA were done from 1990 to 1997, and the last one was decommissioned in 2019. However, several Block III GPS satellites are currently operational.

10. Answer is (c)

   Explanation: Today, there are 6 Air Force and 11 National Geospatial-Intelligence Agency (NGA) monitoring stations, and now each GPS satellite is tracked by at least 3 of them at all times and by an average of almost 5 over a 24-hour period.

11. Answer is (b)

   Explanation: There is a horizontal dilution of precision (HDOP) and vertical dilution of precision (VDOP) where the uncertainty of a solution for positioning has been isolated into its horizontal and vertical components, respectively. When both horizontal and vertical components are combined, the uncertainty of three-dimensional positions is called position dilution of precision (PDOP). There is also time dilution of precision (TDOP), which indicates the uncertainty of the clock. There is a geometric dilution of precision (GDOP), which is the combination of PDOP and TDOP.

12. Answer is (a)

Explanation: The larger the volume of the body defined by the lines from the receiver to the satellites, the better the satellite geometry and the lower the DOP. An ideal arrangement of four satellites would be one directly above the receiver, the others 120° from one another in azimuth near the horizon. With that distribution, the DOP would be nearly 1, the lowest possible value. In practice, the lowest DOPs are generally around 2.

4 Receivers and Methods

COMMON FEATURES OF GPS RECEIVERS

RECEIVERS FOR GNSS SURVEYING

The receivers are important hardware in a GNSS surveying operation. Their characteristics and capabilities influence the techniques available to the user throughout the work. There are many different GNSS receivers on the market. Some of them are appropriate for surveying, and they share some fundamental elements. No level of accuracy is ever guaranteed, but most are also capable of performing differential, real time, precise-point positioning, static, rapid static GNSS, etc., and are usually accompanied by processing and network adjustment software.

A receiver must collect and then convert signals from GNSS satellites into measurements of position, velocity, and time. There is a challenge in that the signal has low power. An orbiting GNSS satellite broadcasts its signal across a cone of arc. From the satellite's point of view and that cone covers a substantial portion of the whole planet. It is instructive to contrast this arrangement with a typical communication satellite that not only has much more power but also broadcasts a very directional signal. Its signals are usually collected by a large dish antenna, but the typical GNSS receiver has relatively nondirectional antennas. Fortunately, antennas used for GNSS receivers do not have to be pointed directly at the signal source. The GNSS signal also intentionally occupies a broader bandwidth than it must to carry its information. This characteristic is used to prevent jamming and mitigate multipath, but most importantly, the GNSS signal itself would be completely obscured by the variety of electromagnetic noise that surrounds us if it were not a spread spectrum coded signal. In fact, when a GNSS signal reaches a receiver, its power is less than the noise level; fortunately, the receiver can still extract the signal and achieve unambiguous satellite tracking using the correlation techniques described earlier. To do this job, the elements of a GNSS receiver function cooperatively and iteratively. That means that the data stream is repeatedly refined by the several components of the device working together as it makes its way through the receiver.

THE ANTENNA

The antenna, *radio frequency* (*RF*) section, filtering, and *intermediate frequency* (*IF*) elements are in the front of a receiver. An L-band antenna, capturing GNSS signals, noise, and possible interference, converts those

DOI: 10.1201/9781003405238-4

electromagnetic waves into electric currents that are appropriate for the RF section of the receiver. Several antenna designs are possible, but the satellite's signal has such a low power density, especially after propagating through the atmosphere, that antenna efficiency is critical. Many receivers have an antenna built in but can also accommodate a separate tripod-mounted or range pole-mounted antenna as well. These separate antennas with their connecting coaxial cables in standard lengths are usually available from the receiver manufacturer. The cables are an important detail. The longer the cable, the more of the GNSS signal is lost traveling through it. In short, GNSS antennas must have high sensitivity, also known as high *gain*. An antenna can be designed to collect one, two, or more frequencies.

Most of the receiver manufacturers use a *microstrip,* aka a patch antenna, which may have a patch for each frequency. With carriers in the 20-cm range, these antenna elements can be as small as 4 or 5 cm and still be a quarter wavelength long, which is where they tend to be the most practical and efficient. Microstrip antennas are compact. They also have a simple construction and a low profile. A ground plane facilitates the use of a microstrip antenna. It not only ameliorates multipath but also tends to increase the antenna's zenith gain. Another commonly used antenna is known as a *dipole*. A dipole antenna has a stable phase center and simple construction with two conductors with a feedline connecting them. It has high directivity and very limited reception on the perpendicular. A *quadrifilar* antenna has two orthogonal bifilar helical loops on a common axis. Quadrifilar antennas have a high profile and are most appropriate for crafts that pitch and roll like boats and airplanes. They are also used in many recreational handheld GNSS receivers. They have a good *gain pattern* and do not require a ground plane but are not azimuthally symmetric.

BANDWIDTH

An antenna ought to have a bandwidth commensurate with its application; in general, the larger the bandwidth, the better the performance. Said another way, the larger the bandwidth the larger the antenna's operating range. However, there is a downside; increased bandwidth degrades the signal-to-noise ratio (SNR) by including more interference. GNSS microstrip antennas usually operate in a range from about 2 to 20 MHz, which corresponds with the null-to-null bandwidth of both new and legacy GPS signals. For example, L2C and the central lobe of the C/A code span 2.046 MHz, whereas L5 and the P(Y) code have a bandwidth of 20.46 MHz. The width of C/A and P(Y) widths are shown in Figure 4.1.

The antenna and front-end of a receiver designed to collect the P(Y) code on L1 and L2 would have a bandwidth of 20.46 MHz, but a system tracking the C/A code or the L2C may have a narrower bandwidth.

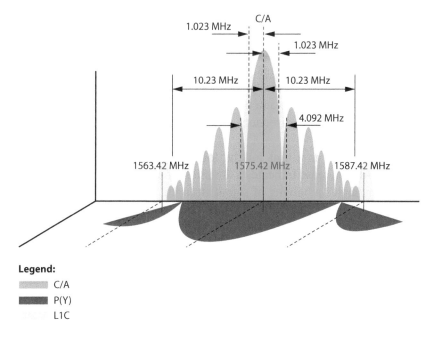

FIGURE 4.1   L1 Carrier with Codes

It would need 2.046 MHz for the central lobe of the C/A code, or if it were designed to track the L1C signal, its bandwidth would need to be 4.092 MHz. A multifrequency microstrip antenna would likely operate in bandwidth from 10 to 20 MHz.

Nearly Hemispheric Coverage

An ideal, theoretical antenna that is able to receive signals equally from all directions is known as an *isotropic* antenna, but that is not fit-for-purpose for GNSS. While a GNSS antenna is designed to have a gain pattern that includes a range of azimuths and elevations, it ought not be perfectly hemispheric. A GNSS antenna is best when it only collects signals from positive elevation angles. Surveying applications filter the signals from very low elevations to reduce the effects of multipath and atmospheric delays. In other words, a portion of the GNSS signal may come into the antenna from below the mask angle; therefore, it is best if the antenna's gain pattern is specifically designed to reject such signals. In fact, research into antenna arrays that promise to facilitate more control of the antenna patterns might contribute advantages in this area.

The narrower an antenna's bandwidth the higher the gain. As things stand, the gain, or gain pattern, describes the success of a GNSS receiver's antenna in collecting more energy from above the mask angle and less

from below the mask angle. A gain of about 3 to 5 decibels (dB) is typical for a GNSS receiver's antenna. Decibels do not indicate the power of
the antenna because the unit is dimensionless. It refers to a comparison.
In this case, the gain of a real GNSS receiver's antenna is measured by
comparing it to that previously mentioned theoretical isotropic antenna.
The hypothetical isotropic antenna is not only supposed to have perfectly
equal capabilities in all directions, but it is also imagined to be a lossless
antenna.

Since a decibel is a unit for the logarithmic measure of the relative
power, a 3-dB increase indicates a doubling of signal strength, and a 3-dB
decrease indicates a halving of signal strength. This means that a typical
omnidirectional antenna with a gain of about 3 dB has about 50% of the
capability of an isotropic antenna. It is important that the GNSS receiver's
antennas and pre-amplifiers be as efficient as possible because the power
received from the GNSS satellites is low. The minimum specified power
received from the GPS C/A code on L1 is −158.5 dBW. The same for L2C
on L2 when broadcast from Block III satellite, but the L2C signal from
Block IIF and Block IIR-M satellites is less powerful at −160 dBW. The
power of L1C is −157 dBW. The most power is assigned to L5 from Block
III satellites at −154 dBW while L1C is −157 dBW.

ANTENNA ORIENTATION

In a perfect GNSS receiver's antenna, the phase center of the gain pattern would be exactly coincident with its actual, physical, center. However,
the contours of equal phase around the antenna's electronic center are not
themselves perfectly spherical. If such a thing were possible, the centering
of the antenna over a point on the Earth would ensure its electronic centering as well, but that absolute certainty remains elusive. It is important to
remember that the position at each end of a GNSS baseline is the position
of the phase center of the receiver's antenna at each end, not their physical
centers. Those phase centers are not immovable points. They move a bit
as the frequency, azimuth, and elevation of the received signals change.
That movement is called the *phase center variation (PCV)*. Most of it
is attributable to changes in satellite elevation. The *phase center offset
(PCO)* is the vector between the antenna's phase center and the Antenna
Reference Point (ARP). As mentioned earlier, the ARP is an accessible
point usually marked on a GNSS receiver antenna where the vertical axis
of the antenna intersects the bottom mount, ground plane, or choke ring.
It is a point to which instrument heights can be measured and from which
the PCO can be measured. Fortunately, the PCOs for many antennas are
available online, i.e., National Geodetic Survey (NGS) antenna calibration
tables. Also, the IGS antenna calibration data, PCO, and PCV patterns are

documented and distributed in their *ANTenna Exchange (ANTEX)* format. It is available for use in positioning and orbit determination software. It is particularly helpful in GNSS precise point positioning (PPP) and high-accuracy carrier-phase observations. Fortunately, the PCOs of specific antenna models are mostly systematic. To compensate for some of this off-set, most receiver manufacturers recommend users take care to orient all their antennas in the same direction when making simultaneous observations on a network of points. Several manufacturers even provide reference marks on the antenna so that each one may be rotated to the same azimuth, usually north, to maintain the same relative position between their physical and phase, electronic, centers. The antenna's configuration also affects another measurement critical to successful GNSS surveying—the height of the instrument. The measurement of the height of the instrument in a GNSS survey is normally made to the ARP. However, it sometimes must include an added correction to bring the total vertical distance to the antenna's phase center.

RECEIVER PROCESSES

Different receiver types use different techniques to process the GNSS signals, but they go through substantially the same steps. The preamplifier increases the signal's power. The gain in the signal coming out of the pre-amplifier needs to be considerably higher than the noise and in a common frequency band. The incoming frequency is combined with the previously mentioned reference signal generated by the receiver's oscillator. The two are multiplied together in a device known as a mixer. They go through a *bandpass filter*, an electronic filter that removes the unwanted frequencies and eliminates some of the residual noise from the signal. This filter will have a bandwidth of about 20 MHz for tracking the P code, but it will be around 2 MHz if the C/A code is required. In any case, the signal that results is known as the IF or beat frequency signal. This beat frequency is the difference between the Doppler-shifted carrier frequency that came from the satellite and the frequency generated by the receiver's own oscillator. The bandwidth of the IF itself can vary from 5 to 10 kHz to ensure that it can accommodate the full range of the Doppler effect on the signals coming in from the satellites. That spread is typically lessened after tracking is achieved. There are usually several IF stages before copies of it are sent into separate channels, each of which extracts the code and carrier information from a particular satellite. As mentioned before, a replica of the C/A or P(Y) code is generated by the receiver's oscillator, that is correlated with the IF signal and the pseudorange can be measured. The receiver also generates a replica of the carrier which is correlated with the IF signal and the shift in phase can be measured.

CHANNELS

A receiver's antenna does not differentiate the signals it collects they all come in together, but they don't stay that way. In a modern receiver, the frequency from each satellite is segregated, acquired, and tracked by its own dedicated channel. It was not that way with early analog receivers. They had a single channel. Satellites were tracked sequentially, one at a time, with about 30 seconds dedicated to each one. There was a bit of a step forward when this slow sequential tracking was replaced with multiplexing. A multiplexing (aka muxing) receiver still gathered some data from one satellite, switched to another, gathered more data, and so on, but the switching was much quicker, about 5 milliseconds per satellite. It could appear to be tracking all of the satellites simultaneously, but a multiplexing receiver still had to dedicate one frequency from one satellite to one channel at a time. It was just that the time was very short.

It was obvious that a receiver that could track all the satellites and all the signals simultaneously but with each in its own channel would be better. Such a parallel architecture came onto the scene about the same time as the advent of the all-digital receivers. The development of the 6-channel receiver was helped along by the appearance of application-specific integrated circuit (ASIC)-type correlator chip. Fast forward to today and the front end of a receiver may still be analog at the antenna, *low noise amplifier,* and down converter, but after that, there is typically an analog digital converter (ADC) followed by a bank of digital channels on an ASIC or other sophisticated chip. In fact, the current all-in-view receivers may have hundreds of channels. Such a multichannel receiver is faster, has better anti-jamming capability, a shorter time to first fix, a more certain phase and code lock, superior SNR, and redundancy, and since the channels operate in parallel, the receiver can maintain accuracy on a moving platform.

The channels are independent from one another. A channel tracks and processes a signal to provide observables and navigation data. Each channel in a multichannel receiver is either busy acquiring a GNSS signal or tracking one. During the acquisition phase, the signal is detected and an initial estimate of its code delay and Doppler frequency is computed. With the Doppler frequency removed, it is correlated with its locally created PRN code replicas, then the channel shifts to tracking. However, maintenance of that tracking requires those estimates to be continuously iterated, updated, and refined as the navigation message is decoded. Everything is fine if lock is maintained, but it can be lost due to wrong or noisy initial estimates, multipath, cycle slips, etc. If it is lost, the channel must go back to the acquisition phase. The tracking loops try to make sure that doesn't happen.

TRACKING LOOPS

A receiver's baseband processor estimates each satellite signal's transmission delay and Doppler shift, but then those estimates need to be refined in tracking loops. There are code tracking loops and *delay lock loops* (*DLL*) which track digressions from the transmission delay estimate. The carrier tracking loops, *phase locking loops* (*PLL*), *frequency-locked loops* (*FLL*) in the receiver track differences from the Doppler frequency estimate. Typically, both the code and the carrier are being tracked in phase lock. The tracking loops are connected to each of the receiver's channels and also work cooperatively with each other. As mentioned, multifrequency receivers have dedicated channels and tracking loops for each frequency.

The tracking loops follow the variations between the incoming signal received from the satellite and the receiver's local oscillator. They vary the input to the numerically controlled oscillator (NCO), a programmable linear frequency generator, to control the generation of the replica signal in the receiver in a feedback loop endeavoring to always ensure that the replica matches the received signals and keeps on matching it. It is challenging, there is noise in the system and correlation fluctuates, but if the tracking loops successfully follow the incoming signal, it achieves maximum correlation. As mentioned in Chapter 1 (see Figure 1.12), that happens when the receiver generated replica exactly matches the code coming in from the GNSS satellite. The same principle applies to carrier tracking. When the receiver's replica matches the incoming signal the correlation output reaches a maximum. The result is an estimate of the carrier signal phase error in the case of the PLL or FLL, and an estimate of the code delay for a DLL.

PSEUDORANGING

In most receivers, the first procedure in processing an incoming GNSS satellite signal is synchronization of the code from the satellite with a replica of that code generated by the receiver itself, i.e., the *code-phase* measurement. When there is no initial match between the satellite's code and the receiver's replica, the receiver time shifts, or slews, the code it is generating until the optimum correlation is found. Then a code tracking loop, the DLL, keeps them aligned. The time shift discovered in that process is a measure of the signal's travel time from the satellite to the phase center of the receiver's antenna. Multiplying this time delay by the speed of light gives a range. The result is called a pseudorange in recognition of the fact that it is contaminated by the errors and biases set out in Chapter 2.

CARRIER-PHASE MEASUREMENT

Once the receiver acquires the code, it has access to a navigation message with its ephemeris, time, and almanac information. Many applications of GNSS can stop there, but pseudoranges alone are not adequate for the majority of surveying applications. Therefore, the next step in signal processing for most receivers involves the carrier-phase observable. As stated earlier, just as they produce a replica of the incoming code, receivers also produce a replica of the incoming carrier wave. The foundation of carrier-phase measurement is the combination of these two frequencies. Remember, the incoming signal from the satellite is subject to an ever-changing Doppler shift while the replica within the receiver is nominally constant.

CARRIER TRACKING LOOP

The process begins after the PRN code has done its job and the code tracking loop is locked. By mixing the satellite's signal with the replica carrier, this process eliminates all the phase modulations, strips the codes from the incoming carrier, and simultaneously creates intermediate signals. The receiver makes its selection with a bandpass filter. Then this signal is sent on to the carrier tracking loop also known as the phase locking loop, PLL, where the NCO is continuously adjusted to follow the beat frequency exactly.

DOPPLER SHIFT

As the satellite passes overhead, the range between the receiver and the satellite changes, that steady change is reflected in a smooth and continuous movement of the phase of the signal coming into the receiver. The rate of that change, known as the *range rate*, is reflected in the constant variation of the signal's Doppler shift. However, if the receiver's oscillator frequency is matching these variations exactly as they are happening, it will duplicate the incoming signal's Doppler shift and phase. As mentioned earlier, this strategy of making measurements using the carrier-phase observable involves counting the elapsed cycles and adding the fractional phase of the receiver's own oscillator.

Doppler information has broad applications in signal processing. It can be used to discriminate between the signals from various GNSS satellites, to determine integer ambiguities in kinematic surveying, as a help in the detection of cycle slips, and as an additional independent observable for autonomous point positioning. Still, the most important application of Doppler data may be the determination of the just mentioned range rate. It is useful because the rate at which the range, the distance, between the satellite and the receiver changes over a particular period of time is approximately equal to the difference between the frequency of the signal at the receiver and the frequency of the signal transmitted by the satellite, a principle demonstrated during the orbits of Sputnik1 and applied with the TRANSIT system.

TYPICAL GPS DOPPLER SHIFT

With respect to the receiver, the satellite is always in motion even if the receiver is static, but the receiver may be in motion in another sense, as it is in kinematic GNSS.

The ability to determine the instantaneous velocity of a moving vehicle has always been a primary application of GNSS and is based on the fact previously mentioned that the Doppler-shift frequency of a satellite's signal is nearly proportional to its range rate (see Figure 4.2).

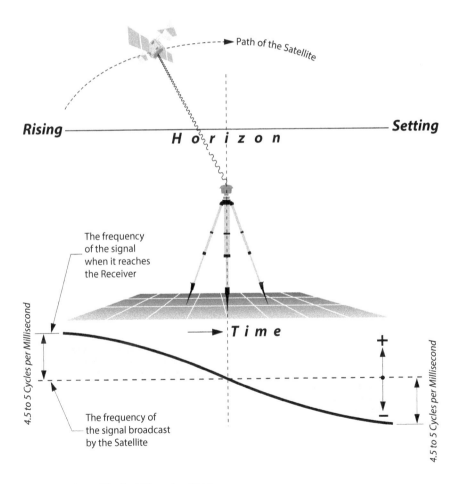

FIGURE 4.2   Typical Doppler Shift

To see how it works, let us look at a static, that is, stationary, GNSS receiver. The signal received would have its maximum Doppler shift, 4.5 to 5 cycles per millisecond, when the satellite is at its maximum range, just as it is rising or setting. The Doppler shift continuously changes throughout the overhead pass. Immediately after the satellite rises, relative to a particular receiver, its Doppler shift gets smaller and smaller until the satellite reaches its closest approach. At the instant its radial velocity with respect to the receiver is zero, the Doppler shift of the signal is zero as well. As the satellite recedes, it grows again, negatively, until the Doppler shift once again reaches its maximum extent just as the satellite sets.

CONTINUOUSLY INTEGRATED DOPPLER

The Doppler shift and the carrier phase are measured by first combining the received frequencies with the nominally constant reference frequency created by the receiver's oscillator. The difference between the two is the often-mentioned *beat frequency*, an IF, and the number of beats over a given time interval is known as the Doppler count for that interval. Since the beats can be counted much more precisely than their continuously changing frequency can be measured, most GNSS receivers keep track of the accumulated cycles, the Doppler count. The sum of consecutive Doppler counts from an entire satellite pass is often stored, and the data can then be treated like a sequential series of biased range differences. Continuously integrated Doppler is such a process. The rate of the change in the continuously integrated Doppler shift of the incoming signal is the same as that of the reconstructed carrier phase. Integration of the Doppler frequency offset can result in an accurate measurement of the advance in carrier-phase between epochs. As mentioned earlier, using double-differences in processing the carrier-phase observables remove most of the error sources other than multipath and receiver noise.

INTEGER AMBIGUITY

The solution of the integer ambiguity, the number of whole cycles on the path from satellite to receiver would be more difficult if it was not preceded by pseudoranges, or code phase measurements in most receivers. This allows the centering of the subsequent double-difference solution. In other words, a pseudorange solution provides an initial estimate of the candidates for the integer ambiguity within a smaller range than would otherwise be the case, and as more measurements become available, it can reduce them even further. After the code-phase measurements narrow the field, there are several methods used to solve the integer ambiguity.

In the geometric method, the carrier-phase data from multiple epochs are processed and the constantly changing satellite geometry is used to find an estimate of the actual position of the receiver. This approach is also used to show the error in the estimate by calculating how its results hold up as the geometry of the constellation changes. This strategy requires a significant amount of satellite motion to succeed and, therefore, takes time to converge on a solution. In the filtering approach, independent measurements are averaged to find the estimate with the lowest noise level. Then there is the method that searches through the range of possible integer combinations and finds the one with the lowest residual. Both the search and filtering methods are *heuristic*. They depend on trial and error. These methods cannot assess the correctness of a particular solution, but they can provide the probability, given certain conditions, that the answer is within given limits. In the end, most GNSS processing software uses some combination of all three ideas. All of these methods narrow the field of the integer ambiguity solution by beginning at an initial position estimate provided by code-phase measurements.

SIGNAL SQUARING

There was a method that did not use the codes carried by the satellite's signal. It was codeless tracking that relied on *signal squaring*. It was first used in the earliest civilian GPS receivers such as the Macrometer Model V-1000. The method supplanted proposals for a TRANSIT-like Doppler solution. It made no use of pseudoranging and relied exclusively on the carrier-phase observable. Like other methods, it also depended on the creation of an intermediate or beat frequency. With signal squaring, the beat frequency was created by multiplying the incoming carrier by itself. The result had double the frequency and half the wavelength of the original. It was squared. There are some drawbacks to the method. For example, in the process of squaring the carrier, it was stripped of all its codes. The chips of the P(Y) code, the C/A code, and the LNAV message normally modulated onto the carrier by 180° phase shifts were eliminated entirely. As discussed earlier, the signals broadcast by the satellites have phase shifts called code states that change from +1 to –1 and vice versa, but squaring the carrier converts them all to exactly 1. The result was that the codes themselves were wiped out. Therefore, this method had to acquire information such as almanac data and clock corrections from other sources. Other drawbacks of squaring the carrier included the deterioration of the SNR because when the carrier was squared the background noise was squared, too, and cycle slips occurred at twice the original carrier frequency. Signal squaring had its upside as well. It reduced susceptibility to multipath. It had no dependence on PRN codes and was not hindered by the encryption of the P(Y) code. The technique worked as well on L2 as it did on L1, and that facilitated ionospheric delay correction. Therefore, signal squaring did provide high accuracy over long baselines.

THE MICROPROCESSOR

The microprocessor controls the entire receiver, managing its collection of data. It controls the digital circuits that, in turn, manage the tracking and measurements, extract the ephemerides and other information from the Navigation message and mitigate multipath and noise among other things. The GNSS receivers used in surveying often send these data to the storage unit. However, more and more they are expected to process the ranging data, do datum conversion, and produce their final positions instantaneously, that is, in real time and then serve up the position through the control and display unit (CDU). There is a two-way street between the microprocessor and the CDU; each can receive information from or send information to the other.

THE CONTROL AND DISPLAY UNIT (CDU)

A GNSS receiver will often have a CDU. From handheld keyboards to soft keys around a screen to digital map displays and interfaces to other instrumentation there are a variety of configurations. Nevertheless, they all have the same fundamental purpose, facilitation of the interaction between the operator and the receiver's microprocessor. A CDU typically displays status, position data, velocity, and time. It may also be used to select different surveying methods, waypoint navigation, and/or set parameters such as epoch interval, mask angle, and antenna height. The CDU can offer a combination of help menus, prompts, datum conversions, readouts of survey results, estimated positional errors, and so forth. The information available from the CDU varies from receiver to receiver. However, when four or more satellites are available, they can generally be expected to display the PRN numbers of the satellites being tracked, the receiver's position in three dimensions, and velocity information. Most of them also display the dilution of precision.

THE STORAGE

GNSS receivers today have internal data logging. The amount of storage required for a particular session depends on several things: the length of the session, the number of satellites above the horizon, the epoch interval, and so forth. For example, presuming the amount of data received from a single GNSS satellite is ~100 bytes per 1-second epoch, a typical multichannel multifrequency receiver observing six satellites and using a 1-second epoch interval over the course of a 1-hour session would require ~2 MB of storage capacity for that session.

THE POWER

BATTERY POWER

Since most receivers in the field operate on battery power, batteries, and their characteristics are fundamental to GNSS surveying. A variety of batteries are used and there are various configurations. In surveying applications, rechargeable batteries are the norm. Lithium, nickel cadmium (NiCad), and nickel metal-hydride (NiMH) may be the most common categories, but lead-acid car batteries still have an application as well. The obvious drawbacks to lead-acid batteries are their size and weight. There are a few others—the corrosive acid, the need to store them charged, and their low cycle life. Nevertheless, lead-acid batteries are especially hard to beat when high power is required. They are economical and long-lasting. NiCad batteries cost more than lead-acid batteries but are small and operate well at low temperatures. Their capacity does decline as the temperature drops. Like lead-acid batteries, NiCad batteries are quite toxic. They self-discharge at the rate of about 10% per month, and even though they do require periodic full discharge, these batteries have an excellent cycle life. NiMH batteries self-discharge a bit more rapidly than NiCad batteries and have a less robust cycle life but are not as toxic. Lithium-ion batteries overcome several of the limitations of the others. They have a relatively low self-discharge rate. They do not require periodic discharging and do not have memory issues as do NiCad batteries. They are light, have a good cycle life, and low toxicity. On the other hand, the others tolerate overcharging and the lithium-ion battery does not. It is best not to charge lithium-ion batteries in temperatures at or below freezing. These batteries require a protection circuit to limit current and voltage but are widely used to power electronic devices, including GNSS receivers. It is fortunate that GNSS receivers operate at low power; from 9 to 36 volts, DC is generally required. This allows longer observations with fewer and lighter batteries than might be otherwise required. It also increases the longevity of the GNSS receivers themselves.

RECEIVERS

There are now so many specialized receivers built it is not possible to be complete in their description; however, all types of GNSS receivers are growing ever smaller. Today's capability to embed hundreds of thousands of transistors on a single silicon semiconductor microchip, known as *very large-scale integration* (*VLSI*), means miniaturized receivers weigh less and consume less power. The trend will undoubtedly continue. There will continue to be more correlators on more channels, better mitigation methods, and utilization of more signals from more constellations than in the past. Over 2 billion GNSS chips are driving the receivers in the civil consumer market including those in smartphones, smartwatches, tablets, notebook PCs, drones, handheld devices, digital cameras, etc. While there are more and more multifrequency and multi-constellation capabilities in such receivers it seems single-frequency devices continue to predominate. Civil consumer receivers typically do not have onboard feature data collection capabilities. They also do not usually have adequate onboard storage for recording the features (coordinates and attributes) required for a mapping project. Such capabilities are not needed for their designed applications. When such a receiver is used to obtain an autonomous or stand-alone position, its precision may be within several meters. Under less optimal field conditions; tree cover and other obstructions, less than favorable GNSS satellite geometry, etc., users can expect the precision of autonomous positions to lessen. Despite the limitations, some of these receivers have capabilities that enhance their systematic precision and achieve a quantifiable accuracy using correction signals available from Earth-orbiting satellites such as the *Wide Area Augmentation System* (*WAAS*) correction, more about that in Chapter 6.

LOCAL AND NETWORK ACCURACY

Network accuracy differs from *local accuracy*. Network accuracy is usually meant to concern the uncertainty of a position relative to a datum or reference frame, local accuracy does not. It represents the uncertainty of a position relative to other positions nearby. In other words, local accuracy would be useful in knowing the accuracy of a line between the two positions at each end. Network accuracy would be less about the accuracy of the positions at each end of the line relative to each other than both relative to the whole datum or reference frame. Local accuracy is also known as relative accuracy and network accuracy is also known as absolute accuracy. The network and local accuracy values provide very different pictures. The local category represents the accuracy of a point with respect to adjacent points. The network category represents the accuracy of a point with respect to the reference system.

Local horizontal and vertical accuracies represent the variance and uncertainty in points relative to adjacent points to which they are directly connected (see Figure 4.3). Local horizontal coordinate accuracy is computed considering the errors between the point in question and other

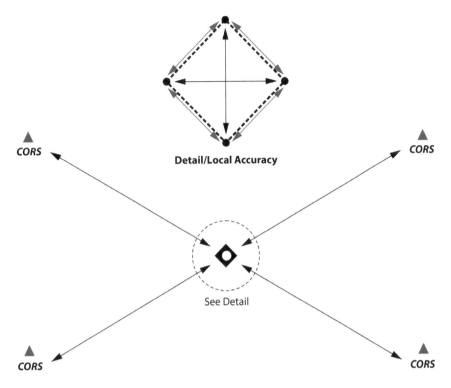

Detail/Local Accuracy

See Detail

FIGURE 4.3   Local and Network Accuracy

adjacent points. Height accuracy is computed using the linear vertical error between the point in question and other adjacent points. Within a well-defined geographical area, local accuracy may be the most immediate concern. However, those tasked with constructing a control network that embraces a wide geographical scope will most often need to know the positions relationship to the realization of the datum or reference frame on which they are working. A point with good local accuracy may not have good network accuracy and vice versa.

Typically, network horizontal and vertical accuracies require that a point's accuracy be specified with respect to an appropriate national geodetic system. In the United States, as a practical matter, this most often means that the work is tied to at least one of the Continuously Operating Reference Stations (CORS), which represent the most accessible realization of the National Spatial Reference System (NSRS) in the nation.

Mapping Receivers

These receivers are generally defined as those that allow the user to configure some settings such as PDOP, SNR, elevation mask, and the logging rate. They most often have an integrated antenna and CDU in the receiver. They generally record pseudoranges and can log data suitable for differential corrections either in real time or for post-processing, many record carrier-phase data, too. Mapping receivers are often capable of storing mapped features (coordinates and attributes) and usually have adequate capacity for mapping applications. This memory is required for differential GNSS, DGNSS receivers, and even those that track only the code phase. For many applications, a receiver must be capable of collecting the same information as is simultaneously collected at a base station, and storing it for post-processing. Receivers typically depend on proprietary post-processing software, which also includes utilities to enable GNSS data to be transferred to a PC and exported in standard *Geographic Information System* (*GIS*) file format(s) either over a cable or a wireless connection. Some mapping receivers for spatial data collection are single-frequency, L1, with code only or both code and carrier. Most mapping receivers of all tracking configurations are WAAS (or other *Satellite-Based Augmentation System* [*SBAS*] enabled) and thereby offer real time results.

SURVEYING RECEIVERS

Receivers designed for surveying strive to use all constellations, all frequencies, and both current and future signals. These code and carrier-phase multisystem, multifrequency receivers have precise oscillators and advanced antennas. They are appropriate for collecting positions on long baselines. They are optimized to run software specific to work in *Real-Time Network* (*RTN*) work, *Real-Time Kinematic* (*RTK*), usually with radio modem integration modules, PPP, etc. They are built for the achievement of consistent network accuracy in static or real time mode. They generally provide options for setting the observational parameters and the components of these receivers can usually be configured in a variety of ways.

Most share some practical characteristics: they have hundreds to many hundreds of independent channels that track the satellites continuously, and they begin acquiring satellite signals from a few seconds to less than a minute from the moment they are switched on. Most acquire all the satellites above their mask angle in a very few minutes, with the time usually lessened by a warm start, and most provide some sort of alert to the user that data is being recorded, and so forth. About three-quarters of them can have their sessions preprogrammed in the office before going to their field sites. Nearly all allow the user to select the logging rate, also known as epoch interval and also known as the sampling interval. While a

1-second interval is often used faster rates of 0.1 seconds (10 Hz) and more increments of tenths of a second are often available. This feature allows the user to stipulate the short period of time between each of the microprocessor's downloads to storage. The faster the data-sampling rate, the larger the volume of data a receiver collects and the larger the amount of storage it needs. A fast rate is helpful in cycle slip detection, and that improves the receiver's performance on baselines longer than 50 km, where the detection and repair of cycle slips can be particularly difficult.

EXERCISES

1. Which of the following is not a consideration in antenna design for GNSS receivers?
   a. Efficient conversion of electromagnetic waves into electric currents
   b. Directional capability
   c. Coincidence of the phase center and the physical center
   d. Reduction of the effects of multipath

2. Which of the ideas listed below is intended to limit the effect of the difference between the phase center and the physical center of GNSS antennas on a baseline measurement?
   a. Ground plane antennas at each end of the baseline.
   b. Choke ring antennas at each end of the baseline.
   c. Rotation of the antenna's reference marks to north at each end of the baseline.
   d. The use of the receiver's built-in antennas at each end of the baseline.

3. Which of the following statements about the utility of the Doppler effect in GNSS is not correct?
   a. The Doppler effect-based change in the frequency collected at a receiver can be used as a measure of the range rate.
   b. The signal received by a static GNSS receiver would see a Doppler shift of ±4.5 to 5 cycles per millisecond during a full pass of a satellite from rising to setting.
   c. Integration of the Doppler frequency offset can result in an accurate measurement of the advance in carrier-phase between epochs.
   d. The Doppler shifted signal was the fundamental observable of the TRANSIT but is no longer used in GNSS.

4. Which of the following statements concerning the channels in an all-in-view multichannel receiver is correct?
   a. The channels are dependent on one another.
   b. Each channel acquires and tracks the signals from several satellites simultaneously.
   c. The channels operating in parallel means the receiver cannot maintain accuracy on a moving platform.
   d. Each channel in an operating multichannel receiver is either busy acquiring a GNSS signal or tracking one.

5. Which of the following is not a drawback of signal squaring?
   a. The effect of multipath is decreased.
   b. The SNR deteriorates.
   c. The codes are stripped from the carrier.
   d. The receiver must acquire information such as almanac data and clock corrections somewhere other than the navigation message.

6. Which of the following is the closest to the length of a C/A code period that is 1023 chips and 1 millisecond duration?
   a. 20,000 km
   b. 300 m
   c. 300 km
   d. 66 m

7. What is an isotropic antenna?
   a. A hypothetical lossless antenna that has equal capabilities in all directions.
   b. A durable, compact antenna with a simple construction and a low profile.
   c. An antenna built with several concentric rings and designed to reduce the effects of multipath.
   d. An antenna that has a stable phase center and simple construction but no reception on the perpendicular and should not be used for navigation.

8. Which of the following statements is correct?
   a. The most powerful GPS signal power is assigned to the L2C signal from Block IIF and Block IIR-M satellites.
   b. The PCO is the movement of an antenna's phase center as the frequency, azimuth, and elevation of the received signal changes.

c. The PCV is an accessible point on an antenna to which instrument heights can be measured.

d. An NCO is used to control the generation of the replica signal in the receiver to ensure that the replica matches the received signals.

9. How does the antenna bandwidth affect an antenna's performance?
   a. It does not have any effect.
   b. The narrower the bandwidth the lower the gain.
   c. Larger bandwidth increases antenna's performance but also increases interference.
   d. The larger the bandwidth the smaller the antenna's operating range.

10. Which of the following describes network accuracy correctly?
    a. Network accuracy and precision are the same.
    b. Network accuracy is also called local accuracy.
    c. Network accuracy represents the location of the point with respect to the datum or reference frame.
    d. Network accuracy is also known as relative accuracy.

ANSWERS AND EXPLANATIONS

1. Answer is (b)

   Explanation: The GNSS signal is quite weak and it is broadcast over a very large area. The efficiency of GNSS antennas is an important consideration. Limiting the effects of multipath is also a very important consideration. The perfect coincidence of the phase center with the physical center in these antennas has yet to be achieved, but it is much sought after and is certainly a design consideration. However, the GPS signal is a spread-spectrum coded signal that occupies a broader frequency bandwidth than it must carry its information. This fact allows GNSS antennas to be omnidirectional. They do not require directional orientation to properly receive the signal.

2. Answer is (c)

   Explanation: Ground planes used with many GNSS antennas and the choke ring antennas are designs that attack the same problem, limiting the effects of multipath. The coincidence of the phase center with the physical center in GNSS antennas is not yet perfected, and since the measurements made by a GNSS receiver are made to the phase center of its antenna, its orientation is

paramount. The rotation of the antennas at each end of a baseline
to the north per an imprinted reference is a strategy to reduce the
effect of the difference between the two points.

3. Answer is (d)

   Explanation: Since the orbit of Sputnik 1 was worked out by
   measuring the Doppler frequency shift of its 20 MHz radio signals received from the satellite, the Doppler shift has been a valuable observable in satellite-based positioning including GNSS.
   Doppler information has broad applications in signal processing,
   but perhaps the most important application of Doppler data is
   the determination of the range rate. It is useful because the rate
   at which the range, the distance, between the satellite, and the
   receiver changes over a particular period of time is approximately
   equal to the difference between the frequency of the signal at the
   receiver and the frequency of the signal transmitted by the satellite, a principle demonstrated with the TRANSIT system and still
   used today.

4. Answer is (d)

   Explanation: The channels are independent from one another.
   Each frequency from each satellite is segregated, acquired, and
   tracked by its own dedicated channel. Since the channels operate
   in parallel, the receiver can maintain accuracy on a moving platform. It is correct that each channel in a multichannel receiver is
   either busy acquiring a GNSS signal or tracking one.

5. Answer is (a)

   Explanation: With signal squaring, the beat or IF is created by
   multiplying the incoming carrier by itself. The result has double
   the frequency and half the wavelength of the original. It is squared.
   The process of squaring the carrier strips off all the codes, which
   means that not only the PRN codes but also the navigation message are not available to the receiver, and with the codes wiped out,
   the receiver must acquire information such as almanac data and
   clock corrections from other sources. Also, the SNR is degraded,
   increasing cycle slips. On the other hand, the effect of multipath
   is decreased, and since squaring works as well on L2 as it does on
   L1, multifrequency ionospheric delay correction is possible.

6. Answer is (c)

   Explanation: The C/A code "chipping rate," the rate at which
   each chip is modulated onto the carrier, is 1.023 Mbps. That means,

at light speed, the chip length is approximately 300 m, but the whole C/A code period is 1023 chips, 1 millisecond long. That is approximately 300 km, and, of course, each satellite repeats its whole 300 km C/A code over and over.

7. Answer is (a)

Explanation: An isotropic antenna is a hypothetical, lossless antenna that has equal capabilities in all directions. This theoretical antenna is used as a basis of comparison when expressing the gain of GNSS and other antennas. A gain of about 3 dB is typical for the usual omnidirectional GNSS antenna. The gain, or gain pattern, describes the success of a GNSS antenna in collecting more energy from above the mask angle and less from below the mask angle. The decibel here does not indicate the power of the antenna; it refers to a comparison. In this case, 3 dB indicates that the GNSS antenna has about 50% of the capability of an isotropic antenna.

8. Answer is (d)

Explanation: The most powerful minimum signal power is assigned to L5 from Block III satellites at −154 dBW. The L2C signal from Block IIF and Block IIR-M satellites is less powerful at −160 dBW. The phase center of the receiver's antenna is not an immovable point. It moves a bit as the frequency, azimuth, and elevation of the received signals change. That change is called the PCV. The PCO is the vector between an antenna's phase center and the ARP. The ARP is an accessible point usually marked on a GNSS receiver antenna where the vertical axis of the antenna intersects the bottom mount, ground plane, or choke ring. It is a point to which instrument heights can be measured. The input to the NCO, the programmable linear frequency generator, is varied to control the generation of the replica signal in the receiver in a feedback loop endeavoring to ensure that the replica matches the received signals and keeps on matching it.

9. Answer is (c)

Explanation: An antenna ought to have a bandwidth commensurate with its application; in general, the larger the bandwidth, the better the performance. However, there is a downside; increased bandwidth degrades the SNR by including more interference. The larger the bandwidth the larger the antenna's operating range. The narrower an antenna's bandwidth the higher the gain.

10. Answer is (c)

Explanation: The use of phrase network accuracy is used to define its difference from local accuracy. Network accuracy here concerns the uncertainty of a position relative to a datum or reference frame. Local accuracy represents the uncertainty of a position relative to other positions nearby. In other words, local accuracy would be useful in knowing the accuracy of a line between the two positions at each end. Network accuracy would not be about the accuracy of the positions at each end of the line relative to each other but rather relative to the whole datum. Local accuracy is also known as relative accuracy and network accuracy is also known as absolute accuracy.

5 Coordinates

A FEW PERTINENT IDEAS ABOUT GEODESY FOR GNSS

PLANE SURVEYING

Plane surveying has traditionally relied on an imaginary flat reference surface, or *datum*, with Cartesian axes. This rectangular system is used to describe measured positions by ordered pairs, usually expressed in northings and eastings, or y- and x-coordinates. Even though surveyors have always known that this assumption of a flat Earth is fundamentally unrealistic, it provided, and continues to provide, an adequate arrangement for small areas. The attachment of elevations to such horizontal coordinates somewhat acknowledges the topographic irregularity of the Earth, but the whole system is always undone by its inherent inaccuracy as surveys grow large.

DEVELOPMENT OF STATE PLANE COORDINATE SYSTEMS

Designed in the 1930s, the purpose of the State Plane Coordinate System (SPCS) was to overcome some of the limitations of the horizontal plane datum when they are applied over large areas while avoiding the imposition of geodetic methods and calculations on local surveyors. Using the conic and cylindrical models of the Lambert and Mercator map projections, respectively, the flat datum was curved but only in one direction. By curving the datums and limiting the area of the zones, the distortion could be limited to a scale ratio of about 1 part in 10,000 without disturbing the traditional system of ordered pairs of Cartesian coordinates. The SPCS was a step ahead at that time. To this day, it provides surveyors with a mechanism for coordination of surveying stations that approximates geodetic accuracy more closely than the commonly used methods of small-scale plane surveying. However, the SPCSs were organized in a time of generally lower accuracy and efficiency in surveying measurement. It was an understandable compromise in an age when such computation required sharp pencils, logarithmic tables, and lots of midnight oil.

The distortion of positions attributable to the transformation of geodetic coordinates into the plane grid coordinates of any one of these projections is generally small. In fact, the state plane system has been recently augmented by Low Distortion Projection (LDP) plane projected coordinate systems that are designed to further minimize linear distortion.

DOI: 10.1201/9781003405238-5

Most GNSS and land surveying software packages provide routines for automatic transformation of latitude and longitude to and from these mapping projections. Similar programs are available from the National Geodetic Survey (NGS). Therefore, for most applications of GNSS, there ought to be no technical compunction about expressing the results in grid coordinates. However, because the results are presented in plane coordinates, it can be easy to lose sight of the geodetic context of the entire process that produced them. The following is an effort to provide some context to that relationship.

GNSS SURVEYORS AND GEODESY

Today, GNSS has thrust surveyors into the thick of geodesy, which is no longer the exclusive realm of distant experts. Thankfully, in the age of the microcomputer, computational drudgery can be handled with off-the-shelf software. Nevertheless, it is unwise to venture into GNSS believing that knowledge of the basics of geodesy is, therefore, unnecessary. It is true that GNSS would be impossible without computers, but blind reliance on the data they generate eventually leads to disaster.

SOME GEODETIC COORDINATE SYSTEMS

THREE-DIMENSIONAL CARTESIAN COORDINATES

A spatial Cartesian system with three axes lends itself to describing the terrestrial positions derived from space-based geodesy. Using three rectangular coordinates instead of two, one can unambiguously define any position on the Earth, or above it for that matter. The three-dimensional Cartesian coordinates (X, Y, Z) derived from this system are known as *earth-centered earth-fixed* (*ECEF*) coordinates. It is a right-handed orthogonal system that rotates with and is attached to the Earth, which is why it is called Earth fixed. A three-dimensional Cartesian coordinate system is right-handed if it can be described by the following model: the extended forefinger of the right hand symbolizes the positive direction of the x-axis. The middle finger of the same hand extended at right angles to the forefinger symbolizes the positive direction of the y-axis. The extended thumb of the right hand, perpendicular to them both, symbolizes the positive direction of the z-axis (see Figure 5.1). However, such a system is only useful if its origin (0,0,0) and its axes (x, y, z) can be fixed to the planet with certainty, something easier said than done.

The usual arrangement is known as the Conventional Terrestrial Reference System (CTRS) and the Conventional Terrestrial System (CTS). The latter name will be used here. The origin is the center of mass of the

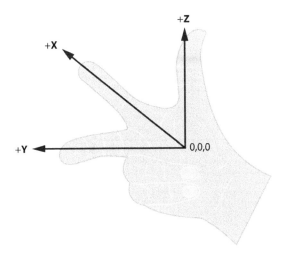

FIGURE 5.1 Right-Handed Three-Dimensional Cartesian Axes

whole Earth including oceans and atmosphere, the *geocenter*. The *x*-axis is a line from that geocenter through its intersection at the zero meridian, also known as the *International Reference Meridian (IRM)*, with the internationally defined conventional equator. The *y*-axis is extended from the geocenter along a line perpendicular from the *x*-axis in the same mean equatorial plane toward 90°E longitude. That means that the positive end of the *y*-axis intersects the actual Earth in the Indian Ocean. In any case, they both rotate with the Earth around the *z*-axis, a line from the geocenter through the internationally defined pole known as the *International Reference Pole (IRP)*.

However, the Earth is constantly moving, of course. While one can say that the Earth has a particular axis of rotation, equator, and zero meridian for an instant, they all change slightly in the next instant. Within all this motion how do you stabilize the origin and direction of the three axes for the long term? One way is to choose a moment in time and consider them fixed to the Earth as they are at that instant.

POLAR MOTION

Here is an example of that process of definition. The Earth's rotational axis wanders slightly with respect to the solid Earth in a very slow oscillation called *polar motion*. The largest component of the movement relative to the Earth's crust has a 430-day cycle known as the *Chandler period*. It was named after American astronomer Seth C. Chandler, who described it in papers in the *Astronomical Journal* in 1891. Another aspect of polar motion is sometimes called polar wander. It is about 0.004 seconds of arc

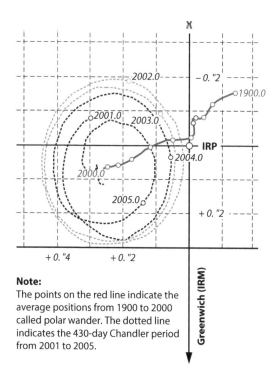

Note:
The points on the red line indicate the average positions from 1900 to 2000 called polar wander. The dotted line indicates the 430-day Chandler period from 2001 to 2005.

FIGURE 5.2   Polar Motion

per year as the pole moves toward Ellesmere Island (see Figure 5.2). The actual displacement caused by the wandering generally does not exceed 12 m. Nevertheless, the CTS of coordinates would be useless if its third axis was constantly wobbling. Originally, an average stable position was chosen for the position of the pole. Between 1900 and 1905, the mean position of the Earth's rotational pole was designated as the *Conventional International Origin* (*CIO*) and the *z*-axis.

This was defined by the *Bureau International de l'Heure* (*BIH*). It has since been refined by the International Earth Rotation and Reference Systems Service (IERS) using very long baseline interferometry (VLBI) and satellite laser ranging (SLR). It is now placed as it was at midnight on New Year's Eve 1983, or January 1, 1984 (UTC). The moment is known as an *epoch* and can be written as 1984.0. So, we now use the axes illustrated in Figure 5.3. The name of the *z*-axis has been changed to the IRP epoch 1984, but it remains within 0.005″ of the previous definition. It provides a geometrically stable and clear definition on the Earth's surface for the *z*-axis.

In this three-dimensional right-handed coordinate system, the *x*-coordinate is a distance from the *y*-*z* plane measured parallel to the *x*-axis. It is always positive from the zero meridian to 90°W longitude and from the

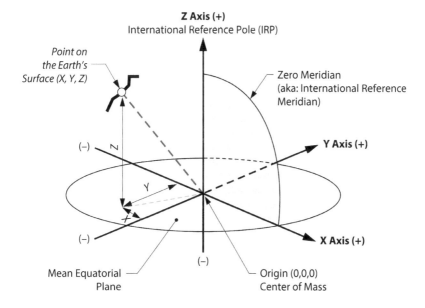

FIGURE 5.3 Three-Dimensional Coordinate (ECEF)

zero meridian to 90°E longitude. In the remaining 180°, the *x*-coordinate is negative. The *y*-coordinate is a perpendicular distance from the plane of the zero meridian. It is always positive in the Eastern Hemisphere and negative in the Western Hemisphere. The *z*-coordinate is a perpendicular distance from the plane of the equator. It is always positive in the Northern Hemisphere and negative in the Southern Hemisphere. Here is an example, the position of the station CTMC expressed in three-dimensional Cartesian coordinates of this type expressed in meters, the native unit of the system (Table 5.1).

Modern GNSS software provide the positions in ECEF. Further, the ends of baselines determined by GNSS observation are typically given in ECEF coordinates so that the vectors themselves become the difference between those *X*, *Y*, and *Z* coordinates. The display of these differences as DX, DY, and DZ is a usual product of these post-processed calculations (see Figure 5.4).

LATITUDE AND LONGITUDE

Despite their utility, such three-dimensional Cartesian coordinates are not the most common method of expressing a geodetic position. Latitude and longitude have been the coordinates of choice for centuries. The designation of these relies on the same two standard lines as three-dimensional

TABLE 5.1
Station CTMC

ITRF2014
CDOT GOLDEN (CTMC), COLORADO

Antenna Reference Point (ARP) : CDOT GOLDEN CORS ARP

PID =DM5962

ITRF2014 POSITION (EPOCH 2010.0)
Computed in Jun 2019 using data through gpswk 1933.

X = -1287788.532 m	latitude = 39 43 17.51335 N
Y = -4742177.053 m	longitude = 105 11 34.38099 W
Z = 4055414.539 m	height = 1818.969 m

ITRF2014 VELOCITY
Computed in Jun 2019 using data through gpswk 1933.

VX = -0.014 m/yr	northward = -0.0061 m/yr
VY = -0.0001 m/yr	eastward = -0.0142 m/yr
VZ = -0.0047 m/yr	upward = 0.0000 m/yr

Occupation Time:	00:30:30:00	
Measurement Epoch Interval (seconds):	15:00	
Solution Time:	Receiver/Satellite double difference Iono free fixed	
Solution Acceptability:	Passed ratio test	
Baseline Slope Distance Std. Dev. (meters):	17044.376	0.000574
	Forward	*Backward*
Normal Section Azimuth:	179° 04' 38.886169"	359° 04' 47.857511"
	0° 0.7' 00.456589"	−0° 16' 12.548396"
Baseline Components (meters):	dn −17042.131 de 274.423 du 34.744	
Standard Deviations:	dx −6552.297 dy −10264.496 dz −11925.530	
	8.437168E-004 1.072974E-003 9.513724E-004	
Aposteriori Covariance Matrix:	7.118580E-007	
	7.389287E-007 1.151273E-006	
	−6.141036E-007 −7.690377E-007 9.051094E-004	
Variance Ratio Cutoff:	62.1	1.5
Reference Variance:	0.556	
Observable Count/Rejected RMS:	Iono free phase 451/0 0.006	

FIGURE 5.4   DX, DY, and DZ from GNSS

Cartesian coordinates: the mean equator and the zero meridian. Unlike them, however, they require some clear representation of the terrestrial surface. In modern practice, latitude and longitude cannot be said to uniquely define a position without a clear definition of the Earth itself.

ELEMENTS OF A GEODETIC DATUM

How can latitude, φ, and longitude, λ, not define a unique position on the Earth? The reference lines—the mean equator and the zero meridian—are clearly defined. The units of degrees, minutes, seconds, and decimals of seconds allow for the finest distinctions of measurement. Finally, the reference surface is the Earth itself. The answer to the question relies, in the first instance, on the fact that there are several categories of latitude and longitude, the geographical coordinates. From the various options, astronomic, geocentric, and geodetic, the discussion here will concern geodetic latitude and longitude as they are typical. Its definition has a good deal to do with the direction of down.

THE DEFLECTION OF THE VERTICAL

Down seems like a pretty straightforward idea. A hanging plumb bob certainly points down. Its string follows the direction of gravity. That is one version of the idea. There are others. Imagine an optical surveying instrument set up over a point. If it is centered precisely with a plumb bob and leveled carefully, the plumb line and the line of the level telescope of the instrument are perpendicular to each other. In other words, the level line, the horizon of the instrument, is perpendicular to gravity. Using an instrument so oriented it is possible to determine the latitude and longitude of the point. Measuring the altitude of a circumpolar star is one good method of finding the latitude of the point from which the measurement is made. The measured altitude would be relative to the horizontal level line of the instrument, of course. A latitude found this way is called *astronomic latitude.*

One might expect that this astronomic latitude would be the same as the geocentric latitude of the point, but they are different. The difference is because a plumb line coincides with the direction of gravity; it does not point to the center of the Earth where the line used to derive geocentric latitude originates. Astronomic latitude also differs from the most widely used version of latitude, geodetic. The line from which geodetic latitude is determined is perpendicular with the surface of the ellipsoidal model of the Earth. That does not match a plumb line either, more about that in the next section. In other words, there are three different versions of down and each with its own latitude. For astronomic latitude down is along a line in the direction of gravity. For geocentric latitude, down is along a line to the center of mass of the Earth, the geocenter. For geodetic latitude down is along a line perpendicular to the ellipsoidal model of the Earth. Often these are three completely different lines, as shown in Figure 5.5.

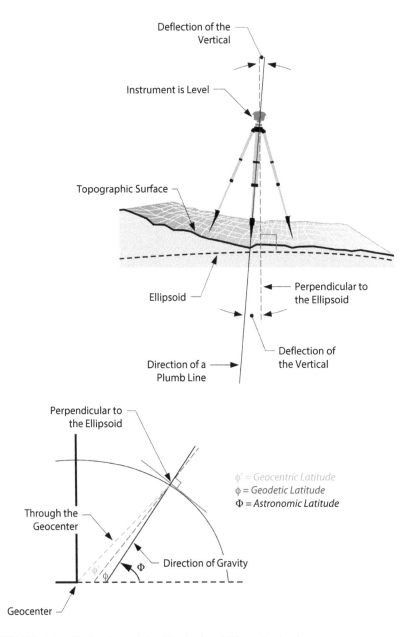

FIGURE 5.5   Deflection of the Vertical and Three Latitudes

GEOCENTRIC, GEODETIC, AND ASTRONOMIC LATITUDE

Each version of down can be extended upward, too, toward the zenith, and there are small angles between them. The angle between the vertical extension of a plumb line and the vertical extension of a line perpendicular to the *ellipsoid* is called the deflection of the vertical. It sounds better than the difference in down. This *deflection of the vertical* defines the actual angular difference between the astronomic latitude and longitude of a point and its geodetic latitude and longitude.

Even though the discussion has so far been limited to latitude, the deflection of the vertical usually has both a north–south and an east–west component. The deflection of the vertical also has an effect on azimuths; for example, there will be a slight difference between the geodetic azimuth of a baseline and the astronomically determined azimuth of the same line.

It is interesting to note that that optical instrument set up so carefully over a point on the Earth cannot be used to measure geodetic latitude and longitude directly because they are relative to an ellipsoidal model of the Earth rather than the Earth itself. Gravity does not even come into the ellipsoidal version of down.

On the *reference ellipsoid* model of the Earth down is a line perpendicular to the ellipsoidal surface at a particular point. On the real Earth down is the direction of gravity at the point. They are most often not the same thing, but fortunately, the difference is usually very small. Since the ellipsoidal model of the Earth is imaginary, it is quite impossible to set up an instrument on it. On the other hand, the measurement of latitude and longitude by astronomic observations has a very long history indeed. Nevertheless, astronomic latitudes and longitudes are not the most used coordinates, geodetic latitudes and longitudes are. Therefore, the conversion from astronomic latitude and longitude to geodetic latitude and longitude has a long history too. Until the advent of GNSS geodetic latitudes and longitudes were often derived from astronomic observations by post-observation calculation. In a sense, that is still true. It is just that calculations can be completed with incredible speed by a modern GNSS receiver's computer. However, a fundamental fact remains unchanged: the instruments by which latitudes and longitudes are measured are oriented to gravity, the ellipsoidal model on which geodetic latitudes and longitudes are determined is not. That is just as true for the antenna of a GNSS receiver, an optical surveying instrument, a camera in an airplane taking aerial photography, or even the GNSS satellite themselves.

DATUMS AND REFERENCE FRAMES

The second part of the answer to the question posed earlier is this: for geographic coordinates to have meaning they must have a context. Despite the certainty of the physical surface of the Earth the lithosphere, it remains notoriously difficult to define in mathematical terms. The dilemma is illustrated by the ancient struggle to represent its curved surface on flat maps. There have been a whole variety of map projections developed over the centuries that rely on mathematical relationships between positions on the Earth's surface and points on the map. Each projection serves a particular application well, but none of them can represent the Earth without distortion. For example, no modern surveyor would presume to promise a client a high-precision control network with data scaled from a paper map. As the technology of measurement has improved, the pressure for greater exactness in the definition of the Earth's shape has increased. Even with electronic tools that widen the scope and increase the precision of the data, perfection is nowhere in sight.

Development of the Ellipsoidal Model

Local topography is obvious to an observer standing on the Earth, but efforts to grasp the more general nature of the planet's shape and size have been occupying scientists for at least 2300 years. There have, of course, been long intervening periods of unmitigated nonsense on the subject. Nevertheless, ever since 200 BC, when Eratosthenes almost calculated the planet's circumference correctly, geodesy has been getting ever closer to expressing the actual shape of the Earth in numerical terms. A leap forward occurred with Newton's thesis that the Earth was an ellipsoid rather than a sphere in the first edition of his *Principia* in 1687. Newton's idea that the actual shape of the Earth was slightly ellipsoidal was not entirely independent. There had already been some other suggestive observations. For example, 15 years earlier, astronomer Jean Richer had found that to maintain the accuracy of the one-second clock he used in his observations in Cayenne, French Guiana, he had to shorten its pendulum significantly. The clock's pendulum, regulated in Paris, tended to swing more slowly as it approached the equator. Newton reasoned that the phenomenon was attributable to a lessening of the force of gravity. Based on his own theoretical work, he explained the weaker gravity by the proposition, "the Earth is higher under the equator than at the poles, and that by an excess of about 17 miles" (*Philosophiae naturalis principia mathematica*, Book III, Proposition XX).

Although Newton's model of the planet bulging along the equator and flattened at the poles was supported by some of his contemporaries, notably Huygens, the inventor of Richer's clock, it was attacked by others. The director of the Paris Observatory, Jean Dominique Cassini, for example, took exception to Newton's concept. Even though the elder Cassini had himself observed the flattening of the poles of Jupiter in 1666, neither he nor his equally learned son Jacques was prepared to accept the same idea when it came to the shape of the Earth. It appeared they had some empirical evidence on their side. Scientists had employed arc measurements at various latitudes for geometric verification of the Earth model, certainly since the early 1500s and perhaps even earlier. They would choose a meridian of longitude, establish on it a beginning and ending point. The two points would have exactly one degree of latitude between them. So prepared, they measured the distance between the two points along that north–south line. The result revealed the length of one degree of latitude along that particular meridian of longitude. Early attempts assumed a spherical Earth, and the results were used to estimate its radius by simple multiplication. In fact, one of the most accurate of the measurements of this type, begun in 1669 by the French Abbé J. Picard, was actually used by Newton in formulating his own law of gravitation. However, Cassini noted that a close analysis of Picard's arc measurement, and others, seemed to show the length of one degree of latitude decreased as one proceeds northward. He concluded that this showed the Earth was not flattened as proposed by Newton but was rather elongated at the poles.

The argument was not resolved until two expeditions between about 1733 and 1744 were completed. They were sponsored by the Paris Académie Royale des Sciences and produced irrefutable proof. One group, which included Clairaut and Maupertuis, was sent to measure the length of one-degree of latitude along a meridian arc near the Arctic Circle, 66°20′ Nφ, in Lapland. Another expedition with Bouguer and Godin, to what is now Ecuador, measured a meridian arc near the equator, 01°31′ Sφ. The work showed that the length of one degree of latitude measured along a meridian increases as you go northward and Newton's conjecture was proved correct. The contradictory evidence of Picard's arc was charged to errors in the latter's measurement of the astronomic latitudes. The ellipsoidal model (see Figure 5.6), bulging at the equator and flattened at the poles, has been used ever since as a representation of the general shape of the Earth's surface. It is called an oblate spheroid. In fact, several reference ellipsoids have been established for various regions of the planet. They are precisely defined by their semi-major axis and flattening. The relationship between these parameters is expressed in the formula:

$$f = 1 - \frac{b}{a}$$

where
  f = flattening
  a = semi-major axis
  b = semi-minor axis

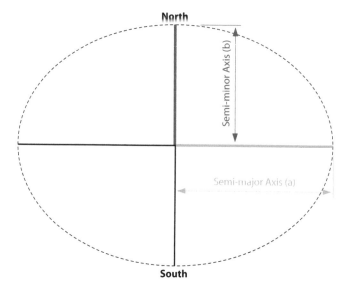

Parameters of a Biaxial Ellipsoid

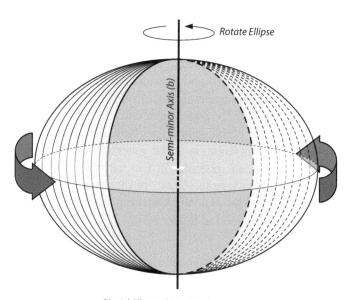

Biaxial Ellipsoid Model of the Earth

FIGURE 5.6   An Ellipsoid

BIAXIAL ELLIPSOIDAL MODEL OF THE EARTH

THE ROLE OF AN ELLIPSOID IN A DATUM

The semi-major axis and flattening can be used to completely define an ellipsoid of revolution. The ellipse is revolved around the minor axis. However, if that ellipsoid is to be used as a geodetic datum, six additional elements are required to describe its orientation with respect to the Earth. In the traditional approach, three specifies its center and three more clearly indicate its orientation around that center. The Clarke 1866 spheroid is one of many reference ellipsoids. Its shape is completely defined by a semi-major axis, a, of 6378.2064 km and a flattening, f, of 1/294.9786982. It is the reference ellipsoid of the datum known to surveyors as the North American Datum of 1927 (NAD27), but it is not the datum itself.

For the Clarke 1866 spheroid to become NAD27, the ellipsoid of reference had to be attached at a point and specifically oriented to the actual surface of the Earth. However, even this ellipsoid, which fits North America well, could not conform to that surface perfectly. Therefore, the initial point was chosen near the center of the anticipated geodetic network to best distribute the inevitable distortion between the ellipsoid and the Earth as the network extended beyond the initial point. The attachment was established at Meades Ranch, Kansas, 39°13′26.686″ Nφ, 98°32′30.506″ Wλ, and *geoid height* was considered to be zero. Those coordinates were not sufficient, however. The establishment of directions from this initial point was required to complete the orientation. The azimuth from Meades Ranch to station Waldo was fixed at 75°28′09.64″ (from south) and the deflection of the vertical set at zero. Once the initial point and directions were fixed, the whole orientation of NAD27 was considered established, including the center of the reference ellipsoid. Its center was imagined to reside somewhere around the center of mass of the Earth. However, the two points were certainly not coincident, nor were they intended to be. In short, NAD27 does not use a geocentric ellipsoid.

REGIONAL ELLIPSOIDS

MEASUREMENT TECHNOLOGY AND DATUM SELECTION

In the period before space-based geodesy was tenable, a regional datum was not unusual. The Australian Geodetic Datum 1966, the European Datum 1950, and the South American Datum 1969, among many others, were also designed as nongeocentric systems. Achievement of the minimum distortion over a particular region was the primary consideration in choosing their ellipsoids and the orientation of those ellipsoids with respect to the surface of the Earth, not their centers position relative to the center

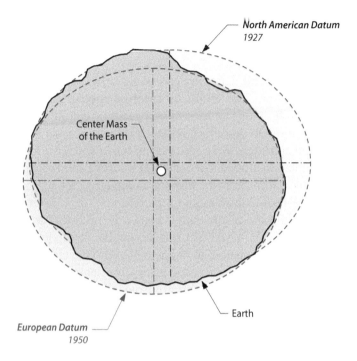

FIGURE 5.7   Nongeocentric Datums

of mass of the Earth (see Figure 5.7). For example, in the CTS, the three-dimensional Cartesian coordinates of the center of the Clarke 1866 spheroid as it was used for NAD27 are about $X = -4$ m, $Y = +166$ m, and $Z = +183$ m.

This approach to the design of datums was bolstered by the fact that the vast majority of geodetic measurements they would be expected to support were of the classical variety. That is, the work was done with theodolites, towers, and tapes. They were Earth-bound. Even after the advent of electronic distance measurement, the general approach involved the determination of horizontal coordinates by measuring from point to point on the Earth's surface and adding heights, otherwise known as *elevations*, through a separate leveling operation. As long as this methodological separation existed between the horizontal and vertical coordinates of a station, the difference between the ellipsoid and the true Earth's surface was not an overriding concern. Such circumstances did not require a geocentric datum. However, as the sophistication of satellite geodesy increased, the need for a truly global, geocentric datum became obvious. The horizontal and vertical information were no longer separate.

Today, the horizontal and vertical components of a position are derived from precisely the same vector, and the choice of the coordinate system used to express them is a matter of convenience. The position vector can be transformed into the three-dimensional Cartesian ECEF system,

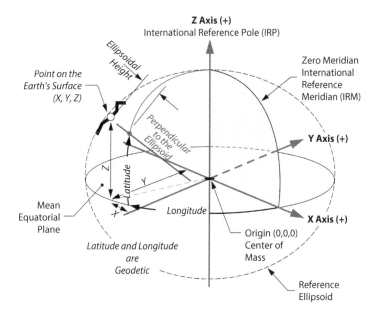

FIGURE 5.8   A Few Fundamentals

the traditional latitude, longitude, and height, or virtually any other well-defined coordinate system. However, since the orbital motion and the subsequent position vector derived from satellite geodesy are themselves Earth-centered, it follows that the most straightforward representations of that data are Earth-centered as well (see Figure 5.8).

POSITION DERIVED FROM GNSS

THE DEVELOPMENT OF A GEOCENTRIC MODEL

Satellites have not only provided the impetus for a geocentric datum, but they have also supplied the means to achieve it. In fact, the orbital per-turbations of man-made near-Earth satellites have probably brought more refinements to the understanding of the shape of the Earth in a shorter span of time than was ever before possible. For example, the analysis of the *precession* of Sputnik 2 in the late 1950s showed researchers that the Earth's semi-minor axis was 85 m shorter than had been previously thought. In 1958, while studying the tracking data from the orbit of Vanguard I, Ann Bailey of the Goddard Spaceflight Center discovered that the planet is shaped a bit like a pear. There is a slight protuberance at the North Pole, a little depression at the South Pole, and a small bulge just south of the equator. These formations and others have been discovered through the observation of small distortions in satellites' otherwise elliptical orbits, little bumps in their road, so to speak. The deviations are caused by the action of Earth's gravity on the satellites as they travel through space. Just as Richer's clock reacted to the lessening of gravity at the equator and thereby revealed one of the largest features of the Earth's shape to Newton, small perturbations in the orbits of satellites, also responding to gravity, reveal details of Earth's shape to today's scientists.

The common aspect of these examples is the direct relationship between direction and magnitude of gravity and the planet's form. In fact, the surface that best fits the Earth's gravity field has been given a name. It is called the geoid.

THE GEOID

An often-used description of the geoidal surface involves idealized oceans. Imagine the oceans of the world utterly still, completely free of currents, tides, friction, variations in temperature, and all other physical forces, except gravity. Reacting to gravity alone, these unattainable calm waters would coincide with the figure known as the geoid (see Figure 5.9). Admitted by small frictionless channels or tubes and allowed to migrate across the land, the water would then, theoretically, define the same geoidal surface across the continents, too.

Of course, 70% of the Earth covered by oceans is not so cooperative, nor is there any such system of channels and tubes. In addition, the physical forces eliminated from the model cannot be avoided in reality. These unavoidable forces actually cause *Mean Sea Level* (*MSL*) to deviate from the geoid. This is one of the reasons that MSL and the surface of the geoid

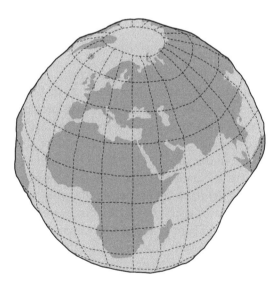

FIGURE 5.9   Exaggerated Representation of the Geoid

are not the same. It is a fact frequently mentioned to emphasize the inconsistency of the original definition of the geoid as it was offered by J.B. Listing in 1872. Listing thought of the geoidal surface as equivalent to MSL. Even though his idea does not stand up to scrutiny today, it can still be instructive.

An Equipotential Surface

Gravity is not consistent across the topographic surface of the Earth. At every point, it has a magnitude and a direction. In other words, anywhere on Earth, gravity can be described by a mathematical vector. Along the solid Earth, such vectors do not have all the same direction or magnitude, but one can imagine a surface of constant gravity potential. Such an *equipotential* surface would be level in the true sense. It would coincide with the top of the hypothetical water in the previous example. MSL does not define such a figure, nevertheless, the geoidal surface is not just a product of imagination. For example, the vertical axis of any properly leveled surveying instrument and the string of any stable plumb bob are perpendicular to the geoid. Just as pendulum clocks and Earth-orbiting satellites, they clearly show that the geoid is real.

Geoidal Undulation

The geoid does not precisely follow MSL, nor does it exactly correspond with the topography of the dry land. It is irregular like the terrestrial surface. It is bumpy. The uneven distribution of the mass of the planet makes it maddeningly. If the solid Earth had no internal density anomalies, the geoid would be smooth and almost exactly ellipsoidal. In that case, the reference ellipsoid could fit the geoid to near perfection and the lives of surveyors would be much simpler, but like the Earth itself, the geoid defies such mathematical consistency and departs from true ellipsoidal form by as much as 100 m in places (see Figure 5.10).

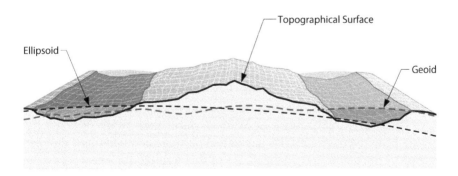

FIGURE 5.10   Three Surfaces

THE MODERN GEOCENTRIC DATUM

Three distinct figures are involved in a geodetic datum for latitude, longitude, and height: the geoid, the reference ellipsoid, and the topographic surface. Due in large measure to the ascendancy of satellite geodesy, it has become highly desirable that they share a common center, the geocenter. While the level surface of the geoid provides a solid foundation for the definitions of heights and the topographic surface of the Earth is necessarily where measurements are made, neither can serve as the reference surface for geodetic positions.

From the continents to the floors of the oceans, the solid Earth's actual surface is too irregular to be represented by a simple mathematical statement. The geoid, which is sometimes under, and sometimes above, the surface of the Earth, has an overall shape that also defies any concise geometrical definition. On the other hand, the ellipsoid not only has the same general shape as the Earth but, unlike the other two figures, can be described simply and completely in mathematical terms. For example, the global *Geodetic Reference System 1980* (*GRS80*) developed by the *International Union of Geodesy and Geophysics* (*IUGG*) in 1979 is based on a reference ellipsoid whose semi-major axis, a, 6378137 m is probably within a few of meters of the Earth's actual equatorial radius. Its flattening, f, is 1/298.257222100 and likely deviates only slightly from the true value. While this is a considerable improvement over Newton's calculation of a flattening ratio of 1/230, he did not have data from Earth orbiting satellites to check his work.

WORLD GEODETIC SYSTEM 1984

GRS80 is not the reference ellipsoid for the GPS system that is WGS84, but it is interesting that the WGS84 reference ellipsoid is actually very similar to GRS80's. Its semi-major axis is identical. The difference between them is chiefly found in its flattening; the flattening of the WGS84 ellipsoid is 1/298.257223563, whereas 1/298.257222101 is the flattening of the GRS80 ellipsoid.

The WGS84 ellipsoid is the reference ellipsoid at the foundation of the coordinate system, known as the *World Geodetic System 1984* (*WGS84*). This datum was established by the U.S. Department of Defense and has been used by the U.S. military since January 21, 1987. However, there have been seven realizations of WGS84 since then. While WGS84 has always been the basis for GPS navigation message computations, the realization of the datum has changed. As of this writing, the latest version of WGS84 is WGS84 (G2139). The number following the letter G is the GPS week number (see Chapter 1). Therefore, single points, aka autonomous or absolute positions provided today by GPS receivers (see Figure 2.8), are based on WGS84 (G2139), which is the seventh update to the realization of the WGS84 Reference Frame.

The original WGS 84 was based on observations from more than 1900 Doppler stations. It was revised to become WGS84 (G730) to incorporate GPS observations. That realization was implemented in GPS by the *Operational Control System, OCS* on June 29, 1994. More GPS-based realizations of WGS84 followed, WGS84 (G873) on January 29, 1997, WGS84 (G1150) on January 20, 2002, WGS84(G1674) on February 8, 2012, and WGS84(G1762) on October 16, 2013. As of January 3, 2021, it became WGS84 (G2139). Most available GPS software can transform WGS84-based coordinates to several other datums. The one that is probably of greatest interest to surveyors in the United States now is the *North American Datum 1983 (NAD83)*, though that will shortly be replaced by the Terrestrial Reference Frames. It is nevertheless instructive to know why and how the relationship between WGS84 and NAD83 changed. The difference between WGS84 as originally rolled out in 1987 and NAD83 as first introduced in 1986 coordinates was so small that transformation was unnecessary. That is no longer the case. The difference between a position expressed in NAD83 (2011) and WGS84 (G2139) can be up to 1 or 2 m. To explain why that is, let us step back a bit.

NORTH AMERICAN DATUM 1983

NAD27

The Clarke 1866 ellipsoid was the foundation of NAD27, and the work that built that foundation was geodetic triangulation surveys. After all, a reference system is only an abstraction until some mode for its practical application is realized, something like physical, identifiable control stations. During the tenure of NAD27, control positions were tied together by tens of thousands of miles of triangulation and some traverses (see Figure 5.11). Its measurements grew into chains of figures with their vertices perpetuated by bronze disks set in stone, concrete, and other permanent media.

These tri-stations, brass caps, known as *passive marks* or *passive control* and their associated coordinates have provided a framework for all types of surveying and mapping projects for many years. They have served to locate international, state, and county boundaries. They have provided geodetic control for the planning of national and local projects, the development of natural resources, national defense, and land management. They have enabled surveys to be fitted together, provided checks, and assisted in the perpetuation of their marks. They have supported scientific inquiry,

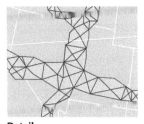

Detail
Idaho, Utah, Colorado, Wyoming,
Nevada, and Arizona

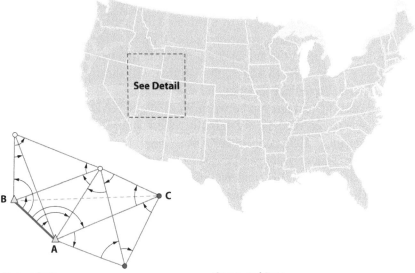

Known Data:
Length of base line AB.
Latitude and longitude of points A and B.
Azimuth of line AB.

Measured Data:
Angles to the new control points.

Computed Data:
Latitude and longitude of point C, and other new points.
Length and azimuth of line AC.
Length and azimuth of all other lines.

FIGURE 5.11   Triangulation

including crustal monitoring studies and other geophysical research, but even as the application of the nationwide control network grew, the revelations of local distortions in NAD27 were reaching unacceptable levels.

Judged by the standards of newer measurement technologies, the quality of some of the observations used in the datum was too low. That, and its lack of an internationally viable geocentric ellipsoid, finally drove its positions to obsolescence. The monuments remain, but it was clear early on that NAD27 had some difficulties. There were problems from too few baselines, Laplace azimuths, and other deficiencies. By the early 1970s, the NAD27 coordinates of the National Geodetic Control Network were no longer adequate.

THE DEVELOPMENT OF THE NORTH AMERICAN DATUM 1983

While a committee of the National Academy of Sciences advocated the need for a new adjustment in its 1971 report, work on the new datum, NAD83, did not really begin until after July 1, 1974. Leading the charge was an old agency with a new name. Called the U.S. Coast & Geodetic Survey in 1878 and then the Coast and Geodetic Survey (C&GS) in 1899, the agency is now known as the *NGS*. It is within the National Oceanic and Atmospheric Administration (NOAA), Department of Commerce. The first ancestor of today's NGS was established back in 1807 and was known as the Survey of the Coast. Its current authority is contained in U.S. Code, Title 33, USC 883a.

NAD83 includes not only the United States but also Central America, Canada, Greenland, and Mexico. The NGS and the Geodetic Survey of Canada set about the task of attaching and orienting its reference ellipsoid, GRS80, to the actual surface of the Earth as it was defined by the best positions available at the time. It took more than ten years to readjust and refine the horizontal coordinate system of North America into what is now NAD83. More than 1.7 million weighted observations derived from classical surveying techniques throughout the Western Hemisphere were involved in the least-squares adjustment. They were supplemented by approximately 30,000 EDM measured baselines, 5000 astronomic azimuths, and more than 650 Doppler stations positioned by the TRANSIT satellite system. Over 100 VLBI vectors were also included, but GPS, in its infancy, contributed little to the realization.

GPS was growing up in the early 1980s, and some of the agencies involved in its development decided to join forces. NOAA, the National Aeronautics and Space Administration (NASA), the U.S. Geological Survey (USGS), and the Department of Defense coordinated their efforts. Each agency was assigned specific responsibilities. NGS was charged with the development of specifications for GPS operations, investigation of related technologies, and the use of GPS for modeling crustal motion. It was also authorized to conduct its subsequent geodetic control surveys with GPS. So, despite an initial sparseness of GPS data in the creation of NAD83, the stage was set for a systematic infusion of its positions as the datum matured. The work was officially completed on July 31, 1986. NAD83 has since gone through several realizations, which have been driven in large part by the constant improvements in surveying and geodesy. The creation of *High Accuracy Reference Networks (HARN)* is a good example.

HIGH ACCURACY REFERENCE NETWORKS

Significant work in the evolution of NAD83 was accomplished in the state-by-state super-net programs. The creation of HARN was the result of cooperative ventures between NGS and the states, and often include other organizations as well. The campaign was originally known as High Precision Geodetic Networks (HPGN). A station spacing of not more than about 62 miles and not less than about 16 miles was the objective in these statewide networks. The accuracy was intended to be 1 part-per-million, or better between stations. In other words, with heavy reliance on GPS observations, these networks were intended to provide extremely accurate, vehicle-accessible, regularly spaced control point monuments with good overhead visibility. The accuracy of the final positions was one of the most important aspects of HARN. Entirely new orders of accuracy were developed for GPS relative positioning techniques by the Federal Geodetic Control Committee (FGCC). Its 1989 provisional standards and specifications for GPS work include Orders AA, A, and B, which were later defined as having minimum geometric accuracies of 3 mm ± 0.01 ppm, 5 mm ± 0.1 ppm, and 8 mm ± 1 ppm, respectively, at the 95%, or 2σ, confidence level. The publication of up-to-date geodetic data has always been one of the most important functions of NGS. The Federal Geodetic Control Subcommittee is within the Federal Geographic Data Committee and has published accuracy standards for geodetic networks in part 2 of the Geospatial Positioning Standards (FGDC-007-1998).

The accuracy of the HARN control was intended to be superior to the vectors derived from the day-to-day GPS observations that are tied to them. In that way, the HARN points significantly reduced the need for user to warp their GPS measurements to fit inferior control. Something that did happen in the early days of GPS. To further ensure such coherence in the HARN, when the GPS measurements were complete, they were submitted to NGS for inclusion in a statewide readjustment of the existing control stations, points of reference covered by the state. Coordinate shifts of 0.3 to 1.0 m from NAD83 values were typical in these readjustments, which were concluded in 1998.

CONTINUOUSLY OPERATING REFERENCE STATIONS

Around this same period, a way to provide GPS surveyors convenient access to NAD83 in the *conterminous* United States and Alaska was maturing. The network of National *Continuously Operating Reference Stations* (*CORS*) was expanding from its beginnings in 1986 when there were only three stations in what was known as the Cooperative International GPS Network (CIGNET). Six years later, in about 1992, the NGS began establishing a network of CORS throughout the United States. Each of these active stations occupied by a GPS receiver constantly tracking the satellites was formed into a network whose most straightforward benefit was, and is, that a user could do relative positioning without operating his own base station.

The original idea was that the system would provide positioning for navigational and marine needs. There were about 50 CORS in 1996. From 1998 to 2004, NGS introduced another series of observations in each state designed to tie the network to the CORS. This work resulted in the Federal Base Network (FBN), a nationwide network of stations. These spatial reference positions were among the most precise available and were particularly dense in crustal motion areas. At that time, the points were spaced at approximately 100 km apart. They were few compared to the much more numerous Cooperative Base Network (CBN), a high-accuracy network of monumented control stations spaced at 25 to 50 km apart throughout the United States and its territories. The CBN was created and maintained by state and private organizations with the help of NGS.

Today, the NOAA CORS Network (NCN) continues to be a cooperative enterprise which includes hundreds of government, academic, and private organizations. There are about 2800 stations of which 1700 or so are operational as of this writing. Of those, about 170 are tracking GPS only and about 1530 are tracking GNSS.

They are independently owned and operated, but each shares their data with NGS, which analyzes and distributes the data free of charge. They are spaced at approximately 70 km. They all must meet NOAA geodetic standards for installation, operation, and data distribution. Among those requirements are stipulations that the station adopt coordinates published by NGS. The receiver at the station must be at least dual frequency (L1 and L2) and capable of tracking at least ten satellites above the horizon at a minimum. It must be capable of both pseudorange and full wavelength carrier-phase tracking so that both the carrier-phase and code range data are recorded on a 30-second or shorter epoch interval that can be analyzed and distributed by NGS through the Internet.

A STEP FORWARD

NAD83 has steadily improved. It has done great service over the years, but a significant, and necessary, step forward is coming, a step that will see NAD83 retired. To understand why it is important to discuss some aspects that limit NAD83's fitness to support GNSS today. One way to illustrate its shortcomings is to compare its current realization with the current ITRF and WGS84.

When NAD83 was created, it was intended to be geocentric. This makes sense as GNSS satellites basically orbit around the center of mass of the Earth. Unfortunately, it is now known that the center of the GRS80 reference ellipsoid for NAD83 is about 2.24 m from the true geocenter, as illustrated in Figure 5.12. However, the *International Terrestrial Reference Frame (ITRF)* is geocentric.

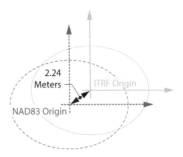

FIGURE 5.12   Reference Ellipsoid Centers

THE INTERNATIONAL TERRESTRIAL REFERENCE FRAME

The ITRF is a realization of the International Terrestrial Reference System (ITRS). The origin of the ITRF is at the center of mass of the whole Earth including the oceans and atmosphere. It is an ECEF reference frame The unit of length is the meter. The orientation of its axes was established as consistent with that of the IERS's predecessor, BIH, at the beginning of 1984.

The IERS was created by the International Astronomical Union (IAU) and the IUGG in 1987. It maintains the ITRS. It monitors Earth Orientation Parameters (EOP) for the scientific community through a global network of hundreds of fiducial stations. The measurements are made at these stations by four space geodetic systems GPS, VLBI, *Lunar Laser Ranging* (*LLR*), SLR and the *Doppler Orbitography and Radiopositioning Integrated by Satellite* (*DORIS*). They track the motion of the approximately 20 tectonic plates with extraordinary accuracy, and thereby they track the motion of the control stations themselves. This provides IERS with the data it needs to publish the coordinates of their stations and the rate of their movement, their velocities. In order to understand how that works, we need to return to a word we have seen before, *epoch*.

In GNSS, the word epoch is used in a couple of different ways. Both are about time; one is a time interval and the other is a moment of time. Epoch was mentioned earlier in the context of a time interval, for example, a one-second epoch being the time elapsed in completing one single GNSS observation. There is another meaning of the word that is important here. It is a moment of time, similar to a date, but expressed in the form of a year and a decimal of that year, typically two decimals. Here is an example of such a moment 2022.51. This is July 4, 2022. It is calculated by considering the days numbered sequentially through the year. In a day-of-year calendar, July 4 is the 185th day of 2022. By dividing the day of the year, 185, by 365, the decimal value .51 is derived. This sort of epoch is an important part of the proper expression of a datum or reference frame.

The Earth is dynamic. Tectonic plates move. As mentioned earlier, a reference frame or datum can be realized by control stations on the ground, but since the ground in which they are monumented moves, the rate of those movements, and the velocities, must be considered. They tell us how fast the control station's positions shift in relation to their reference and how rapidly their coordinates change.

FRAME REFERENCE EPOCH

When a reference frame is established, recognizing that there is constant movement, an epoch is chosen on which that movement is theoretically paused. It is sometimes known as the *frame reference epoch* or just *reference epoch*. It is as if there is an instantaneous snapshot of the station's coordinates. They are caught in a sort of freeze frame. Here is an example. Table 5.2 shows each ITRF realization. The first was in 1988, ITRF88, and there was a new one nearly every year in the beginning of the system. As measurements, procedures and data got more precise, they have been less frequent. In any case, each realization includes a year in its name.

TABLE 5.2
ITRF Realizations and Frame Reference Epochs

International Terrestrial Reference frame realizations	Frame reference epoch
ITRF88	
ITRF89	
ITRF90	1988.00
ITRF91	
ITRF92	
ITRF93	1993.00
ITRF94	1993.00
ITRF96	1997.00
ITRF97	1997.00
ITRF2000	1997.00
ITRF2005	2000.00
ITRF2008	2005.00
ITRF2014	2010.00
ITRF2020	2015.00

i.e., ITRF2020, but it is just a name. The frame reference epochs are the instantaneous moments of the snapshot, the freeze frames of the station's coordinates. At that moment, the frame reference epoch for ITRF2020 was midnight on January 1, 2015. That is the meaning of 2015.00.

Those epochs indicate the moment in time that the motion tracked by the four previously mentioned space geodetic systems is theoretically frozen to define the realization. The tectonic motion does not actually stop, of course, it is constant.

World Geodetic System (WGS84)

In the past, we did not have to be concerned with the shift between the original realizations of NAD83 as it was introduced in 1986 and the original WGS84 introduced in 1987. In those days, WGS84 (original), like NAD83, was not geocentric, so the difference between coordinates of the same point expressed in the two was very close, a centimeter or two. They were so close the difference easily fell within our overall error budget. That is no longer the case. Today, WGS84 is an ECEF reference frame very like ITRF and very closely aligned to it. In their current definitions, coordinates expressed in NAD83 (2011) and coordinates expressed in WGS84 (G2139) of the same point at the same epoch differ up to 1 or 2 m within the conterminous United States. Table 5.3 shows the evolution of WGS84

TABLE 5.3

WGS84 Realizations and Frame Reference Epochs

Date introduced by the U.S. Department of Defense	Realization: G is for GPS Week	Frame reference epoch	Compatible with:
1987	WGS84 (1987), aka WGS84 (original), aka WGS84 (TRANSIT)	1984.00	
06/29/1994	WGS84 (G730)	1994.00	ITRF91
01/29/1997	WGS84 (G873)	1997.00	ITRF94
01/20/2002	WGS84 (G1150)	2001.00	ITRF2000
02/08/2012	WGS84 (G1674)	2005.00	ITRF2008
10/16/2013	WGS84 (G1762)	2005.00	ITRF2008
01/03/2021	WGS84 (G2139)	2016.00	ITRF2014

through several realizations. It shows the date on which each was introduced by the U.S. Department of Defense along with a parenthetical statement that begins with the letter G, standing for a GPS week, as in G2139. Please note these are just part of the name of each realization. Their frame reference epochs are shown in the column labeled, *Frame reference epoch*.

In the right-most column, you see the ITRF realizations with which they are compatible. In other words, those positions of the same point expressed in those pairings of ITRF and WGS84 realizations are within a meter or so of one another. In fact, between the most recent realizations, the proximity of positions of the two reference frames is typically less than a decimeter. Now consider all three reference frames together ITRF, WGS84, and NAD83.

ITRF, WGS84, AND NAD83

As mentioned, if you compare positions of the same point expressed in ITRF2014 and WGS84 (G2139) at the same epoch, they can agree within a few centimeters. However, if you compare positions of the same point expressed in NAD83 and WGS84 they departed from one another a meter or two. Why is that? In answering that, it bears remembering that the current ITRF and WGS84 realizations are dynamic, they are not fixed to any plate. They have a global scope and they are geocentric. On the other hand, NAD83 is static. Its three components are each fixed to a plate, the North American plate—NAD 83(2011), the Pacific plate—NAD 83(PA11), and a small tectonic plate located west of the Mariana Trench, the Mariana plate—NAD 83(MA11). Further, NAD83 is not global and it is not geocentric. In short, the current configuration of NAD83 is not nearly as well suited to GNSS as are ITRF and WGS84. This is one of the reasons NAD83 is going away.

THE NORTH AMERICAN TERRESTRIAL REFERENCE FRAME 2022 (NATRF2022)

NATRF2022 will be one of the four plate-fixed terrestrial reference frames, which NGS will introduce soon after 2022. The tectonic plate for each frame may be inferred from their names; North American Terrestrial Reference Frame of 2022 (NATRF2022), Pacific Terrestrial Reference Frame of 2022 (PTRF2022), Caribbean Terrestrial Reference Frame of 2022 (CTRF2022), and Mariana Terrestrial Reference Frame of 2022 (MTRF2022). They will be truly geocentric, which will remove gross disagreements with the ITRF. In fact, the NGS work will first be in the ITRF2020, and then a mathematical relationship to all four NSRS frames will occur. Finally, even though fixed to their plates, each will rotate at the average rate of the plate bearing its name.

The initial horizontal shift in meters from NAD83 (2011) epoch 2010.0 to NATRF 2022 in the conterminous United States will vary from approximately 0.7 m (2.3 feet) to approximately 2.6 m (8.5 feet) with the largest shift being along the westernmost portion of California. However, the geometric shifts will not be the most fundamental change, that is, the demise of the idea that the coordinates assigned to monuments set in the Earth are static. It was never correct, of course, and the coming reference frames will bring epochs into the prominence they should have. In the past, it was easy to use the official coordinates published by the NGS as if they were unchanging. It made some sense at the time. Measurements made with the systems then available were sufficiently reliable, but they did not draw attention to the constant slow movement of the monuments because the extent of the motion was within their error budgets. With the maturation of GNSS, that is no longer the case. The nearly universal dissemination of GNSS not only reveals the persistent movement, but it is so apparent it cannot be ignored.

In the new paradigm where a point depends on when you ask. Tectonic plate motion and crustal deformation continually change the station's coordinates, so their positions must always be qualified with the epoch at which they are valid. Such official accommodation of time-dependency is nothing new to users of ITRF and WGS84. For example, the frame reference epoch chosen for ITRF2020 was midnight on January 1, 2015, so a station's coordinates in ITRF2020 epoch 2015.00 tell you where that station was at that past moment not where it is now. However, suppose you need to calculate the coordinates of that station as they are right now not a decade or so ago. To do that, you will need the station's velocities in North, East, and Up, or in ECEF X, Y, and Z. With those in hand, you can move the station's coordinates backward or forward in time to the epoch you want. The good news is that as soon as you know those velocities, you can compute the stations' position at any *coordinate epoch*, any moment in time. Such specific coordinate epochs are also known as a *survey epoch* by the NGS. NGS has an online utility called *Horizontal Time-Dependent Positioning* (*HTDP*) that helps with that. It has an underlying crustal dynamics model and given the necessary information can provide estimates of the movement of a mark through time. It also supports 14-parameter Helmert transformations from one NAD 83, ITRF, and WGS84 reference frame to another and from one realization to another. In any case, a station's coordinates are only completely and correctly defined when the reference frame, frame reference epoch, and coordinate epoch, aka survey epoch, are all included. This idea may be somewhat new to some users of NSRS, so to ease the transition to this new paradigm NGS will continue its long-standing policy of publishing static reference frames like NAD83 (2011) epoch 2010.00, but with a twist. They will carry on providing a frame reference epoch for their mark's coordinates starting at 2020.00 in a manner similar to the past protocol, the difference is that there will be a new one every five or ten years. These periodic *reference epoch coordinates (RECs)*, as NGS calls them, will provide consistent static coordinates at one fixed epoch, but they will be so for only five or ten years. Then the REC will change.

There will also be a change in the public's access to NATRF2022. The definitive, up-to-date coordinates will be available from the NOAA CORS Network (NCN) but not necessarily from the passive monuments. The positions of the passive stations, often brass or aluminum disks set in concrete or metal rods, were likely originally derived from conventional surveying as contrasted with active stations of the NCN that constantly track GPS/GNSS satellites. In the past, the coordinates of passive monuments were periodically updated, but after 2022 there will no effort to provide time-dependent coordinates on the passive monuments, so they may be outdated. While this fact will diminish their facility, surveyors may still need to rely on a passive point's old coordinate from time to time. They are encouraged to re-survey such points to help NGS update them. It is fortunate that NGS will also continue to provide users with utilities to process surveyor's data, i.e., Online Positioning User Service (OPUS) and HTDP. These tools will be available to transform coordinates from NAD 83 to the NATRF2022 and assign velocities to them.

STATE PLANE COORDINATE SYSTEMS

Since the advent of GPS, the gulf that once existed between the precision of local surveys and geodetic work has virtually closed and that has changed the relationship between local surveyors in private practice and geodesists. Today's surveyor has relatively easy and direct access to the geodetic coordinate systems through GNSS and CORS networks around the world, but in the past, relatively few engineers and surveyors were employed in geodetic work. Perhaps the greatest importance of the data from the various geodetic surveys then was that they furnished precise points of reference to which the multitude of surveys of lower precision could be tied. This arrangement was clearly illustrated by the design of *SPCS* devised to make the National Control Network accessible to surveyors without geodetic capability.

In 1932, two engineers in North Carolina's highway department, O.B. Bester and George F. Syme, appealed to the then Coast & Geodetic Survey (C&GS, now NGS) for help. They had found that the stretching and compression inevitable in the representation of the curved Earth on a plane was so severe over long route surveys that they could not check into the C&GS geodetic control stations across a state within reasonable limits. The engineers suggested that a plane coordinate grid system be developed that was mathematically related to these control station positions but could be utilized using plane trigonometry.

To mediate the problem, Dr. Oscar Adams of the Division of Geodesy, assisted by Charles Claire, designed the first SPCS. It was based on *map projections* which represent a portion of the Earth on a plane while managing the unavoidable distortion inherent in the process. It has been done for hundreds of years to create paper maps. Using one, Adams knew, it was possible to retain one of the four elements of area, shape, scale, or direction virtually unchanged from its actual value on the Earth, but not all four. A perfect map projection on which all distances, directions, and areas are the same on the ellipsoid and on the map did not, and does not, exist.

The problem is often illustrated by trying to flatten part of an orange peel. The orange peel stands in for the surface of the Earth. A small part, say a square a quarter of an inch on the side, can be pushed flat without much noticeable deformation. However, when the portion gets larger problems appear. Suppose a third of the orange peel is involved, as the center is pushed down, the edges tear and stretch, or both. If the peel gets even bigger the tearing gets more severe. If a map is drawn on the orange before it is peeled, the map gets distorted in unpredictable ways when it is flattened and it is difficult to relate a point on one torn piece with a point on another in any meaningful way.

As it is not possible to retain the correct distances, directions, and areas simultaneously, choices must be made. There is no universal best decision. Adams and Claire had to decide which characteristics would be shown the most correctly, knowing it would be done at the expense of the others. The best solution is the one that gives the most satisfactory results for a particular mapping problem, so Adams decided to use the *Lambert Conformal Conic Projection* for North Carolina. This map projection is known as conic projection because it is based on a *developable* cone. A developable surface is one that may be flattened without all the unpredictable deformation mentioned in the orange peel example. The surface of an ellipsoid, like the orange peel, is *nondevelopable* because flattening it inevitably leads to highly irregular distortion. So, a useful map projection ought to start with a surface that is developable. It happens that a paper cone or cylinder both illustrate this idea nicely. They are illustrations only, models for thinking about the issues involved. As illustrated in Figure 5.13, a right circular cone cut perpendicular up from the base to its apex can then be made completely flat without trouble. The same may be said of a cylinder cut perpendicular from base to base.

On the Lambert Conformal Conic Projection, the parallels of latitude are represented by arcs of concentric circles and meridians of longitude are represented by equally spaced straight radial lines and the meridians and parallels intersect at right angles. When attached to the Earth, the axis of the cone is imagined to be a prolongation of the polar axis. The parallels of latitude are not equally spaced because the scale varies as you move north and south along a meridian of longitude.

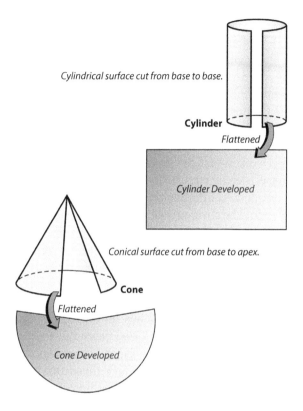

FIGURE 5.13 Development of a Cylinder and a Cone

A cylindrical mapping surface is the basis for the *Transverse Mercator projection*. However, when it is attached to the Earth, the axis of the cylinder is rotated so that it is perpendicular with the polar axis of the ellipsoid. Unlike the Lambert Conic projection, the Transverse Mercator represents meridians of longitude as curves rather than straight lines on the developed grid. Please note that the Transverse Mercator projection used in State Plane Coordinates differs from the Transverse Mercator system used in the Universal Transverse Mercator (UTM) system.

In using these projections as the foundation of the SPCSs, Adams ensured that the results would be *conformal*. Conformal or *orthomorphic* map projections are those in which shape is preserved. Orthomorphic means right shape. In a conformal projection, the angles between intersecting lines and curves retain their original form on the map. In other words, between short lines, a 45° angle on the ellipsoid is a 45° angle on the map. It also means that the scale is the same in all directions from a point; in fact, it is this characteristic that preserves the angles. These aspects were certainly a boon for the North Carolina Highway engineers and benefits that all State Plane Coordinate users have enjoyed since. On long lines, angles on the ellipsoid are not exactly the same on the map projection. Nevertheless, the change is small and systematic.

Eventually, three conformal map projections were used in the designs of the original SPCSs. At that time, each system was based on the North American Datum 1927, NAD27. The two primary projections were the Lambert Conic Conformal Projection and the Transverse Mercator projection. The Oblique Mercator projection was used on the panhandle of Alaska. For North Carolina and other states that are longest east–west, the Lambert Conic projection worked best. SPCSs in states' longest north–south were built on the Transverse Mercator projection. There were exceptions to this general rule. For example, California used the Lambert Conic projection even though the state could be covered with fewer Transverse Mercator zones. The Lambert Conic projection is a bit simpler to use, which may account for the choice. Adams wanted to cover each state with as few zones as possible. A zone in this context is a belt across the state that has one Cartesian coordinate grid with one origin and is projected onto one mapping surface.

Today, every state in the United States, Puerto Rico, and the U.S. Virgin Islands have their own SPCS that relies on an imaginary flat reference surface with Cartesian axes. They describe measured positions by ordered pairs, expressed in northings and eastings, or y- and x-coordinates. In fact, many agencies of government, particularly those that administer state, county, and municipal databases, prefer coordinates in their SPCS. The systems are, as the name implies, state specific. In many states, the system is officially sanctioned by legislation which often allows surveyors to use State Plane Coordinates to legally describe property corners. However, since a completely accurate representation of the Earth on a flat plane is impossible, all plane coordinate systems include distortion. Decreasing distortion is a constant and elusive goal in map projection.

DISTORTION

A typical method for establishing an independent *local coordinate system* can be illustrated by bringing the center of a flat plane tangent to the portion of the Earth to be projected onto it. The resulting simple two-dimensional Cartesian systems is convenient and accommodates the needs of small projects. The method of projection, onto a simple flat plane, is based on the idea that a small area of interest, as with a small section of the orange mentioned previously, conforms so nearly to a plane that distortion is negligible. Subsequently, local tangent planes have been long used by surveyors. Such systems demand little, if any, manipulation of the field observations and the approach has merit when the extent of the work is limited. Nevertheless, the design of such a projection must accommodate some awkward facts. For example, while it would be possible to imagine

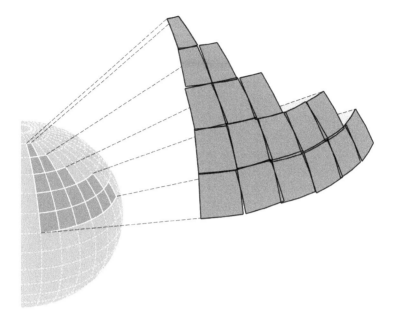

FIGURE 5.14   Local Coordinate Systems Do Not Edge Match

mapping a considerable portion of the Earth using a large number of small individual planes, like facets of a gem, it is seldom done because when these planes are brought together, they cannot be edge-matched properly along their borders (see Figure 5.14).

The problem is unavoidable because the planes, tangent at their centers, inevitably depart more and more from the reference ellipsoid at their edges, and the greater the distance between the ellipsoidal surface and the surface of the map on which it is represented, the greater the distortion on the resulting flat map. This is true of all methods of map projection. Therefore, one is faced with the daunting task of joining together a mosaic of individual maps along their edges where the accuracy of the representation is at its worst, and even if that could be overcome by making the distortion, however large, the same on two adjoining maps, another difficulty would usually remain. Each of these planes often has its own unique coordinate system. When the orientation of the axes, the scale, and the rotation of each one of these individual local systems is not the same as those of its neighbor's local system there are gaps and overlaps between them.

So, the idea of a self-consistent, local map projection based on small, flat planes tangent to the Earth, or the reference ellipsoid, is convenient, but only for small projects that need not be related to adjoining work. When there is no need to venture outside the bounds of a particular local system it can be entirely adequate, but when the area involved grows larger and a significant area needs coverage another strategy is needed.

DECREASING DISTORTION

Distortion can be reduced in several ways. Most involve reducing the distance between the map projection surface and the ellipsoidal surface. One way this is done is to change the elevation of the mapping surface closer to the surface being mapped as is done in some *LDPs*. Another way to shrink the distance between the map projection surface and the ellipsoid is to move the mapping surface from tangency with the ellipsoid and make it cut through it. This produces what is known as a *secant projection*. Thereby the area where distortion is in an acceptable range on the map can be effectively increased (see Figure 5.15).

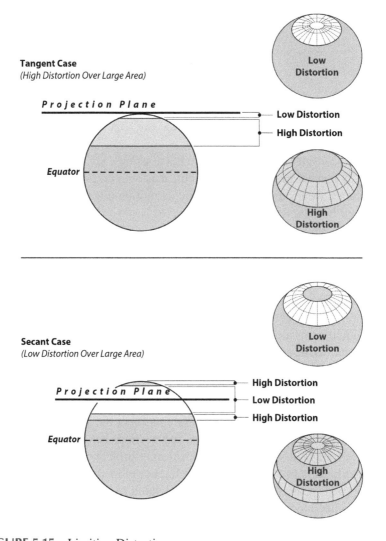

FIGURE 5.15   Limiting Distortion

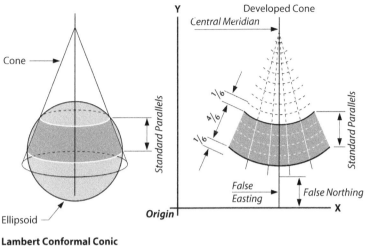

Lambert Conformal Conic Projection

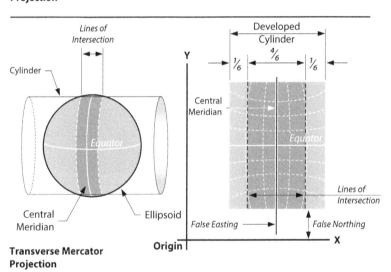

Transverse Mercator Projection

FIGURE 5.16   Two Projections

The distortion can be reduced even further when one of those developable surfaces mentioned earlier is added to the idea of a secant map projection plane (see Figure 5.16).

SECANT AND CYLINDRICAL PROJECTIONS

Both cones and cylinders have an advantage over a flat map projection plane. They are curved in one direction and can be designed to follow the curvature of the area to be mapped in that direction. Also, if a large portion of the ellipsoid is to be mapped, several cones or several cylinders

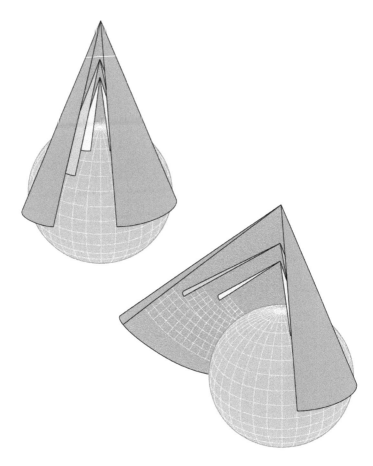

FIGURE 5.17   Several Developable Surfaces

may be used together in the same system to further limit distortion. In that case, each cone or cylinder defines a different zone coverage, as shown in Figure 5.17. This is the approach used in SPCSs.

As mentioned, when a conic or a cylindrical map projection surface is made secant, it intersects the ellipsoid, and the map is brought close to its surface. The map is projected both inward and outward onto it. Two lines of exact scale, standard lines, are created along the small circles where the cone and the cylinder intersect the ellipsoid. Where the ellipsoid and the map projection surface touch, in this case, intersect, there is no distortion. However, between the standard lines, the map is under the ellipsoid and outside of them the map is above it. That means that between the standard lines, the distance from one point to another is longer on the ellipsoid than it is shown on the map (see Figure 5.19), and outside the standard lines, the distance on the ellipsoid is shorter than it is on the map. Any length that is measured along a standard line is the same on the ellipsoid and on the map, which is why another name for the Lambert Conformal Conic standard parallels is lines of exact scale. Ultimately, the goal is very straightforward relating each position on one surface, the reference ellipsoid, to a corresponding position on another surface as faithfully as possible and then flattening that second surface to accommodate two-dimensional Cartesian coordinates. In fact, the whole procedure is in the service of moving from geographic to two-dimensional Cartesian coordinates and back again. These days the complexities of mathematics are handled with computers. Of course, that was not always the case.

STATE PLANE COORDINATE SYSTEM MAP PROJECTIONS

For example, using a single secant cone in the Lambert projection and limiting the extent of a zone, or belt, across a state to about 158 miles, approximately 254 km, Adams limited the distortion of the length of lines. Not only were angles preserved on the final product, but also there were no radical differences between the length of a measured line on the Earth's surface and the length of the same line on the map projection. In other words, the scale of the distortion was small in terms of the surveying technology available at the time. He placed four-sixths of the map projection plane between the standard lines, one-sixth outside at each extremity. The distortion was held to 1 part in 10,000. A maximum distortion in the lengths of lines of 1 part in 10,000 means that the difference between the length of a 2-mile line on the ellipsoid and its representation on the map would only be about 1 foot at the most.

State Plane Coordinates were created to be the basis of a method that approximates geodetic accuracy more closely than the then commonly used transit and tape methods of small-scale plane surveying. Today, surveying methods can easily achieve accuracies beyond 1 part in 100,000 and much more, but the SPCSs were designed at a time of generally lower accuracy and efficiency in surveying measurement. Today, computers easily handle the lengthy and complicated mathematics of geodesy, but the first SPCS was created when such computation required sharp pencils and logarithmic tables. In fact, the original SPCS was so successful in North Carolina that similar systems were devised for all the states in the Union within a year or so. The system was successful because, among other things, it overcame some of the limitations of mapping on a horizontal plane while avoiding the imposition of strict geodetic methods and calculations. It managed to limit distortion and preserved conformality. It did not disturb the familiar system of ordered pairs of two-dimensional Cartesian coordinates, and it covered each state with as few zones as possible. The boundaries of which were generally constructed to follow county lines so that those relying on State Plane Coordinates could work in one zone throughout a jurisdiction.

SPCS27 TO SPCS83

In several instances, the boundaries of State Plane Coordinate Zones today, *SPCS83*, the SPCS based on NAD83, and its reference ellipsoid GRS80, differ from the original zone boundaries. The foundation of the original SPCS, *SPCS27* was NAD27 and its reference ellipsoid Clarke 1866. As mentioned earlier, NAD27 geographical coordinates, latitudes, and longitudes differ significantly from those in NAD83. In fact, conversion from geographic coordinates, latitude, and longitude, to grid coordinates, y and x, and back is one of the three fundamental conversions in the SPCS. It is important because the whole objective of the SPCS is to allow the user to work in plane coordinates but still have the option of expressing any of the points under consideration in either latitude and longitude or State Plane Coordinates without significant loss of accuracy. Therefore, when geodetic control was migrated from NAD27 to NAD83, the SPCS had to go along. When the migration was undertaken in the 1970s, it presented an opportunity for an overhaul of the system. Many options were considered, but, in the end, just a few changes were made. One of the reasons for the conservative approach was the fact that 37 states had passed legislation supporting the use of State Plane Coordinates. Nevertheless, some zones got new numbers and some of the zones changed.

The zones are numbered in the SPCS83 system as shown in Figure 5.18. As mentioned earlier, the original goal was to keep each zone small enough to ensure that the scale distortion was 1 part in 10,000 or less, but when the SPCS83 was designed, that scale was not maintained in some states. In five states, some SPCS27 zones were eliminated altogether and the areas they had covered consolidation into one zone or added to adjoining zones. In three of those states, the result was one single large zone. Those states are South Carolina, Montana, and Nebraska. In SPCS27, South Carolina and Nebraska had two zones, in SPCS83, they have just one zone, 3900, and one zone, 2600, respectively. Montana previously had three zones. It now has one zone, 2500. Therefore, because the area covered by these single zones has become so large, they are not limited by the 1 part in 10,000 standard. California eliminated zone 7 and added that area to zone 0405, formerly zone 5. Two zones previously covered Puerto Rico and the Virgin Islands. They now have one. It is zone 5200. In Michigan, three Transverse Mercator zones were entirely eliminated.

In both the Transverse Mercator and the Lambert projection, the positions of the axes are similar in all SPCS zones. As you can see in Figure 5.16, each zone has a central meridian. These central meridians are true meridians of longitude at the geometric center of the zone. Please note that the central meridian is not the y-axis. If it were, negative coordinates would result. To avoid them, the actual y-axis is moved far to the west of the zone itself. In the old SPCS27 arrangement, the y-axis was 2,000,000 feet west

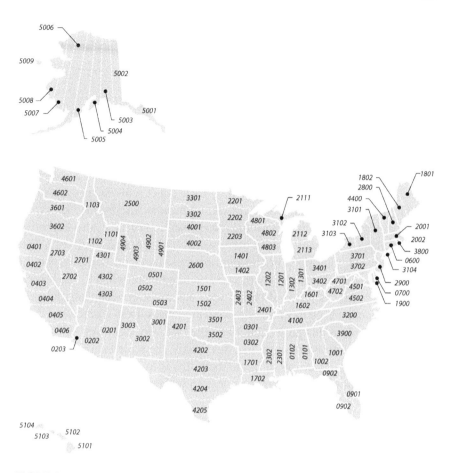

FIGURE 5.18 State Plane Coordinate Systems 1983

from the central meridian in the Lambert Conic projection and 500,000 feet in the Transverse Mercator projection. In the SPCS83 design, those constants have been changed. The most common values are 600,000 m for the Lambert Conic and 200,000 m for the Transverse Mercator. However, there is a good deal of variation in these numbers from state to state and zone to zone. In all cases, however, the y-axis is still far to the west of the zone and there are no negative State Plane Coordinates. No negative coordinates because the x-axis, also known as the baseline, is far to the south of the zone. Where the x-axis and y-axis intersect is the origin of the zone, and that is always south and west of the zone itself. This configuration of the axes ensures that all State Plane Coordinates occur in the first quadrant and are, therefore, always positive.

It is important to note that the fundamental unit SPCS83 is the meter. It was the *U.S. Survey Foot* for the old SPCS27. As of January 1, 2023, the U.S. Survey Foot will be no more. From then on 1 foot will be exactly equal to 0.3048 m, the *International Foot*. This decision was taken under the authority of the National Institute of Standards and Technology (NIST), NGS, the National Ocean Service (NOS), and NOAA.

STATE PLANE COORDINATES SCALE AND DISTANCE

GEODETIC LENGTHS TO GRID LENGTHS

This brings us to the scale factor, also known as the K factor and the projection factor. It was this factor that the original design of the SPCS sought to limit to 1 part in 10,000. As implied by that, effort scale factors are ratios that can be used as multipliers to convert ellipsoidal lengths, also known as geodetic distances, to lengths on the map projection surface, also known as grid distances and vice versa. In other words, the geodetic length of a line, on the ellipsoid, multiplied by the appropriate scale factor will give you the grid length of that line on the map. The grid length multiplied by the inverse of that same scale factor would bring you back to the geodetic length again.

The projection used most on states that are longest from east to west is the Lambert Conic. In this projection, the scale factor for east–west lines along parallels of latitude is constant. In other words, the scale factor is the same all along the line. One way to think about this is to recall that the distance between the ellipsoid and the map projection surface does not change along a parallel of latitude in that projection. On the other hand, along a north–south line, a meridian of longitude, the scale factor is constantly changing on the Lambert Conic. It is no surprise, then, to see that the distance between the ellipsoid and the map projection surface is always changing along a meridian of longitude in that projection. Looking at the Transverse Mercator projection, the projection used most on states longest north to south, the situation is reversed. Another way to say it is the difference in the scale factor at each end of a line is greatest for an east–west line in the Transverse Mercator projection and greatest for a north–south line in the Lambert projection.

Both the Transverse Mercator and the Lambert Conic used a secant projection surface and were generally restricted the width to 158 miles. These were two strategies used to limit scale factors when the SPCSs were designed. Where that was not optimum, the width was sometimes made smaller, which means the distortion was lessened. As the belt of the ellipsoid projected onto the map narrows, the distortion gets smaller. For example, Connecticut is less than 80 miles wide from north to south. It has only one zone. Along its northern and southern boundaries, outside of the standard parallels, the scale factor is 1 part in 40,000, a fourfold improvement over 1 part in 10,000, and in the middle of the state, the scale factor is 1 part in 79,000, nearly an eightfold increase. On the other hand, the scale factor was allowed to get a little bit smaller than 1 part in 10,000 in Texas. By doing that, the state was covered completely with five zones.

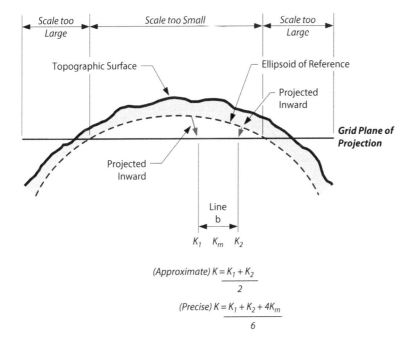

FIGURE 5.19 The Scale Factor

Among the guiding principles in 1933 was covering the states with as few zones as possible and having zone boundaries follow county lines. Still, it requires ten zones and all three projections to cover Alaska.

In Figure 5.19, a typical State Plane Coordinate zone is represented by a grid plane of projection cutting through the ellipsoid of reference. As mentioned earlier, between the intersections of the standard lines, the grid is under the ellipsoid. There, the distance from one point to another is longer on the ellipsoid than on the grid. This means that right in the middle of an SPCS zone, the scale factor is at its minimum. In the middle, a typical minimum SPCS scale factor is not less than 0.9999, though there are exceptions. Outside of the intersections, the grid is above the ellipsoid, where the distance from one point to another is shorter on the ellipsoid than it is on the grid. There at the edge of the zone, a maximum typical SPCS scale factor is generally not more than 1.0001, but again there are now exceptions.

When SPCS 27 was current, scale factors were interpolated from tables published for each state. In the tables for states in which the Lambert Conic projection was used, scale factors change north–south with the changes in latitude. In the tables for states in which the Transverse Mercator projection was used, scale factors change east–west with the changes in the x-coordinate. Today, scale factors are not interpolated from tables for SPCS83. For both the Transverse Mercator and the Lambert Conic projections, they are calculated directly from equations.

There are also several software applications that can be used to automatically calculate scale factors for stations. They can be used to convert latitudes and longitudes to State Plane Coordinates. Given the latitude and longitude of the stations under consideration, part of the available output from these programs is typically the scale factors for those stations.

To illustrate the use of these factors, consider line b to have a length on the ellipsoid of 130,210.44 feet, a bit over 24 miles. That would be its geodetic distance. Suppose that the scale factor for that line was 0.9999536, then the grid distance along line b would be:

$$Geodetic\ Distance * Scale\ Factor = Grid\ Distance$$

$$130,210.44\ ft. * 0.999953617 = 130,204.40\ ft.$$

The difference between the longer geodetic distance and the shorter grid distance here is a little more than 6 feet. That is better than 1 part in 20,000. Distortion lessens and the scale factor approaches 1 as a line nears a standard parallel. Please also recall that on the Lambert projection, an east–west line, that is, a line that follows a parallel of latitude, has the same scale factor at both ends and throughout. However, a line that bears in any other direction will have a different scale factor at each end. A north–south line will have a great difference in the scale factor at its north end compared with the scale factor of its south end. In this vein, here is an approximate formula.

$$K_m = \frac{K_1 + K_2}{2}$$

where K_m is the scale factor for a line, K_1 is the scale factor at one end of the line and K_2 is the scale factor at the other end of the line. The scale factor varies with the latitude in the Lambert projection. For example, suppose the point at the north end of the 24-mile line in Figure 5.20 is called Stormy and has a geographic coordinate of:

37°46′00.7225″

103°46′35.3195″

and at the south end, the point is known as Seven with a geographic coordinate of:

37°30′43.5867″

104°05′26.5420″

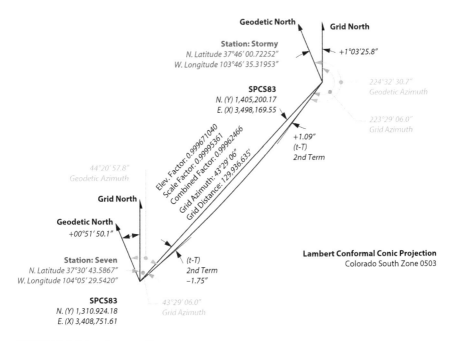

FIGURE 5.20 Stormy-Seven

The scale factor for point Seven is 0.99996113 and the scale factor for point Stormy is 0.99994609. It happens that point Seven is further south and closer to the standard parallel than point Stormy, and it, therefore, follows that the scale factor at Seven is closer to 1. It would be exactly 1 if it were on the standard parallel, which is why the standard parallels are called lines of exact scale. The typical scale factor for the line is the average of the scale factors at the two end points:

$$K_m = \frac{K_1 + K_2}{2}$$

$$0.99995361 = \frac{0.99996113 + 0.99994609}{2}$$

Deriving the scale factor at each end and averaging them is the usual method for calculating the scale factor of a line.

That is not the whole story when it comes to reducing the distance to the State Plane Coordinate grid. Measurement of lines must always be done on the topographic surface of the Earth and not on the ellipsoid. Therefore, the first step in deriving a grid distance must be moving a measured line from the Earth to the ellipsoid. In other words, converting a distance measured on the topographic surface to a geodetic distance on the reference ellipsoid. This is done with another ratio that is also used as a multiplier.

Originally, this factor had a rather unfortunate name. It was known as the sea level factor in SPCS27. It was given that name because, as you may recall that when NAD27 was established using the Clarke 1866 reference ellipsoid, the distance between the ellipsoid and the geoid was declared to be zero at Meades Ranch in Kansas. That meant that in the middle of the country, the sea level surface, the geoid, and the ellipsoid were defined as being coincident. Since the Clarke 1866 ellipsoid fit the United States quite well, the separation between the two surfaces, the ellipsoid and geoid, only grew to about 12 m anywhere in the conterminous United States. With such a small distance between them, many practitioners at the time took the point of view that, for all practical purposes, the ellipsoid and the geoid were in the same place, and that place was called sea level. Hence reducing the distance measure on the surface of the Earth to the ellipsoid was said to be reducing it to sea level. Today, that idea and that name for the factor are misleading because, of course, the GRS80 ellipsoid on which NAD83 is based is certainly not the same as MSL. Now the separation between the geoid and ellipsoid can grow as large as 53 m and technology by which lines are measured has improved dramatically. Therefore, in SPCS83, the factor for reducing a measured distance to the ellipsoid is known as the ellipsoid factor. In any case, both the old and the new name can be covered under the name the *elevation factor*. Regardless of the name applied to the factor, it is a ratio. The ratio is the relationship between an approximation of the Earth's radius and that same approximation with the mean ellipsoidal height of the measured line added to it. For example, consider station Boulder and station Peak illustrated in Figure 5.21.

Boulder
  39°59′29.1299″
  105°15′39.6758″

Geocentric Radius 20,897,577.640

Peak
  40°01′19.1582″
  105°30′55.1283″

Geocentric Radius 20,896,795.555

The distance between these two stations is 72,126.21 feet. This distance is sometimes called the *ground distance* or the *horizontal distance at mean elevation*. In other words, it is not the slope distance but rather the distance between them corrected to an averaged horizontal plane, as is common practice. For practical purposes, then this is the distance between the two stations on the topographic surface of the Earth. On the way to finding the

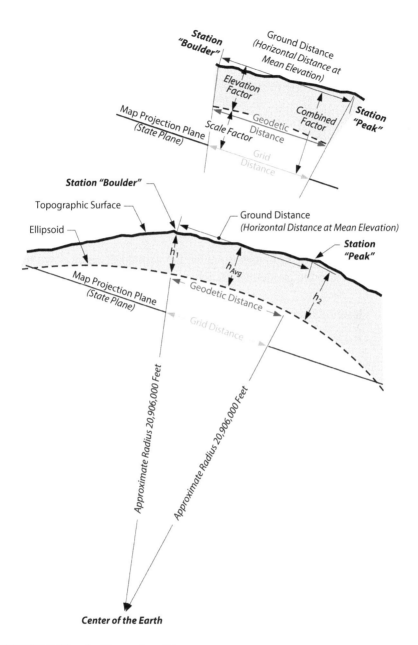

FIGURE 5.21 An Example—Distances

grid distance from Boulder to Peak, there is the interim step, calculating the geodetic distance between them, that is, the distance on the ellipsoid. We need the elevation factor and here is how it is determined. The ellipsoidal height of Boulder, h_1, is 5437 feet. The ellipsoidal height of Peak, h_2, is 9099 feet. The average ellipsoidal height of the two is 7268 feet.

The approximate radius of the ellipsoid, traditionally used in this work, is 20,906,000 feet. The elevation factor is calculated:

$$\text{Elevation Factor} = \frac{R}{R + h_{avg}}$$

$$\text{Elevation Factor} = \frac{20,906,000 \text{ ft.}}{20,906,000 \text{ ft.} + 7268 \text{ ft.}}$$

$$\text{Elevation Factor} = \frac{20,906,000 \text{ ft.}}{20,913,268 \text{ ft.}}$$

$$\text{Elevation Factor} = 0.99965247$$

This factor then is the ratio used to move the ground distance down to the ellipsoid, down to the geodetic distance.

Ground Distance Boulder to Peak = 72,126.21 ft.

Geodetic Distance = Ground Distance × Elevation Factor

Geodetic Distance = 72,126.21 × 0.99965247

Geodetic Distance = 72,101.14 ft.

It is possible to refine the calculation of the elevation factor by using an average of the actual *geocentric radius* rather than the approximate 20,906,000 feet. In the area of stations Boulder and Peak, that is the average radius 20,897,186.597 feet. It is a bit shorter, but it is worth noting that within the conterminous United States, such variation will not cause a calculated geodetic distance to differ significantly. The difference is one-hundredth of a foot in this case. However, it is worthwhile to take care to use the ellipsoidal heights of the stations when calculating the elevation factor rather than the *orthometric* heights.

In calculating the elevation factor in SPCS27, no real distinction was made between ellipsoid height and orthometric height. However, in SPCS83, the averages of the ellipsoidal heights at each end of the line are used. If the ellipsoid height is not directly available, it can be calculated from the formula (see Figure 5.30):

$$h = H + N$$

where
h = ellipsoid height
H = orthometric height
N = geoid height

As mentioned previously, converting a geodetic distance to a grid distance is done with an averaged scale factor:

$$K = \frac{K_1 + K_2}{2}$$

In this instance, the scale factor at Boulder is 0.99996703, and at Peak, it is 0.99996477.

$$0.99996590 = \frac{0.99996703 + 0.99996477}{2}$$

Using the scale factor, it is possible to reduce the geodetic distance 72,101.14 feet to a grid distance:

$$\text{Geodetic Distance} \times \text{Scale Factor} = \text{Grid Distance}$$
$$72,101.14 \text{ ft.} \times 0.99996590 = 72,098.68 \text{ ft.}$$

There are two steps, first from ground distance to geodetic distance using the elevation factor and second from geodetic distance to grid distance using the scale factor. They can be combined into one. Multiplying the elevation factor and the scale factor produces a single ratio that is usually known as the *combined factor* or the *grid* factor. Using this grid factor, the measured line is converted from a ground distance to a grid distance in one jump. Here is how it works. In the example above, the elevation factor for the line from Boulder to Peak is 0.99965247, and the scale factor is 0.99996590:

$$\text{Grid Factor} = \text{Scale Factor} \times \text{Elevation Factor}$$
$$0.99961838 = 0.99996590 \times 0.99965247$$

Then, using the grid factor, the ground distance is converted to a grid distance.

$$\text{Grid Distance} = \text{Grid Factor} \times \text{Ground Distance}$$
$$72,098.68 \text{ ft.} = 0.99961838 \times 72,126.21 \text{ ft.}$$

Also, the grid factor can be used to go the other way. If the grid distance is divided by the grid factor, it is converted to a ground distance.

$$\text{Ground Distance} = \text{Grid Distance}/\text{Grid Factor}$$
$$72,126.21 \text{ ft.} = \frac{72,098.68 \text{ ft.}}{0.99961838}$$

The NGS published data for passive stations includes state plane northing, mapping angle, and grid azimuth in the appropriate zone. A scale factor is also included for easy conversions. They also include *UTM* coordinates.

UNIVERSAL TRANSVERSE MERCATOR COORDINATES

The UTM system is far from an anachronism. UTM, with the Universal Polar Stereographic (UPS) system, covers the whole world in one consistent system. It can be four times less precise than typical SPCSs with a scale factor that typically reaches 0.9996. Yet the ease of using UTM and its worldwide coverage make it useful to those planning work that embraces large areas, especially when such work would have to cross many different SPCS zones. The UTM projection has been adopted by the IUGG, the same organization that reached the international agreement to use GRS80 as the reference ellipsoid for the modern geocentric datum. The U.S. NASA and other military and civilian organizations worldwide also use UTM coordinates for various mapping needs. Nearly all National Geospatial-Intelligence Agency (NGA) topographic maps, *United States Geological Survey* (*USGS*) quad sheets, and many aeronautical charts show the UTM grid lines.

It is often said that UTM is a military system created by the U.S. Army, but several nations, and the North Atlantic Treaty Organization (NATO), played roles in its creation after World War II. At that time, the goal was to design a consistent coordinate system that could promote cooperation between the military organizations of several nations. Before the introduction of UTM, allies found that their differing systems hindered the synchronization of military operations. Conferences were held on the subject from 1945 to 1951 with representatives from Belgium, Portugal, France, and Britain, and the outlines of the present UTM system were developed. By 1951 the U.S. Army introduced a system that was very similar to that currently used. The UTM projection divides the world into 60 zones that begin at longitude 180°, the International Date Line. Zone 1 is from 180° to 174°W longitude. The conterminous United States are within UTM zones 10 to 19.

Here is a convenient way to find the zone number for a particular longitude. Consider west longitude negative and east longitude positive, add 180° and divide by 6. Any answer greater than an integer is rounded to the next highest integer and you have the zone. For example, Denver, Colorado is near 105°W longitude, −105°.

$$-105° + 180° = 75°$$

$$75°/6 = 12.50$$

Round up to 13

Therefore, Denver, Colorado, is in UTM Zone 13, as shown in Figure 5.22.

All UTM zones have a width of 6° of longitude. From north to south, the zones extend from 84°N latitude to 80°S latitude. Originally the northern limit was at 80°N latitude and the southern 80°S latitude. On the south, the latitude is a small circle that conveniently traverses the ocean well south of

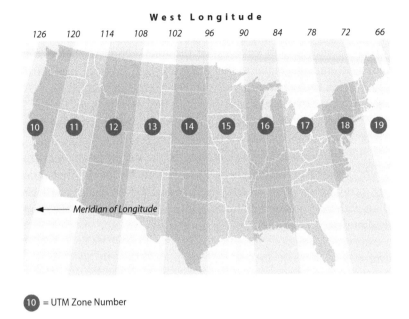

FIGURE 5.22   UTM Zones in Conterminous United States

Africa, Australia, and South America. However, 80°N latitude was found to exclude parts of Russia and Greenland and was extended to 84°N latitude.

UNIVERSAL TRANSVERSE MERCATOR ZONES OF THE WORLD

The foundation of the 60 UTM zones is a secant Transverse Mercator projection. The central meridian of the zones is exactly in the middle. For example, in Zone 1 from 180°W to 174°W longitude, the central meridian is 177°W longitude, so each zone extends 3° east and west from its central meridian. The UTM secant projection gives approximately 180 km between the lines of exact scale where the cylinder intersects the ellipsoid. The scale factor grows from 0.9996 along the central meridian of a UTM zone to 1.00000 at 180 km to the east and west. Please recall that SPCS zones are usually more limited in width, ~158 miles, and, therefore, have a smaller range of scale factors than do the UTM zones. In State Plane Coordinates, the scale factor is usually no more than 1 part in 10,000. In UTM coordinates, it can be as large as 1 part in 2500. The reference ellipsoids for UTM coordinates vary among five different figures. One can obtain 1983 UTM coordinates by referencing the UTM zone constants to the GRS80 ellipsoid of NAD83, then 1983 UTM coordinates will be obtained.

Unlike any of the systems previously discussed, every coordinate in a UTM zone occurs twice, once in the Northern Hemisphere and once in the

Southern Hemisphere. This is a consequence of the fact that there are two origins in each UTM zone. The origin for the portion of the zone north of the equator is moved 500 km west of the intersection of the zone's central meridian and the equator. This arrangement ensures that all of the coordinates for that zone in the Northern Hemisphere will be positive. The origin for the coordinates in the Southern Hemisphere for the same zone is 500 km west of the central meridian as well, but in the Southern Hemisphere, the origin is not on the equator, it is 10,000 km south of it, close to the South Pole. This orientation of the origin guarantees that all of the coordinates in the Southern Hemisphere are in the first quadrant and are positive. In other words, the intersection of each zone's central meridian with the equator defines its origin of coordinates. In the Southern Hemisphere, each zone's origin is given the coordinates:

$$\text{easting} = X_0 = 500,000 \text{ meters, and northing} = Y_0 = 10,000,000 \text{ meters}$$

In the Northern Hemisphere, the zone's origin's values are:

$$\text{easting} = X_0 = 500,000 \text{ meters, and northing} = Y_0 = 0 \text{ meters}$$

In fact, in the official version of the UTM system, there are more divisions in each UTM zone than the north–south demarcation at the equator. As shown in Figure 5.23, each zone is divided into 20 subzones. Each of

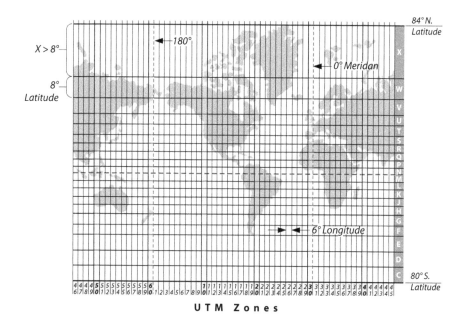

UTM Zones

FIGURE 5.23   UTM Zones Around the World

the subzones covers 8° of latitude and is lettered from C on the south to X on the north. Actually, subzone X is a bit longer than 8°; remember the extension of the system from 80°N latitude to 84°N latitude. That all went into subzone X. It is also interesting that I and O are not included. They resemble one and zero too closely. In any case, these subzones are little used outside of military applications of the system.

The developed UTM grid is defined in meters. Each zone is projected onto the cylinder that is oriented in the same way as that used in the Transverse Mercator SPCS described earlier. The radius of the cylinder is chosen to keep the scale errors within acceptable limits (see Figure 5.24).

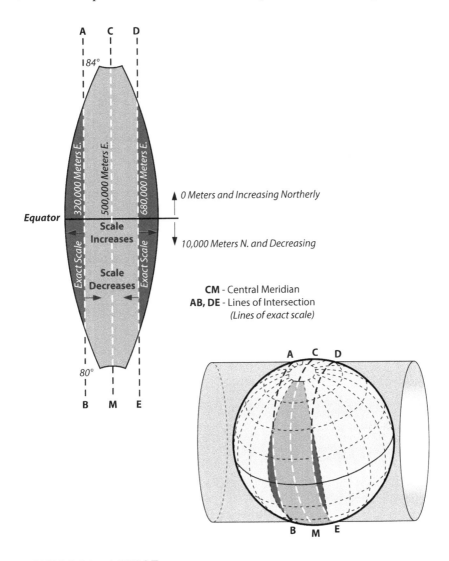

FIGURE 5.24 A UTM Zone

UTM zones nearly cover the Earth, except the Polar Regions, which are covered by two polar zones called the UPS projection one on the north, whose central point is at the North Pole; one on the south, whose central point is at the South Pole (see Figure 5.25). The UPS are azimuthal stereographic projections like those mentioned earlier. The North zone covers latitudes 84°N to 90°N. The South zone covers latitudes 80°S to 90°S. The scale factor is 0.994, and the false easting and northing are both 2000 km.

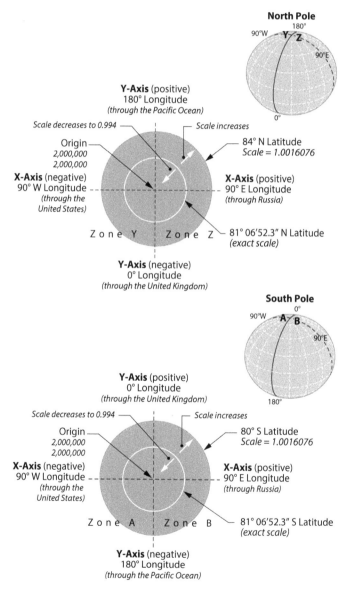

FIGURE 5.25 A Universal Polar Stereographic (UPS) Projection

HEIGHTS

ELLIPSOIDAL HEIGHTS

A point on the Earth's surface is not completely defined by its latitude and longitude. In such a context there is, of course, a third element, that of height. Surveyors have traditionally referred to this component of a position as its elevation. One classical method of determining elevations is spirit leveling. A level, correctly oriented at a point on the surface of the Earth, defines a line parallel to the geoid there. The elevations determined by differential level circuits are orthometric; that is, they are defined by their vertical distance above the geoid as it would be measured along a plumb line. However, orthometric elevations are not directly available from the ECEF X, Y, and Z position vectors derived from GNSS measurements. As mentioned before, modern geodetic datums rely on the surfaces of geocentric ellipsoids to approximate the surface of the Earth. The actual surface of the Earth does not coincide with these nice smooth surfaces, even though that is where the points represented by the coordinate pairs lay. Abstract points may be on the ellipsoid, but the physical features those coordinates intend to represent are on the topographic surface of the Earth. The distance from a coordinate pair on the reference ellipsoid to the point on the surface of the Earth, when measured along a line perpendicular to the reference ellipsoid, is known by more than one name. It is called the *ellipsoidal height,* and it is also called the *geodetic height* and is usually symbolized by h.

In Figure 5.26, the ellipsoidal height of a station is illustrated. The concept of an ellipsoidal height is straightforward. A reference ellipsoid may be above or below the surface of the Earth at a particular place.

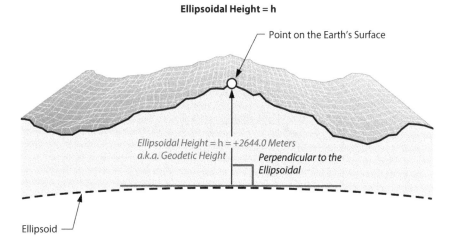

Ellipsoidal Height = h

Point on the Earth's Surface

Ellipsoidal Height = h = +2644.0 Meters
a.k.a. Geodetic Height

Perpendicular to the Ellipsoidal

Ellipsoid

FIGURE 5.26   An Ellipsoidal Height

If the ellipsoid's surface is below the surface of the Earth at the point, the ellipsoidal height has a positive sign, if the ellipsoid's surface is above the surface of the Earth at the point, the ellipsoidal height has a negative sign. It is important to remember that the measurement of an ellipsoidal height is along a line perpendicular to the ellipsoid, not along a plumb line. Most often, they are not the same, and since a reference ellipsoid is a geometric imagining, it is quite impossible to actually set up an instrument on it. That makes it tough to measure ellipsoidal height using surveying instruments. In other words, ellipsoidal height is not what most people think of as an elevation. Said another way, an ellipsoidal height is not measured in the direction of gravity. It is not measured in the conventional sense of down or up.

Nevertheless, the ellipsoidal height of a point is readily determined using a GNSS receiver. In other words, it has the capability of determining the three-dimensional coordinates of a point in a short time. If the system has the parameters of the reference ellipsoid in its software, it can calculate latitude, longitude, and the ellipsoidal height. Actually, in a manner of speaking, ellipsoidal heights are new, at least in common usage, since they could not be easily determined until GNSS became a practical tool in the 1980s.

However, ellipsoidal heights are not all the same because reference ellipsoids, or sometimes just their origins, can differ. For example, an ellipsoidal height expressed in ITRF would be based on an ellipsoid with exactly the same shape as the NAD83 ellipsoid, GRS80. Nevertheless, the heights would be different because the origin has a different relationship with the Earth's surface (see Figure 5.27).

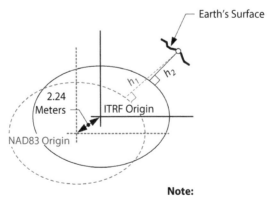

Note:
h_1 is NAD83 Ellipsoidal Height
h_2 is ITRF Ellipsoidal Height

FIGURE 5.27   A Shift

ORTHOMETRIC HEIGHTS

Spirit Leveling

Long before ellipsoidal heights were so conveniently available, knowing the elevation of a point was critical to the complete definition of a position. In fact, there are more than 200 different vertical datums in use in the world today. They were, and still are, determined by spirit leveling.

It is difficult to overstate the amount of effort devoted to differential spirit level work that has carried vertical control across the United States. The transcontinental precision leveling surveys done by the C&GS from coast to coast were followed by thousands of miles of spirit leveling work of varying precision. When the 39th parallel survey reached the west coast in 1907, there were approximately 19,700 miles, 31,789 km, of geodetic leveling in the national network. That was more than doubled 22 years later in 1929 to approximately 46,700 miles, 75,159 km. Though the work was excellent for the time, as the quantity of leveling information grew, so did the errors and inconsistencies. The foundation of the work was ultimately intended to be MSL as measured by tide station gauges. Inevitably this growth in leveling information and benchmarks made a new general adjustment of the network necessary to bring the resulting elevations closer to their true values relative to MSL.

There had already been four previous general adjustments to the vertical network across the United States by 1929. They were done in 1900, 1903, 1907, and 1912. The adjustment in 1900 was based upon elevations held to MSL as determined at five tide stations. The adjustments in 1907 and 1912 left the eastern half of the United States fixed as adjusted in 1903. In 1927, there was a special adjustment of the leveling network. This adjustment was not fixed to MSL at all tide stations and after it was completed because it became apparent that the MSL surface, as defined by tidal observations, tended to slope upwards to the north along both the Pacific and Atlantic Coasts, with the Pacific being higher than the Atlantic. However, in the adjustment that established the Sea Level Datum of 1929, the determinations of MSL at 26 tide stations, 21 in the United States, and 5 in Canada, were held fixed. Sea level was the intended foundation of these adjustments, and it might make sense to say a few words about the forces that shaped it.

EVOLUTION OF A VERTICAL DATUM

SEA LEVEL

Both the sun and the moon exert tidal forces on the Earth, but the moon's force is greater. The sun's tidal force is about half of that exerted on the Earth by the moon. The moon makes a complete elliptical orbit around the

Earth every 27.3 days. There is a gravitational force between the moon and the Earth. Each pulls on the other, and at any moment, the gravitational pull is greatest on the portion of the planet that happens to be closest to the moon. That tidal force produces a bulge in the Earth's waters. On the other side of the Earth, opposite the bulge, centrifugal force exceeds the gravitational force, and waters in this area are forced out away from the surface, creating another bulge. The two bulges are not stationary, they move. They move because the moon moving slowly relative to the Earth as it proceeds along its orbit and because the Earth is rotating beneath the moon. That rotation is relatively rapid compared to the moon's movement. Therefore, a coastal area in the high middle latitudes may find itself with a high tide early in the day when it is close to the moon and a low tide in the middle of the day when it has rotated away from it. This cycle will begin again with another high tide a bit more than 24 hours after the first high tide. A bit more than 24 hours because from the moment the moon reaches a particular meridian to the next time it is there is actually about 24 hours and 50 minutes, a period called a lunar day.

This sort of tide with one high water and one low water in a lunar day is known as a *diurnal tide*. This characteristic tide would be most likely to occur in the middle latitudes to the high latitudes when the moon is near its maximum declination, as you can see from Figure 5.28. The declination of a celestial body is like the latitude of a point on the Earth. It is an angle

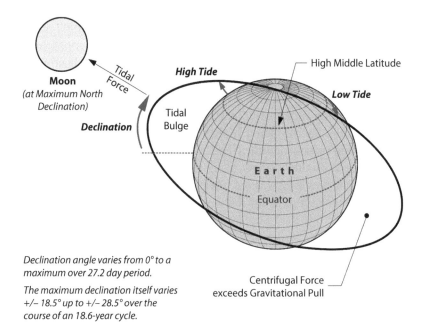

Declination angle varies from 0° to a maximum over 27.2 day period.

The maximum declination itself varies +/– 18.5° up to +/– 28.5° over the course of an 18.6-year cycle.

FIGURE 5.28   Tides

measured at the center of the Earth from the plane of the equator, positive to the north and negative to the south to the subject, which is, in this case, the moon. The moon's declination varies from its minimum of 0° at the equator to its maximum over a 27.2-day period, and that maximum declination oscillates too. It goes from +/–18.5° up to +/–28.5° over the course of an 18.6-year cycle.

Another factor that contributes to the behavior of tides is the elliptical nature of the moon's orbit around the Earth. When the moon is closest to the Earth, that is its perigee, the gravitational force between the Earth and the moon is 20% greater than usual. At apogee, when the moon is farthest from the Earth, the force is 20% less than usual. The variations in the force have exactly the affect you would expect on the tides, making them higher and lower than usual. It is about 27.5 days from perigee to perigee.

To summarize, the moon's orbital period is 27.3 days. It also takes 27.2 days for the moon to move from its maximum declinations back to 0° directly over the equator. There are 27.5 days from one perigee to the next. You can see that these cycles are almost the same, almost, but not quite. They are just different enough that it takes from 18 to 19 years for the moon to go through all the possible combinations of its cycles with respect to the sun and the moon. Therefore, if you want to be certain that you have recorded the full range of tidal variation at a place, you must observe and record the tides at that location for 19 years. This 19-year period, sometimes called the Metonic cycle, is the foundation of the definition of MSL. MSL can be defined as the arithmetic mean of hourly heights of the sea at a primary-control tide station observed over a period of 19 years. The mean in MSL refers to the average of these observations over time at one place. It is important to note that it does not refer to an average calculation made from measurements at several different places. Therefore, when the Sea Level Datum of 1929 was fixed to MSL at 26 tide stations, that meant an attempt was made to fit it to 26 different and distinct Local MSLs. In other words, it was warped to coincide with 26 different elevations.

The topography of the sea changes from place to place, and that means, for example, MSL in Florida is not the same as MSL in California. The fact is MSL varies, and the water's temperature, salinity, currents, density, wind, and other physical forces all cause changes in the sea surface's topography. For example, the Atlantic Ocean north of the Gulf Stream's strong current is around 1 m lower than it is farther south, and the denser water of the Atlantic is generally about 40 cm lower than the Pacific. At the Panama Canal, the actual difference is about 20 cm from the east end to the west end.

A Different Approach

After it was formally established, thousands of miles of leveling were added to the Sea Level Datum of 1929 (SLD29). The Canadian network also contributed data to the Sea Level Datum of 1929, but Canada did not ultimately use what eventually came to be known as the *National Geodetic Vertical Datum of 1929* (*NGVD 29*). The name was changed on May 10, 1973, because, in the end, the result did not and really could not coincide with MSL. It became apparent that the precise leveling done to produce the fundamental data had great internal consistency, but when the network was warped to fit so many tidal station determinations of MSL, that consistency suffered. By the time the name was changed to NGVD29 in 1973, there were more than 400,000 miles of new leveling work included. There were distortions in the network. Original benchmarks had been disturbed, destroyed, or lost. The NGS thought it time to consider a new adjustment. This time there was a different approach. Instead of fixing the adjustment to tidal stations, the new adjustment would be minimally constrained. That means that it would be fixed to only one station, not 26. That station turned out to be Father Point (Pointe-Au-Père)/Rimouski, an International Great Lakes Datum of 1985 (IGLD 85) station near the mouth of the St. Lawrence River and on its southern bank. In other words, for all practical purposes, the new adjustment of the huge network was not intended to be a sea level datum at all. It was a change in thinking that was eminently practical. While it is relatively straightforward to determine MSL at a station in coastal areas, carrying that reference reliably to the middle of a continent is quite another matter. Therefore, the new datum would not be subject to the variations in sea surface topography. It was unimportant whether the new adjustment's zero elevation and MSL were the same things or not.

The Zero Point

Precise leveling proceeded from the zero-reference established at Pointe-au-Père, Quebec in 1953. The resulting benchmark elevations were originally published in September 1961. The result of this effort was International Great Lakes Datum 1955. After nearly 30 years, the work was revised. The revision effort began in 1976, and the result was IGLD 1985. It was motivated by several developments including deterioration of the zero-reference point gauge location and improved surveying methods. However, one of the major reasons for the revision was the movement of previously established benchmarks due to *isostatic rebound*. This effect is literally the Earth's crust rising slowly, rebounding, from the removal of the weight and subsurface fluids caused by the retreat of the glaciers from the last ice age. The choice of the tide gauge at Pointe-au-Père, Quebec as the zero reference for IGLD was logical in 1955. It was reliable. It had already been connected to the network with precise leveling. It was at the outlet of the Great Lakes, but by 1984 the wharf at Pointe-au-Père had deteriorated and the gauge was moved. It was subsequently moved about 3 miles to Rimouski, Quebec, and precise levels were run between the two. It was there that the zero reference for IGLD 1985 and what became a new adjustment called *North American Vertical Datum of 1988 (NAVD88)* was established. The re-adjustment, known as NAVD88, was begun in the 1970s. It addressed the elevations of benchmarks all across the nation. The effort also included fieldwork. Destroyed and disturbed benchmarks were replaced, and over 50,000 miles of leveling were actually redone before NAVD88 was ready in June 1991. The differences between elevations of benchmarks determined in NGVD29 compared with the elevations of the same benchmarks in NAVD88 vary from approximately –1.3 feet in the east to approximately +4.9 feet in the west in the 48 conterminous states of the United States. The larger differences tend to be on the coasts, as one would expect, since NGVD29 was forced to fit MSL at many tidal stations and NAVD88 was held to just one.

THE GEOID

Any object in the Earth's gravitational field has *potential energy* derived from being pulled toward the Earth. Quantifying this potential energy is one way to talk about height because the amount of potential energy an object derives from the force of gravity is related to its height. As previously mentioned, a surface across which the potential of gravity the same is known as an equipotential surface. MSL itself is not an equipotential surface, of course, because it is affected by forces other than gravity, temperature, salinity, currents, wind, and so forth. The geoid, on the other hand, is defined by gravity alone. The geoid is the equipotential surface arranged to fit MSL as well as possible in a least-squares sense (see Figure 5.29). So, while there is a relationship between MSL and the geoid, they are not the same. They could be the same if the oceans of the world could be utterly still, completely free of currents, tides, friction, variations in temperature, and all other physical forces, except gravity.

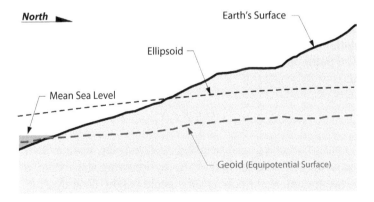

FIGURE 5.29   Ellipsoid–Geoid–Mean Sea Level

Reacting to gravity alone, these unattainable calm waters would coincide with the geoid. If the water was then directed by small frictionless channels or tubes and allowed to migrate across the land, the water would then, theoretically, define the same geoidal surface across the continents, too. However, the physical forces cannot be eliminated, so they cause MSL to deviate up to 1, even 2, meters from the geoid.

Since the geoid is completely defined by gravity, it is not smooth and continuous. It is lumpy because gravity is not consistent across the surface of the Earth. At every point, gravity has a magnitude and a direction, but these vectors do not all have the same direction or magnitude. Some parts of the Earth are denser than others. Where the Earth is denser, there is more gravity, and the fact that the Earth is not a sphere also affects gravity. It follows then that defining the geoid precisely involves measuring the direction and magnitude of gravity at many places. The geoid undulates with the uneven distribution of the mass of the Earth. It has all the irregularity that the attendant variation in gravity implies. In fact, the separation between the lumpy surface of the geoid and the smooth GRS80 ellipsoid worldwide varies from about +85 m west of Ireland to about −106 m in the area south of India near Ceylon. In the conterminous United States, sometimes-abbreviated *CONUS*, the distances between the geoid and the GRS80 ellipsoid, known as *geoid heights,* are less. They vary from about −8 m to about −53 m.

A geoid height is the distance measured along a line perpendicular to the ellipsoid of reference to the geoid. They are usually symbolized, *N*. If the geoid is above the ellipsoid, *N* is positive; if the geoid is below the ellipsoid, *N* is negative. It is negative in Figure 5.30 because the geoid is underneath the ellipsoid as it is throughout the conterminous United States. In Alaska, it is the other way around; the ellipsoid is underneath the geoid, and *N* is positive.

Please recall that an ellipsoid height is symbolized, *h*. The ellipsoid height is also measured along a line perpendicular to the ellipsoid of reference but to a point on the surface of the Earth. However, an orthometric height, symbolized, *H*, is measured along a plumb line from the geoid to a point on the surface of the Earth. In either case, by using the formula,

$$H = h - N$$

one can convert an ellipsoidal height, h, derived, say, from a GNSS observation, into an orthometric height, *H*, by knowing the extent of geoid-ellipsoid separation, also known as the geoid height, *N*, at that point.

The formula $H = h - N$ does not account for the fact that the plumb line along which an orthometric height is measured is curved, as you see

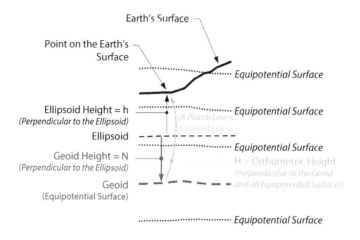

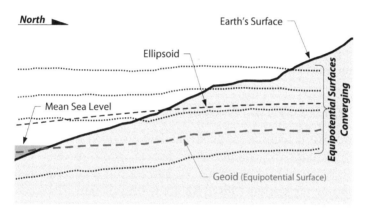

FIGURE 5.30 Height Conversion

in Figure 5.30. Curved because it is perpendicular to every equipotential surface through which it passes, and since equipotential surfaces are not parallel with each other, the plumb line must curve to maintain perpendicularity with them. This deviation of a plumb line from the perpendicular to the ellipsoid reaches about 1 minute of arc in only the most extreme cases. Therefore, any height difference that is caused by the curvature is negligible. It would take a height of over 6 miles for the curvature to amount to even 1 mm of difference in height.

GEOID MODELS

Major improvements have been made in the mapping the geoid on both national and global scales, and because there are large complex variations in the geoid related to both the density and relief of the Earth, geoid models and interpolation software have been developed to support the conversion of ellipsoidal elevations to orthometric elevations. For example, in early 1991, NGS presented a program known as GEOID90. This program allowed a user to find N, the geoidal height, in meters for any NAD83 latitude and longitude in the United States. The GEOID90 model was computed at the end of 1990, using over a million gravity observations. It was followed by the GEOID93 model at the beginning of 1993, using more than five times the number of gravity values used to create GEOID90. Both provided a grid of geoid height values in 3 minutes of latitude by 3 minutes of longitude grid with an accuracy of about 10 cm. Next, the GEOID96 model resulted in a height grid in 2 minutes of latitude by 2 minutes of longitude grid. GEOID99 covered the conterminous United States, and the U.S. Virgin Islands, Puerto Rico, Hawaii, and Alaska, and it was the first to combine gravity values with GPS ellipsoid heights on previously leveled benchmarks. GEOID03 superseded the previous models. It was followed by GEOID09, and the current geoid model was released in June 2020 and is known as GEOID18. However, a significant change is coming. The NGS is working on a completely new vertical datum that will replace the North American Vertical Datum 1988, NAVD88.

North American-Pacific Geopotential Datum of 2022

There are two categories of geoid models provided by NGS. The first category includes the composite, or hybrid, geoid models. GEOID96 was NGS's first hybrid geoid model, and the subsequent models up to and including the current GEOID18 are hybrid models. The second category of geoid models include the gravimetric geoids. These are derived entirely from gravity measurements alone. The two are related because the hybrid geoid models are biased versions of the underlying gravimetric geoid models from which they were built. In other words, hybrid models start with gravimetric models, which are then warped to fit the official vertical datum, NAVD88. To understand why, please recall that in order to calculate an NAVD88 orthometric height from a GNSS measured ellipsoidal height you must have the geoid height N, the separation between the geoid and the ellipsoid. The origin of N is at the heart of the matter.

There are about a half million orthometric heights on benchmarks established by differential leveling across the United States. About 26,000 of those have also been measured by GPS too. Those benchmarks are known as *GPSBM*s. The many years of level work have provided orthometric heights, H, at these benchmarks, and GPS measurements have provided ellipsoidal heights, h on them too. As you know, it is the difference between H and h that yields N. So, it is on a GPSBM. The difference between H from the leveling and h from GPS yields a version of N. I say a version of *N* because there is another. In the second version, GPS still provides the height above the ellipsoid *h*, but in this case, the orthometric height *H* does not come from leveling but from the gravimetric geoid. It is not surprising that the first and the second *N* values are not the same, but which one is used? At present, it is necessary to use the first version. For now, NAVD88 is the zero surface of the official vertical datum in the United States today, and the half million orthometric NAVD88 heights derived by leveling across the nation cannot be contradicted at this point. That means the first version of *N* mentioned here is used to create orthometric heights from GPS measurements. Therefore, a hybrid geoid model may start with a gravimetric geoid model, but then it must be warped to fit NAVD88, the zero surface that was, as you recall, adjusted to the single station at Father Point (Pointe-Au-Père)/Rimouski. However, this situation will change with the implementation of the *North American-Pacific Geopotential Datum of 2022* (*NAPGD2022*).

A gravimetric geoid model served as the basis for the current hybrid model GEOID18. It was defined entirely by remarkably sophisticated gravity measurements, terrestrial, airborne, and space-based (i.e., GRACE satellite). Like its predecessors, GEOID18 does not coincide to the gravimetric geoid. It has been constrained to the NAVD88 surface and it turns out that the NAVD88 zero surface is, on average, ~1.9 feet, ~58 cm, below the gravimetric geoid, as shown in Figure 5.31. Therefore, it follows that the mean difference between an estimated NAPGD2022 orthometric height derived from the most up-to-date gravimetric geoid and NAVD 88 orthometric heights is ~1.9 feet, ~58 cm. However, the average does not tell the whole story.

The comparison across the breadth of coverage shows that the NAVD88 zero surface is tilted across the conterminous United States. The difference

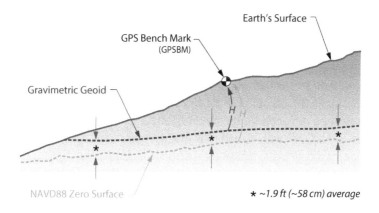

FIGURE 5.31   Gravimetric Geoid

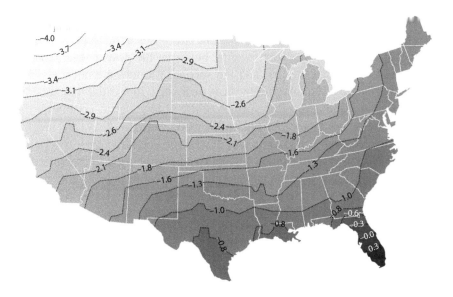

FIGURE 5.32   NAPD2022 – NAVD88 Orthometric Heights **(Estimated difference in feet)**

varies from near-zero in Florida to approximately −4.4 feet, −1.35 m in the Pacific Northwest (see Figure 5.32). There is also a tilt across Alaska from approximately −0.36 feet, −0.11 m in the southwest to −6.9 feet, ~2.10 m in the northeast.

Also, the centers of the hybrid and the gravimetric models differ. The hybrid geoids are tied to the GRS80 reference ellipsoid as used for NAD83, and as you know, that is not geocentric. The gravimetric geoid refers to the same GRS80 reference ellipsoid but as utilized in ITRF, where it is geocentric. Remember that comparison between the NAD83 and the ITRF centers reveals a difference of approximately 2.24 m. That shift contributes to the difference in the heights from the two frames. It would not be appropriate to use a gravimetric geoid to convert between the NAD 83 ellipsoid heights and NAVD 88 orthometric heights. On the other hand, GEOID18 is built for that purpose exactly. However, please note that it is to be the last hybrid geoid model that NGS creates. The current vertical datums will be replaced by the NAPGD2022.

Since NAVD88 was rolled out many years ago, information from the space-based gravity missions such as the European Space Agency's (ESAs) Gravity field and steady-state Ocean Circulation Explorer (GOCE) and National Aeronautics and Space Administration's (NASA) Gravity Recovery And Climate Experiment (GRACE), not to mention technological advances in GNSS have systematically improved measurement. Those improvements have revealed inadequacies in the official vertical datum, which have led to plans to replace NAVD88 with the new NAPGD2022 geopotential reference frame. It is called a geopotential reference frame rather than a vertical datum because the determination of orthometric heights is only part of its purpose. It will also be the model from which the deflection of the vertical, surface gravity, and other elements of the Earth's gravity field will be calculated.

Preparations for the new geopotential reference frame include the NGS plan known as GRAV-D, Gravity for the Redefinition of the American Vertical Datum. One objective of the program is to build an unconstrained gravimetric geoid model that will support both orthometric and *dynamic heights,* but this geoid will not only be modeled it will be monitored. The masses of the planet are in a constant flux, and thereby gravity changes. As gravity changes, heights change. Clearly, NAVD88, a datum based on a static gravity model cannot account for that.

The realization of NAVD88 included conventional level work and the establishment of orthometric heights on hundreds of thousands of passive benchmarks. The benchmarks were frequently set along railroads and highways because they offered flat continuous pathways for differential levels. This approach makes the distribution of benchmarks uneven. It also makes them vulnerable. When tracks and lanes are re-constructed or improved thousands are destroyed or disturbed. It happens every year. Even when they survive, regional distortions are inevitable, their heights can be invalidated without notice by crustal movements such as iso-static rebound, freezing and thawing, uplift, and subsidence. Subsidence is particularly troublesome in low-lying coastal areas along the Gulf of Mexico, Chesapeake Bay, and California agricultural regions. Some of NAVD88's original benchmark heights have been updated or super-seded with local re-leveling or GPS/GNSS. However, there has been no nationwide effort to re-adjust NAVD 88 since its implementation and NGS does not have sufficient resources to check and/or replace exist-ing passive benchmarks. This is particularly significant because users currently rely on those benchmarks for access to the datum. In the new paradigm, NGS will constantly monitor the varying gravity field mak-ing it available to users at an accuracy of 10 microGals, anytime. In other words, the foundation of the new geopotential reference frame will be a temporally tracked gravimetric geoid model. Clearly, such a system cannot be served to the users via passive monuments. Therefore, NGS will likely not maintain or check benchmarks. It follows that as their reliability decays, reliance on passive benchmarks will decline. However, users with a GPS/GNSS receiver and the gravimetric model

will always have direct, immediate, and accurate access to orthometric heights. Not only will the constant monitoring ensure that crustal
movement will be known and accounted for, the target accuracy for the
system is 2 cm, 0.06 feet.

As you probably intuit from this discussion, the long running separation
of horizontal and vertical aspects of surveying is falling away. It is being
replaced by a 4D system. A geocentric system that not only includes three-
dimensional Cartesian coordinates and time too.

EXERCISES

1. What is the datum used for the GPS navigation message?
   a. NAD83 (CORS96)
   b. NAD27
   c. GRS80
   d. WGS84 (G2139)

2. What technology contributed the least number of positions to the
   original adjustment of NAD83?
   a. EDM baselines
   b. TRANSIT Doppler positions
   c. Conventional optical surveying
   d. GPS

3. In the State Plane Coordinate systems in the United States, which
   mapping projection listed below is not used?
   a. The Transverse Mercator projection
   b. The Oblique Mercator projection
   c. The Lambert Conformal Conic projection
   d. The Universal Transverse Mercator projection

4. What information is necessary to convert an ellipsoidal height to
   an orthometric height?
   a. The geoid height
   b. The State Plane coordinate
   c. The semi-major axis of the ellipsoid
   d. GPS Time

5. Which statement about the geoid is correct?
   a. The geoid's surface is always perpendicular to gravity.
   b. The geoid's surface is the same as mean sea level.
   c. The geoid's surface is always parallel with an ellipsoid.
   d. The geoid's surface is the same as the topographic surface.

6. Which one of the four plate-fixed terrestrial references that NGS will introduce soon after 2022 is on a plate not included in the three components of NAD83?
   a. North American Terrestrial Reference Frame of 2022 (NATRF2022)
   b. Pacific Terrestrial Reference Frame of 2022 (PTRF2022)
   c. Caribbean Terrestrial Reference Frame of 2022 (CTRF2022)
   d. Mariana Terrestrial Reference Frame of 2022 (MTRF2022)

7. Which UTM zones cover the conterminous United States?
   a. Zones 10 North to 19 North
   b. Zones 1 North to 12 North
   c. Zones 6 North to 30 North
   d. Zones 20 North to 30 North

8. Which of the following organizations currently maintains the International Terrestrial Reference System (ITRS)?
   a. NGS
   b. IERS
   c. BIH
   d. C&GS

9. The combined factor is used in SPCS conversion. How is the combined factor calculated?
   a. The scale factor is multiplied by the elevation factor.
   b. The scale factor is divided by the grid factor.
   c. The grid factor added to the elevation factor and the sum is divided by two.
   d. The scale factor is multiplied by the grid factor.

10. Which of the following units of measurement will be eliminated as of January 1, 2023?
    a. Meter
    b. International foot
    c. U.S. survey foot
    d. Decibel

11. Deflection of the vertical is:
    a. The angle between the gravity vector and plumb line
    b. The angle between the plumb line and the line perpendicular to the reference ellipsoid
    c. The deviation of the instrument from level
    d. The angle between the plumb line and the vector to the geocenter

12. Orthomorphic map projection preserves:
  a. Azimuth and distance
  b. Angles and shapes
  c. Areas and distances
  d. Directions and areas

ANSWERS AND EXPLANATIONS

1. Answer is (d)

   Explanation: The WGS84 datum has been used by the U.S. military since January 21, 1987, as the basis for the GPS Navigation message computations. Therefore, coordinates provided directly by GPS receivers are based in WGS84. The newest realization of WGS84 is WGS84 (G2139). It was implemented by GPS Operational Control System in 2021.

2. Answer is (d)

   Explanation: It took more than ten years to readjust and redefine the horizontal coordinate system of North America into what is now NAD83. More than 1.7 million positions derived from classical surveying techniques throughout the Western Hemisphere were involved in the adjustment. They were supplemented by approximately 30,000 EDM measured baselines, 5,000 astronomic azimuths, and 650 Doppler stations positioned by the TRANSIT satellite system. Over 100 Very Long Baseline Interferometry (VLBI) vectors were also included. GPS, in its infancy, contributed very little.

3. Answer is (d)

   Explanation: In the United States, State Plane systems are based on the Transverse Mercator projection, an oblique Mercator projection, and the Lambert Conic map projection grid. Every state, Puerto Rico, and the U.S. Virgin Islands has its own plane rectangular coordinate system. The Universal Transverse Mercator projection is not used in the U.S. State Plane coordinate system.

4. Answer is (a)

   Explanation: The geoid undulates with the uneven distribution of the mass of the Earth and has all the irregularity that implies. In fact, the separation between the bumpy surface of the geoid and the smooth GRS80 ellipsoid varies from up to 100 m. Therefore, the only way a surveyor can convert an ellipsoidal height from a GNSS observation on a particular station into a useable orthometric elevation is to know the extent of geoid-ellipsoid separation at that point. That separation is known as the geoid height and is symbolized by N. Toward that end, major improvements have been made over the past quarter century or so in the mapping of the geoid on both national and global scales. This work has gone a long way toward the accurate determination of N. The formula for transforming ellipsoidal heights, h, into orthometric elevations, H, is $H = h - N$.

5. Answer is (a)

Explanation: The geoid is a representation of the Earth's gravity field. It is an equipotential surface that is everywhere perpendicular to the direction of gravity. In other words, it is perpendicular to a plumb line at every point. As there are many equipotential surfaces, the geoid is that which best fits mean sea level in a least-squares sense. Mean sea level is the average height of the surface of the sea for all stages of the tide. It was and sometimes still is used as a reference for elevations. However, it is not the same as the geoid. Mean sea level departs from the surface of the geoid. These displacements are known as the sea surface topography. Neither is the ellipsoid, a smooth mathematically defined surface, always parallel to the bumpy geoid. Finally, the geoid is certainly not coincident with the topographic surface of the Earth.

6. Answer is (c)

Explanation: The three components of NAD83 are each fixed to a plate, the North American plate—NAD 83 (2011), the Pacific plate—NAD 83 (PA11), and a small tectonic plate located west of the Mariana Trench, the Mariana plate—NAD 83 (MA11). The North American Terrestrial Reference Frame 2022 (NATRF2022) will be one of the four plate-fixed terrestrial reference which NGS will introduce soon after 2022. The tectonic plate for each frame may be inferred from their names; North American Terrestrial Reference Frame of 2022 (NATRF2022), Pacific Terrestrial Reference Frame of 2022 (PTRF2022), Mariana Terrestrial Reference Frame of 2022 (MTRF2022), and Caribbean Terrestrial Reference Frame of 2022 (CTRF2022).

7. Answer is (a)

Explanation: The UTM projection divides the world into 60 zones that begin at λ 180°, each with a width of 6° of longitude, extending from 84° Nφ and 80° Sφ. Its coverage is completed by the addition of two polar zones. The conterminous United States are within UTM zones 10 to 19. The UTM grid is defined in meters. Each zone is projected onto a cylinder that is oriented in the same way as that used in the Transverse Mercator State Plane Coordinates. The radius of the cylinder is chosen to keep the scale errors within acceptable limits. Coordinates of points from the reference ellipsoid within a particular zone are projected onto the UTM grid. The intersection of each zone's central meridian with the equator defines its origin of coordinates. In the Southern Hemisphere, each origin is given the coordinates: easting $= X_0 = 500,000$ m, and northing $= Y_0 = 10,000,000$ m, to ensure that all points have positive coordinates. In the Northern Hemisphere, the values are: easting $= X_0 = 500,000$ m, and northing $= Y_0 = 0$ m, at the origin. The scale factor grows from 0.9996 along the central meridian of a UTM zone to 1.00000 at 180,000 m to the east and west.

8. Answer is (b)

Explanation: The best geocentric reference frame currently available is the International Terrestrial Reference Frame, ITRF. Its origin is at the center of mass of the whole Earth including the oceans and atmosphere. The unit of length is the meter. The orientation of its axes was established as consistent with that of the IERS's predecessor, Bureau International de l'Heure, BIH, at the beginning of 1984. Today, the ITRF is maintained by the International Earth Rotation and Reference Systems Service, IERS, which monitors Earth Orientation Parameters (EOP) for the scientific community through a global network of fiducial stations. This is done with GNSS, Very Long Baseline Interferometry (VLBI), Lunar Laser Ranging (LLR), Satellite Laser Ranging (SLR), Doppler Orbitography and Radiopositioning Integrated by Satellite (DORIS), and the positions of the observing stations are now considered to be highly accurate. The International Terrestrial Reference Frame (ITRF) is a series of realizations. In other words, it is revised and published on a regular basis.

9. Answer is (a)

Explanation: The grid factor changes with the ellipsoidal height. It also changes with location in relation to the standard lines of its SPCS zone. The grid factor is derived by multiplying the scale factor by the elevation factor. The product is nearly 1 and is known as either the grid factor or the combination factor. There is a different combination factor for every line in SPCS.

10. Answer: (c)

Explanation: The fundamental unit for SPCS27 was the U.S. Survey Foot. It is very important to note that as of January 1, 2023, the U.S. Survey Foot is no more. From then on 1 foot will be exactly equal to 0.3048 m; in other words, it will be the International Foot. This decision was taken under the authority of the National Institute of Standards and Technology (NIST), the National Geodetic Survey (NGS), the National Ocean Service (NOS), and the National Oceanic and Atmospheric Administration (NOAA).

11. Answer: (b)

Explanation: The angle between the plumb line and the line perpendicular to the ellipsoid is called the deflection of the vertical. The deflection of the vertical usually has both a north–south and an east–west component.

12. Answer: (b)

Explanation: Map projections in which shape is preserved are known as conformal or orthomorphic. Orthomorphic means right shape. In a conformal projection, the angles between intersecting lines and curves around limited areas retain their original form on the map. In other words, between short lines, meaning lines under about 10 miles or so, a 45° angle on the ellipsoid is a 45° angle on the map. It also means that the scale is the same in all directions from a point; in fact, it is this characteristic that preserves the angles.

6 Static GNSS Surveying

Static GNSS surveying has been used on control surveys and will probably continue to be the preferred technique in that category for some time. The technology has virtually conquered two stumbling blocks that have defeated the plans of conventional surveyors for generations. Inclement weather does not disrupt GNSS observations, and a lack of intervisibility between stations is of no concern whatsoever, at least in post-processed GNSS. Still, it is far from so independent of conditions in the sky and on the ground that the process of designing a survey can now be reduced to points-per-day formulas, as some would like. Even with falling costs, the initial investment in the technology remains large by most surveyors' standards. However, there is seldom anything more expensive in a project than a surprise.

Static was the first GNSS method of surveying used in the field. Relative static positioning involves several stationary receivers simultaneously collecting data from at least four satellites during observation sessions that usually last from 30 minutes to 2 hours or longer. A typical application of this method would be the determination of vectors, or baselines, between several static receivers, often one or more of them being Continuously Operating Reference Stations (CORS). The accuracies achieved can be from 1 ppm to 0.1 ppm over tens of kilometers. There are few absolute requirements for relative static positioning. The requisites include more than one receiver, four or more satellites, and a mostly unobstructed sky above the stations to be occupied, but as in most surveying, the rest of the elements of the system are dependent on several other considerations, which sometimes implies planning.

A FEW WORDS ABOUT ACCURACY

When planning a GNSS survey, one of the most important parameters is the accuracy specification. A clear accuracy goal avoids ambiguity both during and after the work is done. First, it is important to remember that there is a difference between precision and accuracy. One aspect of precision can be visualized as the tightness of the clustering of measurements; the closer the grouping the more precise the measurement. Accuracy, on the other hand, requires one more element.

In Figure 6.1, the true positions for A, B, and C are the centers of the targets. We must have those standard values because it is the comparison

DOI: 10.1201/9781003405238-6

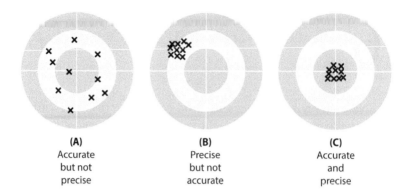

(A)
Accurate
but not
precise

(B)
Precise
but not
accurate

(C)
Accurate
and
precise

FIGURE 6.1   Precision and Accuracy 1

of the measurements with the standard values that the outcome of the work can be found to be sufficiently near the ideal or true value or not.

For example, on the left in Figure 6.2, it may seem at first that the average of the measurements in the GPS-A group is more accurate than the average of those in GPS-B because the GPS-A group is more precise. However, when the true position is introduced on the right, it is revealed that the GPS-B group's average is the more accurate of the two because accuracy and precision are not the same. When it comes to accuracy, there are other important details too. Local accuracy and network accuracy are not the same. As mentioned earlier, local accuracy, also known as relative accuracy, represents the uncertainty in the positions relative to the other adjacent points to which they should be directly connected. Network accuracy, also known as absolute accuracy, requires that a position's accuracy be specified with respect to an appropriate truth set

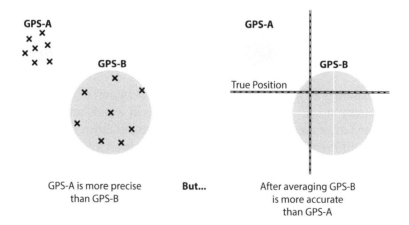

GPS-A is more precise
than GPS-B

But...

After averaging GPS-B
is more accurate
than GPS-A

FIGURE 6.2   Precision and Accuracy 2

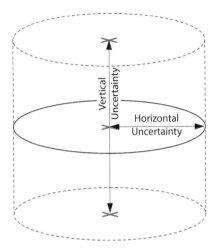

FIGURE 6.3   Horizontal and Vertical Accuracy

such as a national geodetic datum. When differentially corrected GNSS surveys are tied to the National Spatial Reference System (NSRS) in the United States via CORS network accuracy can be derived. It is typical for uncertainty in horizontal accuracies to be expressed in a number that is radial. The uncertainties in vertical accuracies are given similarly, but they are linear, not radial. In both cases, the limits are always plus or minus (±).

In other words, the reporting standard in the horizontal component is the radius of a circle of uncertainty, such that the true location of the point falls within that circle at some level of reliability, i.e., 95% of the time. Also, the reporting standard in the vertical component is a linear uncertainty value, such that the true location of the point falls within ± of that linear uncertainty to some degree of reliability (see Figure 6.3). In GNSS positioning, it is reasonable to expect that the vertical uncertainty will be about 1/3 larger than the horizontal uncertainty. In other words, if the absolute horizontal accuracy of a GNSS position is ~ ±1 m, then the estimate of the absolute vertical accuracy of the same GNSS position would be ~ ±3 m.

Figure 6.4 shows a spread of positions around the center of the range. As the radius of the error circle grows larger the certainty that the true position is inside it increases (it never reaches 100%).

Standards of Accuracy

The Federal Geodetic Control Committee (FGCC) wrote accuracy standards for GPS relative positioning techniques. These were preceded by older standards of first, second, and third order that then became subsumed under group C in the newer scheme. Until the last decades of the

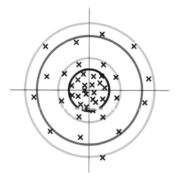

Circular Error Probable (CEP) = 50% = 0.5 m Error Circle Radius = ± 0.5 m
1-Sigma (1σ) = 68% = 0.6 m Error Circle Radius = ± 0.6 m
E90 = 90% = 0.9 m Error Circle Radius = ± 0.9 m
E95 = 95% = 1.0 m Error Circle Radius = ± 1.0 m

FIGURE 6.4   Horizontal Point Accuracy

twentieth century, the cost of achieving first-order accuracy was considered beyond the reach of most conventional surveyors. Besides, surveyors often said that such results were far in excess of their needs anyway. The burden of the equipment, techniques and planning that is required to reach its 2σ relative error ratio of 1 part in 100,000 of the old first-order was something most surveyors were happy to leave to government agencies. The FGCC's standards of B, A, and AA, are respectively 10, 100, and 1000 times more accurate than first-order. With the advent of GPS, the attainment of these accuracies did not require corresponding 10-, 100-, and 1000-fold increases in equipment, training, personnel, or effort. They were now well within the reach of private GPS surveyors both economically and technically. These accuracy standards are now superseded. In 1998, the Federal Geodetic Control Subcommittee under the Federal Geographic Data committee published the *Geospatial Positioning Accuracy Standards Part 2: Standards for Geodetic Networks*. These standards supplanted the earlier standards of 1984 and 1989. The FGDC has also issued the *Geographic Information Framework Data Content Standard Part 4: Geodetic Control*, May 2008.

NEW DESIGN CRITERIA

These upgrades in accuracy standards not only accommodate control by static GNSS, but they also have cast survey design into a new light for many surveyors. Nevertheless, it is not correct to say that every job suddenly requires the highest achievable accuracy, nor is it correct to say that every survey now demands a design. In some situations, a crew of two, or even one surveyor on-site may carry a survey from start to finish with no more planning than minute-to-minute decisions can provide, even though the basis and the content of those decisions may be quite different from those made in a conventional survey. In areas that are not heavily treed and generally free of overhead obstructions, sufficient accuracy may be possible without a prior design. It is unlikely that a survey of photocontrol or work on a cleared construction site would present overhead obstructions problems comparable with a static GNSS control survey among the trees in the Rocky Mountains or at the bottom of the Grand Canyon. However, work in a wide-open area may demand preliminary attention. For example, just the location of appropriate vertical and horizontal control stations or obtaining permits for access across privately owned property or government installations can be critical to the success of the work.

The Lay of the Land

An initial visit to the site of the survey is not always possible. Today, online mapping, if it is up to date, can make virtual site evaluation possible. Topography, as it affects the line of sight between stations, is of no concern on a static GNSS project, but its influence on transportation from station to station is a primary consideration. Perhaps some areas are only accessible by helicopter or other special vehicle. Initial inquiries can be made. Roads may be excellent in one area of the project and poor in another. The general density of vegetation, buildings, or fences may introduce questions of overhead obstruction or multipath. The pattern of land ownership relative to the location of project points may raise or lower the level of concern about obtaining permission to cross property.

NATIONAL GEODETIC SURVEY (NGS) CONTROL

NGS Control Data Sheets

NGS monuments that can be occupied with survey equipment are known as passive mark. Their utility will decline as NGS introduces the previously mentioned changes, but they can still provide control when properly utilized. That utilization should be informed by an understanding of the datasheet that is easily available online for each passive station. There is a good deal of information about the passive survey monuments on each individual sheet (see Figure 6.5). In addition to the latitude and longitude, the published data include the State Plane Coordinates in the appropriate zones. The first line of each data sheet includes the retrieval date. Then the station's category is indicated. There are several, and among them are

The NGS Data Sheet

See file dsdata.pdf for more information about the datasheet.

```
PROGRAM = datasheet95, VERSION = 8.12.5.14
Starting Datasheet Retrieval...
1      National Geodetic Survey, Retrieval Date = AUGUST 4, 2022


KK1696  ************************************************************************************************
KK1696  CBN - This is a Cooperative Base Network Control Station.
KK1696  DESIGNATION - JOG
KK1696  PID - KK1696
KK1696  STATE/COUNTY - CO/ARAPAHOE
KK1696  COUNTRY - US
KK1696  USGS QUAD - PARKER (2019)
KK1696
KK1696                       *CURRENT SURVEY CONTROL
KK1696  -------------------------------------------------------------------------------
KK1696* NAD 83 (2011) POSITION - 39 34 05.17554(N) 104 52 18.24483(W) ADJUSTED
KK1696* NAD 83 (2011) ELLIP HT - 1779.027 (meters) (01/16/18) ADJUSTED
KK1696* NAD 83 (2011) EPOCH - 2010.00
KK1696* NAVD88 ORTHO HEIGHT - 1796.2 (meters)  5893. (feet) GPS OBS
KK1696
KK1696  -------------------------------------------------------------------------------
KK1696  NAVD 88 orthometric height was determined with geoid model GEOID09
KK1696  GEOID HEIGHT -    -17.150 (meters)     GEOID09
KK1696  GEOID HEIGHT -    -17.125 (meters)     GEOID18
KK1696  NAD 83 (2011) X -  -1,263,970.417 (meters)    COMP
KK1696  NAD 83 (2011) Y -  - 4,759,798,468 (meters)   COMP
KK1696  NAD 83 (2011) Z -  - 4,042,268.398 (meters)   COMP
KK1696  LAPLACE CORR -    - 5.33 (seconds)    DEFLEC18
KK1696
KK1696  Network accuracy estimates per FGDC Geospatial Positioning Accuracy
KK1696  Standards:
KK1696  FGDC (95% conf, cm)  Standard deviation (cm)  CoorNE
KK1696  Horiz Ellip  SD_N   SD_E   SD_h (unitless)
KK1696  -------------------------------------------------------------------------------
KK1696  NETWORK  0.65  1.71  0.30  0.22  0.87  -0.04303572
KK1696  -------------------------------------------------------------------------------
KK1696  Click here for local accuracies and other accuracy information.
KK1696
KK1696
KK1696.The horizontal coordinates were establilshed by GPS observations
KK1696.and adjusted by the National Geodetic Survey in June 2012.
KK1696
KK1696.NAD 83 (2011) refers to NAD 83 coordinates where the reference frame has
KK1696.been affixed to the stable North American tectonic plate. See
KK1696.NA2011 for more information.
KK1696
KK1696.The horizontal coordinates are valid at the epoch date displayed above
KK1696.which is a decimal equivalence of Year/Month/Day.
KK1696
KK1696.The ortometric height was determined by GPS observations and a
KK1696.high-resolution geoid model.
KK1696.Significant digits in the geoid height do not necessarily refelect accuracy.
KK1696.GEOID18 height accuracy estimates available here.
KK1696
KK1696.Click photographs - Photos may exist for this station.
```

FIGURE 6.5a   A NGS Control Data Sheet *(Continued)*

KK1696.The X, Y, and Z were computed from the position and the ellipsoidal ht.
KK1696
KK1696.The Laplace correction was computed from DEFLEC18 derived deflections.
KK1696
KK1696.The ellipsoidal height was determined by GPS observations
KK1696.and is referenced to NAD 83.
KK1696
KK1696 The following values were computed from the NAD 83 (2011) position.
KK1696
KK1696;   North   East   Units Scale Factor Converg.
KK1696;SPC CO C - 497,563.467   968,386.201   MT 0.99996908 +0 23 46.5
KK1696;SPC CO C - 1,632,422.81   3,177,113.73 sFT 0.99996908 +0 23 46.5
KK1696;UTM 13   - 4,379,830.668 511,017.357   MT 0.99960149 +0 04054.1
KK1696
KK1696!   - Elev Factor x Scale Factor + Combined Factor
KK1696!SPC CO C   - 0.99972097 x 0.99996908 = 0.99969006
KK1696!UTM 13   - 0.99972097 x 0.99960149 = 0.99932257
KK1696
KK1696:   Primary Azimuth Mark   Grid Az
KK1696:SPC CO C   - JOG AZ MK   175 36 49.2
KK1696:UTM 13   - JOG AZ MK   175 55 41.6
KK1696
KK1696_U.S. NATIONAL GRID SPATIAL ADDRESS: 13SED1101779830(NAD 83)
KK1696
KK1696|--|
KK1696| PID Reference Object   Distance   Geod. Az |
KK1696| dddmmss.s |
KK1696| KK1695 DENVER INVERNESS TANK   66.921 METERS 11309 |
KK1696| CP8337 JOG RM 1   13.428 METERS 13410 |
KK1696| KK1699 JOG AZ MK   APPROX. 0.7 KM 1760035.7 |
KK1696| KK1701 LITTLETON HONEYWELL CORP TANK   APPROX. 5.2 KM 2840836.7 |
KK1696|--|
KK1696
KK1696   SUPERSEDED SURVEY CONTROL
KK1696
KK1696 ELLIP (06/27/12) 1779.178 (m)   GP(2010.00)
KK1696 NAD 83(2007) - 39 34 05.17592(N)   104 52 18.245229(W)   AD(2002.00) 0
KK1696 ELLIP H (02/10/02) 1779.167 (m)   GP(2002.00)
KK1696 ELLIP H (10/21/02) 1779.200 (m)   GP(   )5 1
KK1696 NAD 83(1992) - 39 34 05.17575 (N)   104 52 18.24592(W)   AD(   ) B
KK1696 ELLIP H (05/26/92) 1779.261 (m)   GP(   )4 1
KK1696 NAD 83(1986) - 39 34 05.17172(N)   104 52 18.24385(W)   AD(   ) 1
KK1696 NAD 27 - 39 34 05.21252(N)   104052016.31971(W)   AD(   ) 1
KK1696 NAVD 88 (06/08/05) 1796.4 (m) GEOID03 model used   GPS OBS
KK1696 NAVD 88 (12/14/94) 1796.3 (m) UNKNOWN model used   GPS OBS
KK1696 NAVD 88 (11/02/93) 1796.4 (m) GEOID93 model used   GPS OBS
KK1696 NAVD 88 (05/26/92) 1796.3 (m) UNKNOWN model used   GPS OBS
KK1696 NGVD 29 (07/19/86) 1795.6 (m)   5891.   (f) VERT ANG
KK1696
KK1696.Superseded values are not recommended for survey control.
KK1696
KK1696.NGS no longer agjusts projects to the NAD 27 or NGVD 29 datums,
KK1696.See file dsdata.pdf to determine how the superseded data were derived.
KK1696

FIGURE 6.5b   *(Continued)*

KK1696_MARKER; DH = HORIZONTAL CONTROL DISK
KK1696_SETTING: 7 = SET IN TOP OF CONCRETE MONUMENT
KK1696_STAMPING: JOG 1977
KK1696_MARK LOGO: NGS
KK1696_PROJECTION: RECESSED 10CENTIMETERS
KK1696_MAGNETIC: A = STEEL ROD ADJACENT TO MONUMENT
KK1696_STABILITY: C = MAY Hold, BUT OF TYPE COMMONLY SUBJECT TO
KK1696+STABILITY: SURFACE MOTION
KK1696_SATELLITE: THE SITE LOCATION WAS REPORTED AS SUITABLE FOR
KK1696+SATELLITE: SATELLITE OBSERVATIONS - November 14, 2017
KK1696

KK1696 HISTORY	- Date	Condition	Report By
KK1696 HISTORY	- 1977	MONUMENTED	NGS
KK1696 HISTORY	- 1977	GOOD	NGS
KK1696 HISTORY	- 19910515	GOOD	NGS
KK1696 HISTORY	- 19921214	GOOD	CODOT
KK1696 HISTORY	- 19940307	GOOD	MSAM
KK1696 HISTORY	- 19950206	GOOD	CODOT
KK1696 HISTORY	- 19960415	GOOD	MSAM
KK1696 HISTORY	- 19990405	GOOD	MSAM
KK1696 HISTORY	- 20040228	GOOD	INDIV
KK1696 HISTORY	- 20050516	GOOD	INDIV
KK1696 HISTORY	- 20050516	GOOD	WSSUR
KK1696 HISTORY	- 20060605	GOOD	INDIV
KK1696 HISTORY	- 20080203	GOOD	METRSC
KK1696 HISTORY	- 20100805	GOOD	WOOLPT
KK1696 HISTORY	- 20140226	GOOD	KII
KK1696 HISTORY	- 20171114	GOOD	WOOLPT

KK1696
KK1696   STATION DESCRIPTION
KK1696
KK1696'DESCRIBED BY NATIONAL GEODETIC SURVEY 1977 (LHW)
KK1696'THE STATION IS LOCATED ON THE EAST SIDE OF INTERSTATE HIGHWAY
KK1696'25 AND NORTHEAST OF THE JUNCTION OF STATE HIGHWAY 470 AND
KK1696'INTERSTATE HIGHWAY 25, ON ROAD RIGHT-OF-WAY, ON PROPERTY
KK1696'OWNED BY THE STATE OF COLORADO. IT IS JUST SOUTHEAST OF
KK1696'DENVER CITY LIMITS AND IN THE SOUTHEAST QUARTER OF SECTION
KK1696'43,T. 5S., R. 67 W.
KK1696'
KK1696'TO REACH THE STATION FROM THE JUNCTION OF INTERSTATE HIGHWAY
KK1696'25 AND STATE HIGHWAY 88 (ARAPAHOE ROAD) GO SOUTH-SOUTHEAST
KK1696'ON INTERSTATE HIGHWAY 25 FOR 2.5 MILES TO COUNTY LINE ROAD
KK1696'EXIT. TAKE EXIT RAMP TO A STOP SIGN, TURN LEFT ON COUNTY LINE
KK1696'ROAD FOR 0.3 MILE TO A FRONTAGE ROAD ON THE RIGHT AND ENTRANCE
KK1696'RAMP FOR NORTHBOUND INTERSTATE 2. (TO REACH AZIMUTH
KK1696'MARK FROM HERE TURN RIGHT ON FRONTAGE ROAD AND GO WESTERLY AND
KK1696'SOUTH FOR 0.4 MILE TO AZIMUTH MARK ON THE RIGHT). TURN
KK1696'LEFT ONTO ENTRANCE RAMP AND GO NORTHERLY FOR 0.1
KK1696'MILE TO THE STATION ON THE RIGHT NEXT TO A CHAIN LINK FENCE.
KK1696
KK1696'STATION MARKS, STAMPED---JOG 1977---,ARE STANDARD DISKS. THE
KK1696'SURFACE DISK IS SET IN THE TOP OF A 12-INCH CYLINDRICAL
KK1696'CONCRETE MONUMENT THAT IS FLUSH WITH THE GROUND. IT IS 5.5
KK1696'FEET WEST OF A CHAIN LINK FENCE AND 85 FEET EAST OF THE CENTER
KK1696'OF ENTRANCE RAMP FOR NORTHBOUND INTERSTATE HIGHWAY 25. THE
KK1696'UNDERGROUND DISK IS SET IN AN IRREGULAR MASS OF CONCRETE ABOUT
KK1696'42 INCHES BELOW THE GROUND.

FIGURE 6.5c   *(Continued)*

Continuously Operating Reference Station, Federal Base Network Control Station, and Cooperative Base Network Control Station. This is followed by the station's designation, which is its name, and its Point Identification, PID. Either of these may be used to search for the station in the NGS database. The PID is also found all along the left side of each data sheet record and is always two upper case letters followed by four numbers. The state, county, country, and U.S. Geological Survey (USGS)-7.5-minute quad name follows. Even though the station is in the area covered by the quad sheet, it may not actually appear on the map. Under the heading "Current Survey Control," you will find the latitude and longitude of the station in NAD83, which is fixed to the North American plate, currently in NAD83 (2011) and its height in NAVD88. The orthometric height in meters is listed as "ORTHO HEIGHT" and followed by the same in feet. When the height is derived from GPS observation, a geoid model must be used to determine the orthometric height. The model used is given.

Adjustments to NAD27 and NGVD29 datums are a thing of the past. However, these old values may be shown under Superseded Survey Control along with earlier realizations of NAD83 and NAVD88. Horizontal values may be either Scaled, if the station is a benchmark or adjusted, if the station is indeed a horizontal control point. There are 13 sources of vertical control values shown on NGS data sheets. Here are a few of the categories. There is *Adjusted*, which are given to 3 decimal places and are derived from the least-squares adjustment of precise leveling. Another category is *Posted*, which indicates that the station was adjusted after the general NAVD adjustment in 1991. When a station's elevation has been found by precise leveling but nonrigorous adjustment, it is called *Computed*. Stations' vertical values are given to 1 decimal place if they are from GPS observation (Obs) or vertical angle measurements (Vert Ang). They have no decimal places if they were scaled from topographic map, *Scaled*, or found by conversion from NGVD29 values using the program known as VERTCON. When they are available, Earth-Centered Earth-fixed (ECEF) coordinates are shown in X, Y, and Z. These are right-handed system, three-dimensional Cartesian coordinates and are computed from the position and the ellipsoidal height. They are the same type of X, Y, and Z coordinates presented in Chapter 5. These values are followed by the quantity which when added to an astronomic azimuth yields a geodetic azimuth. It is known as the Laplace correction. Please note that NGS uses a clockwise rotation regarding the Laplace correction. The ellipsoid height per the NAD83 ellipsoid is shown followed by the geoid height, where the position is covered by NGS's GEOID program. The FGDC network accuracy is shown at a 95% reliability level. Photographs of the station may also be available in some cases. When the data sheet is retrieved online one can use the link provided to bring them up.

COORDINATES

NGS data sheets provide State Plane and Universal Transverse Mercator (UTM) coordinates, the latter only for horizontal control stations. Scale factors for conversion from ellipsoidal distances to grid distances are given. While coordinates for azimuth marks or reference objects are not shown, azimuths to the primary azimuth mark are. They are clockwise from north. This information may be followed by distances to reference objects.

THE STATION MARK

Along with mark setting information, the type of monument and the history of mark recovery, the NGS data sheets provide a valuable to-reach description. It begins with the general location of the station. Then starting at a well-known location, the route is described with right and left turns, directions, road names, and the distance traveled along each leg in kilometers. When the mark is reached, the monument is described and horizontal and vertical ties are shown. Finally, there may be notes about obstructions to GNSS visibility and so forth.

SIGNIFICANCE OF THE INFORMATION

The value of the description of the monument's location and the route used to reach it is directly proportional to the date it was prepared and the remoteness of its location. The conditions around older stations often change dramatically when the area has become accessible to the public. If the age and location of a station increase the probability that it has been disturbed or destroyed then reference monuments can be noted as alternatives worthy of on-site investigation. However, special care ought to be taken to ensure that the reference monuments are not confused with the station marks themselves.

CONTROL FROM CONTINUOUSLY OPERATING NETWORKS

As NGS implements NATRF2022 and the other three TRFs mentioned in Chapter 5, the utility of occupying passive geodetic monuments in the field will undoubtedly lessen. It will often be supplanted by downloading the data available online from appropriate CORS or using online processing utilities such as OPUS. The CORS, also known as active stations, comprise fiducial networks that support a variety of GNSS applications. While they are frequently administered by governmental organizations, some are managed by public–private organizations, and some are commercial ventures. The most straightforward benefit of CORS is the user's ability to do relative positioning without operating his own base station. While CORS can be configured to support differential GNSS and Real-Time Kinematic (RTK) applications, as in Real-Time Networks, most networks constantly collect GNSS tracking data from known positions and archive the observations for subsequent download by users from the Internet.

In many instances, the original impetus of a network of CORS was geodynamic monitoring, as illustrated by the GEONET established by the Geographical Survey Institute (GSI) in Japan after the Kobe Earthquake. Networks that support the monitoring of the International Terrestrial Reference System (ITRS) have been created around the world by the International GNSS Service (IGS), which is a service of the *International Association of Geodesy (IAG)* and the Federation of Astronomical and Geophysical Data Analysis Services originally established in 1993. Despite the original motivation for the establishment of CORS networks, the result has been a boon for high-accuracy GNSS positioning. The data collected by these networks is quite valuable to GNSS surveyors around the world. Surveyors in the United States can take advantage of the CORS network administered by the National Geodetic Survey, NGS. The NGS system has two components, the Cooperative CORS and the National CORS. Together they comprise a network of hundreds of stations which constantly log multifrequency GNSS data and make the data available in the Receiver Independent Exchange (RINEX) format.

NGS CONTINUOUSLY OPERATING REFERENCE STATIONS

NGS manages the NOAA Continuously Operating Reference Station (CORS) Network to support post-processing GNSS data. Information is available in both code and carrier-phase GNSS data from receivers at these stations throughout the United States and its territories. There is a static continuous GNSS antenna and related equipment at each station. These active stations provide a direct connection to the U.S. NSRS. That data can be conveniently downloaded in its original form from the Internet free of charge for up to 30 days after its collection. It is also available later, but after it has been decimated to a 30-second format. Currently, nearly all coordinates provided by NGS for the CORS sites are available in NAD83 (2011) epoch 2010.00. The coordinates of CORS stations are also published in ITRF2014 epoch 2010.00. These positions differ from NAD83 (2011), of course. Since the points are moving, the coordinates in both NAD83 (2011) and ITRF2014 are accompanied by velocities. These velocities can be used to calculate the stations position at different epochs using NGS's Horizontal Time-Dependent Positioning (HTDP) utility.

NGS CORS Reference Points

At a CORS site, NGS provides the coordinates of the L1 phase center and the Antenna Reference Point (ARP). Generally speaking, it is best to adopt the position that can be physically measured, that means the coordinates given for the ARP. It is the coordinate of the part of the antenna from which the phase center offsets (PCOs) are calculated that is usually where the vertical axis of the antenna intersects the bottom mount, ground plane, or choke ring. The phase centers of antennas are not immovable points. They change slightly as the frequency, azimuth, and elevation of the satellite's signals change. In any case, the vectors between the phase centers for L1, L2, and L5 and the ARP differ a bit both vertically and horizontally. NGS provides the position of the phase center on average at a particular CORS site. As most post-processing software will, given the ARP, provide the correction for the phase center of an antenna, based on antenna type.

International CORS

Like NGS, IGS also provides CORS data. However, it has a global scope. The information on the individual stations can be accessed including the ITRF three-dimensional Cartesian coordinates and velocities for the IGS sites, but not all the sites are available from IGS servers. The Scripps Orbit and Permanent Array Center (SOPAC) and the ITRF web sites are also convenient access points for international CORS data. There are several others.

STATIC SURVEY PROJECT DESIGN

The selection of satellites to track, start and stop times, mask elevation angle, assignment of data file names, reference position, bandwidth, and sampling rate are some options useful in the static mode, as well as other GNSS surveying methods. These features may appear to be prosaic, but their practicality is not always obvious. A good survey project design can pay dividends by limiting lost time and maximizing productivity.

Horizontal Control

When geodetic surveying was more dependent on optics than electronic signals from space, horizontal control stations were set with station intervisibility in mind, not ease of access. Therefore, it is not surprising that passive stations established in that way are frequently difficult to reach. Not only are they found on the tops of buildings and mountains, but they are also in woods, beside transmission towers, near fences, and places generally obstructed from GNSS signals. The geodetic surveyors that established them could hardly have foreseen a time when a clear view of the sky above their heads would be crucial to high-quality control. In fact, it is only recently that most private surveyors have had any routine use for passive NGS stations. Many station marks have not been occupied for quite a long time. Since the primary monuments are often found deteriorated, overgrown, unstable, or destroyed, it is important that surveyors be well acquainted with the underground marks, reference marks (R.M.'s), and other methods used to perpetuate control stations.

When passive stations, whether NGS or other marks, are part of the survey plan it is a good idea to propose reconnaissance of several more than the absolute minimum of three horizontal control stations. Fewer than three makes any check of their positions virtually impossible. Many more are usually required in a GNSS route survey. In general, in GNSS networks, the more well-chosen horizontal control stations that are available, the better. Some stations will almost certainly prove unsuitable unless they have been used previously in GNSS work.

STATION LOCATION

The location of the stations, relative to the project itself, is also an important consideration in choosing horizontal control. For work other than route surveys, a handy rule of thumb is to divide the project into four quadrants and to choose at least one horizontal control station in each. The actual survey should have at least one horizontal control station in three of the four quadrants. Each of them ought to be as near as possible to the project boundary. Supplementary control in the interior of the network can then be used to add more stability to the network (Figure 6.6).

Route surveys require horizontal control at the beginning, the end, and the middle at a minimum. Long routes should be bridged with control on both sides of the line at appropriate intervals. The standard symbol for indicating horizontal control on the project map is a triangle.

VERTICAL CONTROL

Those stations with a published accuracy high enough for consideration as vertical control are symbolized by an open square or circle on the map. Those stations that are sufficient for both horizontal and vertical control are particularly helpful and are designated by a combination of the triangle and square. A minimum of four vertical control stations are needed to anchor a GNSS static network in which passive marks are used. A large project should have more. In general, the more benchmarks available

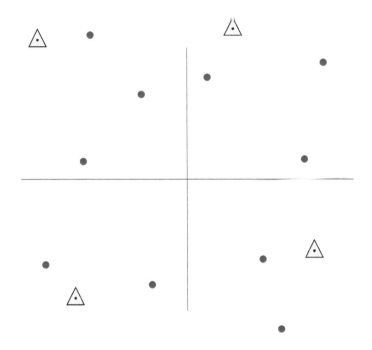

△ HORIZONTAL CONTROL STATION

● PROJECT POINT

FIGURE 6.6   Station Location 1

the better. Vertical control is best located at the four corners of a project
(Figure 6.7).

Orthometric elevations are best transferred by means of classic spirit
leveling, such work should be built into the project plan when it is necessary. When spirit levels are planned to provide vertical control positions,
special care may be necessary to ensure that the precision of such conventional work is as consistent as possible with the rest of the survey. Route
surveys require vertical control at the beginning and the end. They should
be bridged with benchmarks on both sides of the line at intervals from 5
to 10 km. When the distances involved are too long for spirit leveling to
be used effectively, two independent GNSS measurements may suffice to
connect a benchmark to the project. However, it is important to recall the
difference between the ellipsoidal heights available from a GNSS observation and the orthometric elevations yielded by a level circuit.

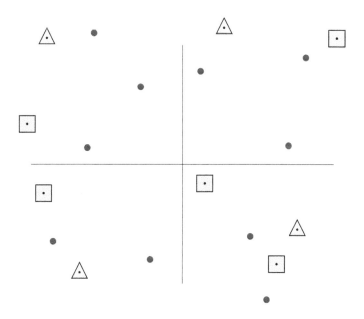

△ HORIZONTAL CONTROL STATION

▣ VERTICAL CONTROL STATION

● PROJECT POINT

FIGURE 6.7 Station Location 2

PLOTTING PROJECT POINTS

A solid dot is the standard symbol used to indicate the position of project points. Some variation is used when a distinction must be drawn between those points that are in place and those that must be set (Figure 6.8).

When its location is appropriate, it is always a good idea to have a vertical or horizontal control station serve double duty as a project point. While the precision of their plotting may vary, it is important that project points be located as precisely as possible even at this preliminary stage.

The subsequent observation schedule will depend to some degree on the arrangement of the baselines. Also, the preliminary evaluation of access, obstructions, and other information depends on the position of the project point relative to these features.

EVALUATING ACCESS

When all potential control and project positions have been plotted and given a unique identifier, some aspects of the survey can be addressed

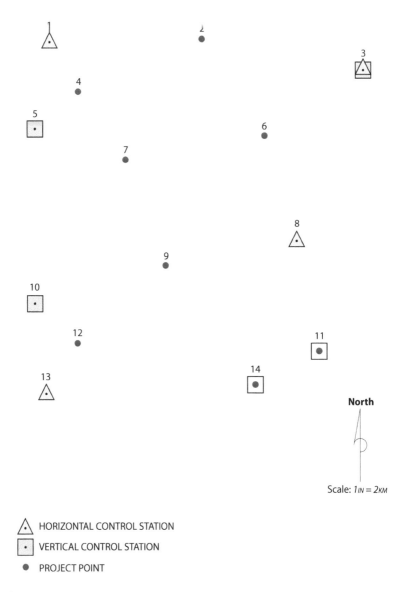

FIGURE 6.8   Control and Project Points

a bit more specifically. If good roads are favorably located, if open areas are indicated around the stations, and if no station falls in an area where special permission will be required for its occupation, then the preliminary plan of the survey ought to be remarkably trouble-free. However, it is likely that one or more of these conditions will not be so fortunately arranged. It is wise to remember that while inclement weather does not disturb GNSS observations, without sufficient preparation, it can play havoc with survey-ors' ability to reach points over difficult roads or by aircraft.

PLANNING OFFSETS

If control stations or project points are located in areas where the map indicates that topography or vegetation will obstruct the satellite's signals, alternatives may be considered. A shift of the position of a project point into a clear area may be possible where the change does not have a significant effect on the overall network. A control station may also be the basis for a less obstructed position, transferred with a short-level circuit or traverse. Of course, such a transfer requires the availability of conventional surveying equipment on the project. In situations where such movement is not possible, careful consideration of the actual paths of the satellites at the station itself during on-site reconnaissance may reveal enough windows in the gaps between obstructions to collect sufficient data by strictly defining the observation sessions.

PLANNING AZIMUTH MARKS

Azimuth marks are a common requirement in GNSS projects. They are an accompaniment to static GNSS stations when a client intends to use them to control subsequent conventional surveying work. Of course, the line between the station and the azimuth mark should be as long as convenience and the preservation of line-of-sight allows. It is wise to take care that short baselines do not degrade the overall integrity of the project. Occupations of the station and its azimuth mark should be simultaneous for a direct measurement of the baseline between them. Both should also be tied to the larger network as independent stations. There should be two or more occupations of each station when the distance between them is less than 2 km.

While an alternative approach may be to derive the azimuth between a GNSS station and its azimuth mark with an astronomic observation, it is important to remember that a small error, attributable to the deflection of the vertical, will be present in such an observation. The small angle between the plumb line and a normal to the ellipsoid at the station can either be ignored because they are likely to be quite similar at both ends or removed with a Laplace correction.

OBTAINING PERMISSIONS

Another aspect of access can be considered when the project map finally shows all the pertinent points. Nothing can bring a well-planned survey to a halt faster than a locked gate, an irate landowner, or a government official that is convinced he should have been consulted previously. To the extent that it is possible from the available mapping, affected private landowners and government jurisdictions should be identified and contacted. Taking this precaution at the earliest stage of the survey planning can increase the chance that the sometimes-long process of obtaining permissions, gate keys, badges, or other credentials has a better chance of completion before the survey begins. Any aspect of a GNSS survey plan derived from examining mapping, virtual or hardcopy, must be considered preliminary. Most features change with time, and even those that are relatively constant cannot be portrayed on a map with complete exactitude. Nevertheless, steps toward a coherent, workable design can be taken using the information they provide.

SOME GNSS SURVEY DESIGN FACTS

Though much of the preliminary work in producing the plan of a GNSS survey is a matter of estimation, some hard facts must be considered, too. For example, the number of GNSS receivers available for the work and the number of satellites above the observer's horizon at a given time in each place is two ingredients that can be determined with some certainty. Most GNSS software packages and online utilities provide users with routines that help them determine the satellite windows, the periods of time when the largest numbers of satellites are simultaneously available. Today, observers are assured of 24-hour coverage; however, the mere presence of adequate satellites above an observer's horizon does not guarantee collection of sufficient data. Therefore, despite the near certainty that adequate numbers of satellites will be available, evaluation of their configuration as expressed in the position dilution of precision (PDOP) is still crucial in planning a GNSS survey.

POSITION DILUTION OF PRECISION

The assessment of the productivity of a GNSS survey almost always hinges, in part at least, on the length of the observation sessions required to satisfy the survey specifications. The determination of the session's duration depends on several particulars, such as the length of the baseline and the relative position, that is, the geometry, of the satellites, among others. When a large constellation of satellites is in an optimal geometry and the PDOP is low, the length of the session needed to achieve the required accuracy can be shorter. Stated another way, the GNSS receiver's position is derived from the simultaneous solution of vectors between it and the constellation of satellites above the observer's horizon. The quality of that solution depends, in large part, on the distribution of those vectors. For example, any position determined when the satellites are crowded together in one part of the sky will be unreliable because all the vectors will have virtually the same direction. Fortunately, a computer can predict such an unfavorable configuration if it is given the ephemeris of each satellite, the approximate position of the receiver, and the time of the planned observation.

Provided with a forecast of a large PDOP, the GNSS survey planner should consider an alternate observation plan. On the other hand, when some satellites are nearly directly above the receiver and several others are evenly distributed near the horizon, the arrangement is nearly ideal. The planner of the survey would be likely to consider such a window. More satellites improve the resulting position if they are well distributed in the sky above the receiver. For example, if the planner finds 15 satellites will be above the horizon in the region where the work is to be done and the PDOP is below 2, that window would be a likely candidate for observation.

There are other important considerations. The satellites are constantly moving in relation to the receiver and to each other. Because satellites rise and set, the PDOP is constantly changing. Within all this movement, the GNSS survey designer must have some way of correlating the longest and most important baselines with the longest windows, the most satellites, and the lowest PDOP. Most GNSS software packages, given a particular location and period of time, can provide illustrations of the satellite configuration.

POLAR PLOT

One such diagram is a plot of the satellite's tracks drawn on a graphical representation of the upper half of the celestial sphere with the observer's zenith at the center and perimeter circle as the horizon. The azimuths and elevations of the satellites above the specified mask angle are connected into arcs that represent the paths of all available satellites. The utility of this sort of drawing has lessened with so many satellites available. The picture can become quite crowded and difficult to decipher (see Figure 3.7). Another printout is a tabular list of the elevation and azimuth of each satellite at time intervals selected by the user.

AN EXAMPLE

The geographical coordinates of the position of point Morant in Table 6.1 needed expression to the nearest minute only, a sufficient approximation for the purpose. The ephemeris data were 5 days old when the chart was generated by the computer, but the data were still an adequate representation of the satellite's movements to use in planning. The mask angle was specified at 15°, so the program would consider a satellite set when it moved below that elevation angle. The zone time was Pacific Daylight Time, 7 hours behind Coordinated Universal Time, UTC. The full constellation provided 24 healthy satellites, and the sampling rate indicated that

TABLE 6.1
Satellite Azimuth and Elevation Table 1

Point: *Morant* Lat 36:45:0 N Lon 121:45:0W Ephemeris: 9/24/2014

Date: *Wed., Sept. 29, 2014* Mask Angle: 15 (deg) Zone: *Time Pacific Day (−7)*

24 Satellites: *1 2 3 7 9 12 13 14 15 16 17 18 19 20 21 22 23 24 25 26 27 28 29 31*

Sampling Rate: *10 minutes*

Time	El	Az	El	Az	El	Az	El	Az	El	Az	El	Az	El	Az	El	Az	PDOP
SV	2		16		18		19		27		28		29		31		
								constellation of 8 SVs									
0:00	16	219	15	317	77	121	66	330	41	287	23	65	36	129	30	109	1.7
0:10	20	221	18	314	73	131	67	341	44	292	22	60	32	132	33	104	1.8
0:20	24	223	20	310	68	137	68	353	47	297	21	56	28	135	35	99	1.8
0:30	28	226	22	306	64	142	68	5	50	302	20	51	24	138	36	93	1.9
0:40	32	229	23	302	59	146	67	17	52	308	18	48	20	140	37	88	1.8
0:50	36	232	24	297	54	148	66	28	55	314	16	44	16	142	38	82	1.8
SV	2		16		18		19		27		31						
								constellation of 6 SVs									
1:00	40	235	24	293	49	151	65	39	58	320	38	76					3.0
1:10	43	239	24	288	44	153	63	49	61	328	37	70					3.0
1:20	47	244	24	283	40	155	61	57	64	336	36	64					2.8
1:30	51	249	23	278	35	156	59	65	66	345	34	60					2.6
SV	2		7		16		18		19		27		31				
								constellation of 7 SVs									
1:40	54	254	16	186	22	273	30	157	56	73	68	356	32	55			2.3
1:50	57	260	21	186	20	269	26	158	53	79	70	9	29	51			2.2
2:00	60	268	25	186	19	264	22	159	50	85	71	23	26	48			2.0
SV	2		7		16		18		19		27				31		
								constellation of 8 SVs									
2:10	66	276	30	185	16	260	17	160	47	91	15	319	71	38	23	45	1.7

the azimuth and elevation of those above the mask angle would be shown every 10 minutes. At 0:00 hour, satellite pseudorandom noise (PRN) 2 could be found on an azimuth of 219° and an elevation of 16° above the horizon by an observer at 36°45′ Nφ and 121°45′ Wλ. The table indicates that PRN 2 was rising and got continually higher in the sky for the 2 hours and 10 minutes covered by the chart. The satellite PRN 16 was also rising at 0:00 but reached its maximum altitude at about 1:10 and began to set. Unlike PRN 2, PRN 16 was not tabulated in the same row throughout the chart. It was supplanted when PRN 7 rose above the mask angle and PRN 16 shifted one column to the right. The same may be said of PRN 18 and PRN 19. Both of these satellites began high in the sky, unlike PRN 28 and PRN 29. They were just above 15° and setting when the table began and set after approximately 1 hour of availability. They would not have been seen again at this location for about 12 hours.

This chart indicated changes in the available constellation from eight space vehicles (SVs) between 0:00 and 0:50, six between 1:00 and 1:30, seven from 1:40 to 2:00, and back to eight at 2:10. The constellation never dipped below the minimum of four satellites, and the PDOP was good throughout. The PDOP varied between a low of 1.7 and a high of 3.0. Over the interval covered by the table, the PDOP never reached the unsatisfactory level of 5 or 6, which is when a planner should avoid observation.

Choosing the Window

Using this chart, the GNSS survey designer might well have concluded that the best available window was the first. There was nearly an hour of 8-satellite data with a PDOP below 2. However, the data indicated that good observations could be made at any time covered here, except for one thing: it was the middle of the night. It is worth noting that the ionospheric error is usually smaller after sundown. In fact, the FGDC specified multifrequency receivers for daylight observations for the achievement of the highest accuracies, due, in part, to the increased ionospheric delay during those hours.

Table 6.2 shows data from later in the day. It covers a period of 2 hours when a constellation of 5 and 6 satellites was always available. However, through the first hour, from 6:30 to 7:30, the PDOP hovered around 5 and 6. For the first half of that hour, four of the satellites—PRN 9, PRN 12, PRN 13, and PRN 24—were all near the same elevation. During the same period, PRN 9 and PRN 12 were only approximately 50° apart in azimuth, as well. Even though a sufficient constellation of satellites was constantly available, the survey designer may well have considered only the last 30 to 50 minutes of the time covered by this chart as suitable for observation.

There is one caution, however. Azimuth-elevation tables are a convenient tool in the division of the observing day into sessions, but it should

TABLE 6.2
Satellite Azimuth and Elevation Table 2

Point: *Morant* Lat *36:45:0 N Lon 121:45:0W* Ephemeris: *9/24/2014*

Date: *Wed., Sept. 29, 2014* Mask Angle: *15 (deg)* Zone: *Time Pacific Day (-7)*

24 Satellites: *1 2 3 7 9 12 13 14 15 16 17 18 19 20 21 22 23 24 25 26 27 28 29 31*

Sampling Rate: *10 minutes*

Time	El	Az	El	Az	El	Az	El	Az	El	Az	El	Az	El	Az	PDOP
SV	7		9		12		13		24						
					constellation of 5 SVs										
6:30	28	54	60	271	61	319	62	15	48	177					6.3
6:40	24	57	60	261	66	314	57	19	53	176					6.0
6:50	21	60	59	252	70	305	53	22	58	175					5.3
7:00	18	62	57	243	73	292	49	25	63	172					4.6
SV	9		12		13		20		24						
					constellation of 5 SVs										
7:10	54	235	74	274	44	28	16	308	68	169					4.8
7:20	51	229	74	255	40	32	20	310	72	163					5.7
7:30	47	224	72	238	37	35	23	311	77	153					5.1
7:40	43	219	68	226	33	38	27	313	80	134					4.0
SV	9		12		13		16		20		24				
					constellation of 6 SVs										
7:50	39	215	64	218	29	41	16	149	31	314	81	102			2.1
8:00	35	212	59	213	26	45	19	146	36	314	80	73			2.3
8:10	31	209	54	209	23	48	23	143	40	315	76	57			2.4
8:20	27	207	49	206	19	52	27	140	44	314	72	49			2.5
8:30	23	204	44	204	16	55	30	137	48	314	67	45			2.5

not be taken for granted that every satellite listed is healthy and in service. In the planning stage, diligence can prevent creation of a design dependent on satellites that prove unavailable. Similarly, after the field work is completed, it can prevent inclusion of unhealthy data in the post-processing.

Supposing that the period from 7:40 to 8:30 was found to be a good window, the planner may have regarded it as a single 50-minute session or divided it into shorter sessions. One aspect of that decision was probably the length of the baseline in question. In static GNSS, a long line of 30 km may require 50 minutes of 6 or more satellite data, but a short line of 3 km may not. Another consideration was probably the approximation of the time necessary to move from one station to another.

NAMING THE VARIABLES

The next step in the static GNSS survey design is drawing the preliminary plan of the baselines on the project map. Once some idea of the configuration of the baselines has been established, an observation schedule can be organized. Toward that end, the FGCC developed a set of formulas, those formulas will be used here. For illustration, suppose that the project map (see Figure 6.9) includes horizontal control, vertical control, and project points for a planned GNSS network. They will be symbolized by m. There are four multifrequency GNSS receivers available for this project. They will be symbolized by r. There will be five observation sessions each day during the project. They will be symbolized by d. To summarize:

m = total number of stations (existing and new) = 14
d = number of possible observing sessions per observing day = 5
r = number of receivers = 4 multifrequency

The design developed from this map must be preliminary. The session for each day of observation will depend on the success of the work the day before. Please recall that the plan must be provisional until the baseline lengths, the obstructions at the observation sites, the transportation difficulties, the ionospheric disturbances, and the satellite geometry are actually known. Those questions can only be answered during the reconnaissance and the observations that follow. Even though these equivocations apply, the next step is to draw the baseline's measurement plan.

RECEIVERS

Relative static positioning, just as all the subsequent surveying methods discussed here, involves several receivers occupying many sites. Problems can be avoided as long as the receivers on a project are compatible.

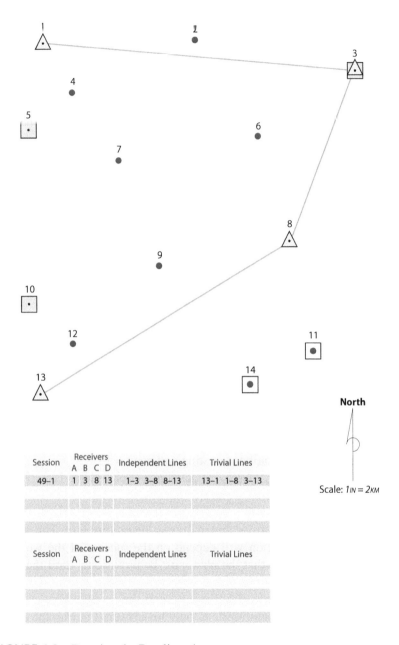

Session	Receivers A B C D	Independent Lines	Trivial Lines
49–1	1 3 8 13	1–3 3–8 8–13	13–1 1–8 3–13
Session	Receivers A B C D	Independent Lines	Trivial Lines

North

Scale: $1_{IN} = 2_{KM}$

FIGURE 6.9　Drawing the Baselines 1

For example, it is helpful if they have the same number of channels and signal processing techniques. However, the RINEX format, developed by the Astronomical Institute of the University of Bern, allows different receivers and post-processing software to work together. Almost all GNSS processing software will output RINEX files. The number and type of channels available to a receiver is a consideration because the more satellites the receiver can track continuously the better. Another factor that ought to be weighed is whether a receiver has single or multifrequency capability. Single-frequency receivers are best applied to relatively short baselines, say, under 25 km. The biases at one end of such a vector are likely to be similar to those at the other. Multifrequency receivers, on the other hand, have the capability to nearly eliminate the effects of ionospheric refraction and can handle longer baselines.

DRAWING THE BASELINES

HORIZONTAL CONTROL

A good rule of thumb is to verify the integrity of the horizontal control by observing baselines between these stations first. The vectors can be used to both corroborate the accuracy of the published coordinates and later to resolve the scale, shift, and rotation parameters between the control positions and the new network that will be determined by GNSS. These baselines are frequently the longest in the project, and there is an added benefit to measuring them first. By processing a portion of the data collected on the longest baselines early in the project, the most appropriate length of the subsequent sessions can be found. This test may allow improvement in productivity on the job without erosion of the final positions.

JULIAN DAY IN NAMING SESSIONS

The table at the bottom of Figure 6.9 indicates that the name of the first session connecting the horizontal control is 49-1. The date of the planned session is given in the Julian system. Taken most literally, as mentioned earlier, Julian dates are counted from January 1, 4713 BC. However, most practitioners of GNSS use the term to mean the day of the current year measured consecutively from January 1.

Under this construction, since there are 31 days in January, Julian day 49 is February 18 of the current year. The designation 49-1 means that this is to be the first session on that day. Some prefer to use letters to distinguish the session. In that case, the label would be 49-A.

INDEPENDENT LINES

This project will be done with four receivers. The table shows that receiver A will occupy point 1; receiver B, point 3; receiver C, point 8; and receiver D, point 13 in the first session. However, the illustration shows only three of the possible six baselines that will be produced by this arrangement.

Only the independent, also known as nontrivial, lines are shown on the map. The three lines that are not drawn are called trivial and are also known as dependent lines. This idea is based on restricting the use of the lines created in each observing session to the absolute minimum needed to produce a unique solution. Whenever four receivers are used, six lines are created. However, any three of those lines will fully define the position of each occupied station in relation to the others in the session. Therefore, the user can consider any three of the six lines independent, but once the decision is made only those three baselines are included in the network. The remaining baselines are then considered trivial and discarded. In practice, the three shortest lines in a four-receiver session are almost always deemed the independent vectors, and the three longest lines are eliminated as trivial or dependent. That is the case with the session illustrated.

Where r is the number of receivers, every session yields $r - 1$ independent baselines. For example, four receivers used in 10 sessions would produce 30 independent baselines. It cannot be said that the shortest lines are always chosen to be independent lines. Sometimes there are reasons to reject one of the shorter vectors due to incomplete data, cycle slips, multipath, or some other weakness in the measurements. Before such decisions can be made, each session will require analysis after the data has been collected. In the planning stage, it is best to consider the shortest vectors as independent lines.

Another aspect of the distinction between independent and trivial lines involves the concept of error of closure or loop closure. Loop closure is a procedure by which the internal consistency of a GNSS network is discovered. A series of baseline vector components from more than one GNSS session, forming a loop or closed figure, is added together. The closure error is the ratio of the length of the line representing the combined errors of all the vector's components to the length of the perimeter of the figure. Any loop closures that only use baselines derived from a single common GNSS session will yield an apparent error of zero because they are derived from the same simultaneous observations. For example, all the baselines between the four receivers in session 49-1 of the illustrated project will be based on ranges to the same constellation of GNSS satellites over the same period. Therefore, the trivial lines of 13-1, 1-8, and 3-13 will be derived from the same information used to determine the independent lines of 1-3, 3-8, and 8-13. It follows that if the fourth line from station 13 to station 1 were included to close the figure of the illustrated session, the error of closure would be zero. The same may be said of the inclusion of any of the trivial lines. Their addition cannot add any redundancy or any geometric strength to the lines of the session because they are all derived from the same data. If redundancy cannot be added to a GNSS session by including any more than the minimum number of independent lines, how can the baselines be checked? Where does redundancy in GNSS work come from?

REDUNDANCY

If only two receivers were used to complete the illustrated project, there would be no trivial lines and it might seem there would be no redundancy at all. However, to connect every station with its closest neighbor, each station would have to be occupied at least twice, and each time during a different session. For example, with receiver A on station 1 and receiver B on station 2, the first session could establish the baseline between them. The second session could then be used to measure the baseline between station 1 and station 4. It would certainly be possible to simply move receiver B to station 4 and leave receiver A undisturbed on station 1. However, some redundancy could be added to the work if receiver A were reset. If it were re-centered, re-plumbed, and its H.I. re-measured, some check on both of its occupations on station 1 would be possible when the network was completed. Under this scheme, a loop closure at the end of the project would have some meaning. If one were to use such a scheme on the illustrated project and connect into one loop all 14 baselines determined by the 14 two-receiver sessions, the resulting error of closure would be useful. It could be used to detect blunders in work, such as mismeasured heights of instruments, H.I.s. Such a loop would include many different sessions. The ranges between the satellites and the receivers defining the baselines in such a circuit would be from different constellations at different times. On the other hand, if it were possible to occupy all 14 stations in the illustrated project with 14 different receivers simultaneously and do the entire survey in one session, a loop closure would be meaningless.

In the real world, such a project is not done with 14 receivers nor with 2 receivers but with 3, 4, or 5. The achievement of redundancy takes a middle road. The number of independent occupations is still an important source of redundancy. In the two-receiver arrangement, every line can be independent, but that is not the case when a project is done with any larger number of receivers. As soon as three or more receivers are considered, the discussion of redundant measurement must be restricted to independent baselines, excluding trivial lines. Redundancy is then partly defined by the number of independent baselines that are measured more than once, as well as by the percentage of stations that are occupied more than once. While it is not possible to repeat a baseline without reoccupying its endpoints, it is possible to reoccupy a large percentage of the stations in a project without repeating a single baseline. These two aspects of redundancy in GNSS—the repetition of independent baselines and the reoccupation of stations—are somewhat separate.

Figure 6.10 shows one of the many possible approaches to setting up the baselines for this GNSS project. The survey design calls for the horizontal control to be occupied in session 49-1. It is to be followed by measurements

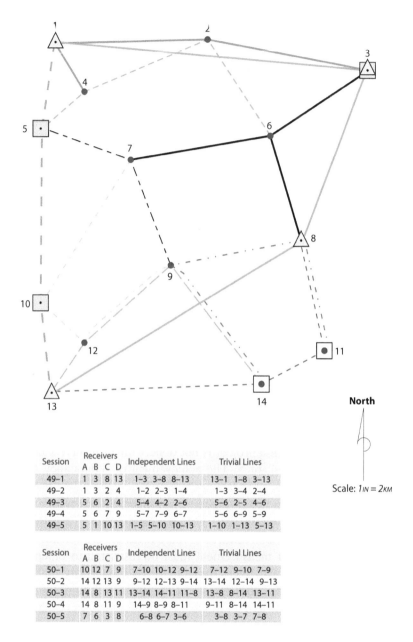

Session	Receivers A B C D	Independent Lines	Trivial Lines
49–1	1 3 8 13	1–3 3–8 8–13	13–1 1–8 3–13
49–2	1 3 2 4	1–2 2–3 1–4	1–3 3–4 2–4
49–3	5 6 2 4	5–4 4–2 2–6	5–6 2–5 4–6
49–4	5 6 7 9	5–7 7–9 6–7	5–6 6–9 5–9
49–5	5 1 10 13	1–5 5–10 10–13	1–10 1–13 5–13

Session	Receivers A B C D	Independent Lines	Trivial Lines
50–1	10 12 7 9	7–10 10–12 9–12	7–12 9–10 7–9
50–2	14 12 13 9	9–12 12–13 9–14	13–14 12–14 9–13
50–3	14 8 13 11	13–14 14–11 11–8	13–8 8–14 13–11
50–4	14 8 11 9	14–9 8–9 8–11	9–11 8–14 14–11
50–5	7 6 3 8	6–8 6–7 3–6	3–8 3–7 7–8

North

Scale: 1IN = 2KM

FIGURE 6.10   Drawing the Baselines 2

between two control stations and the nearest adjacent project points in session 49-2. As shown in the table at the bottom of the figure, there will be
redundant occupations on stations 1 and 3. Even though the same receivers
will occupy those points, their operators will be instructed to reset them at
a different H.I.s for the new session. A better, but probably less efficient,

plan would be to occupy these stations with different receivers than were used in the first session.

FORMING LOOPS

As the baselines are drawn on the project map for a static GNSS survey or any GNSS work where accuracy is the primary consideration, the designer should remember that part of their effectiveness depends on the formation of complete geometric figures. When the project is completed, these independent vectors should be capable of formation into closed loops that incorporate baselines from two to four different sessions. In the illustrated baseline plan, no loop contains more than ten vectors, no loop is more than 100 km long, and every observed baseline will have a place in a closed loop.

FINDING THE NUMBER OF SESSIONS

The illustrated survey design calls for ten sessions, but the calculation does not include human error, equipment breakdown, and other unforeseeable difficulties. It would be impractical to presume a completely trouble-free project. The FGCC proposed the following formula for arriving at a more realistic estimate:

$$s = \frac{(m \cdot n)}{r} + \frac{(m \cdot n)(p-1)}{r} + k \cdot m$$

where
  s = the number of observing sessions
  r = the number of receivers
  m = the total number of stations involved

But n, p, and k require a bit more explanation. The variable n is a representation of the level of redundancy that has been built into the network based on the number of occupations on each station. The illustrated survey design includes more than two occupations on all but 4 of the 14 stations in the network. In fact, 10 of the 14 positions will be visited three or four times during the survey. There are a total of 40 occupations by the 4 receivers in the 10 planned sessions. By dividing 40 occupations by 14 stations, it can be found that each station will be visited an average of 2.857 times. Therefore, the planned redundancy represented by factor n is equal to 2.857 in this project. The experience of a firm is symbolized by the variable p in the formula. The division of the final number of actual sessions required to complete past projects by the initial estimation yields a ratio that can be used to improve future predictions. That ratio is the production factor, p. A typical production factor is 1.1. A safety factor of 0.1, known as k, is recommended for GNSS projects within 100 km of a company's home base. Beyond that radius, an increase to 0.2 is advised.

The substitution of the appropriate quantities for the illustrated project increases the prediction of the number of observation sessions required for its completion:

$$s = \frac{(m \cdot n)}{r} + \frac{(m \cdot n)(p-1)}{r} + k \cdot m$$

$$s = \frac{(14)(2.857)}{4} + \frac{(14)(2.857)(1.1-1)}{4} + (0.2)(14)$$

$$s = \frac{(40)}{4} + \frac{(4)}{4} + 2.8$$

$$s = 10 + 1 + 2.8$$

$$s = 14 \text{ sessions (rounded to the nearest integer)}$$

In other words, the 2-day, 10-session schedule is a minimum period for the baseline plan drawn on the project map. A more realistic estimate of the observation schedule includes 14 sessions. It is also important to keep in mind that the observation schedule does not include time for on-site reconnaissance.

TIES TO THE VERTICAL CONTROL

The ties from the vertical control stations to the overall network are usually not handled by the same methods used with the horizontal control. The first session of the illustrated project was devoted to occupation of all the horizontal control stations. There is no similar method with the vertical control stations. First, the geoidal undulation would be indistinguishable from baseline measurement error. Second, the primary objective in vertical control is for each station to be adequately tied to its closest neighbor in the network. If a benchmark can serve as a project point, it is nearly always advisable to use it, as was done with stations 11 and 14 in the illustrated project. A conventional level circuit can often be used to transfer a reliable orthometric elevation from a vertical control station to a nearby project point.

STATIC GNSS CONTROL OBSERVATIONS

The prospects for the success of a GNSS project are directly proportional to the quality and training of the people doing it. The handling of the equipment, the on-site reconnaissance, the creation of field logs, and the inevitable last-minute adjustments to the survey design all depend on the training of the personnel involved for their success. There are those who say the operation of GNSS receivers no longer requires highly qualified survey personnel. That might be true if effective GNSS surveying needed only the pushing of the appropriate buttons at the appropriate time. In fact, when all goes as planned, it may appear to the uninitiated that GNSS has made experienced field surveyors obsolete, but when the unavoidable breakdowns in planning or equipment occur, the capable people, who seemed so superfluous moments before, suddenly become indispensable.

EQUIPMENT

Conventional Equipment

Most GNSS projects require conventional surveying equipment for spirit-leveling circuits, offsetting horizontal control stations, and monumenting project points, among other things. It is perhaps a bit ironic that this most advanced surveying method also frequently has a need of the most basic equipment. The use of brush hooks, machetes, axes, and so forth, can sometimes salvage an otherwise unusable position by removing overhead obstacles. Another strategy for overcoming such hindrances has been developed using various types of survey masts to elevate a separate GNSS antenna above the obstructing canopy. Flagging, paint, and the various techniques of marking that surveyors have developed over the years are still a necessity in GNSS work. The pressure of working in unfamiliar terrain is often combined with urgency. Even though there is usually not a moment to spare in moving from station to station, a GNSS surveyor frequently does not have the benefit of having visited the points before. In such situations, the clear marking of both the route and the station during reconnaissance is vital.

Despite the best route marking, a surveyor may not be able to reach the planned station or, having arrived, finds some new obstacle or unanticipated problem that can only be solved by marking and occupying an impromptu offset position for a session. A hammer, nails, shiners, paint, and so forth, are essential in such situations. In short, the full range of conventional surveying equipment and expertise have a place in GNSS. For some, their role may be more abbreviated than it was formerly, but one element that can never be outdated is good judgment.

Safety Equipment

The high-visibility vests, cones, lights, flagmen, and signs needed for traffic control cannot be neglected in GNSS work. Unlike conventional surveying operations, GNSS observations are not deterred by harsh weather.

Occupying a control station in a highway is dangerous enough under the best of conditions, but in a rainstorm, fog, or blizzard, it can be absolute folly without the proper precautions. Any time and trouble taken to avoid infraction of the local regulations regarding traffic management will be compensated by an uninterrupted observation schedule. Weather conditions also affect travel between the stations of the survey, both in vehicles and on foot. Equipment and plans to deal with emergencies should be part of any GNSS project. First aid kits, fire extinguishers, and the usual safety equipment are necessary. Training in safety procedures can be an extraordinary benefit, but perhaps the most important capability in an emergency is communication.

Communications

Whether the equipment is handheld or vehicle mounted, two-way radios and cell phones are used in most GNSS operations. The use of cell phones may eliminate the communication problem in some areas, but probably not in remote locations where connectivity is unavailable. In any case, the line of sight that is no longer necessary for the surveying measurements in GNSS is sorely missed in the effort to maintain clear radio contact between the receiver operators. A radio link between surveyors can increase the efficiency and safety of a GNSS project, but it is particularly valuable when last-minute changes in the observation schedule are necessary. When an observer is unable to reach a station or a receiver suddenly becomes inoperable, unless adjustments to the schedule can be made quickly, each end of all lines into the missed station will require re-observation. The success of static GNSS control survey may hinge on all receivers collecting their data simultaneously. High-wattage, private-line FM radios are quite useful when line of sight is available between them or when a repeater is available. However, it is more and more difficult to ensure reliable communication between receiver operators in geodetic surveys, especially as their lines grow longer. Despite the limitations of the systems available achievement of the best possible communication between surveyors on a GNSS project pays dividends in the long run.

GNSS Equipment

Most GNSS receivers capable of geodetic accuracy are designed to be mounted on a tripod, usually with a tribrach and adaptor. However, there is a trend toward bipod- or range pole-mounted antennas. An advantage of these devices is that they ensure a constant height of the antenna above the station. The mismeasured height of the antenna above the mark is probably the most pervasive and frequent blunder in GNSS control surveying. The tape or rod used to measure the height of the antenna is sometimes built into the receiver and, sometimes, a separate device. It is important that the height of instrument (H.I.) be measured accurately and consistently in both feet and meters without merely converting from one to the other mathematically. It is also important that the value be recorded in the field notes and, where possible, also entered into the receiver itself.

Where tribrachs are used to mount the antenna, the tribrach's optical centering should be checked and calibrated. It is critical that the effort to perform GNSS surveys to an accuracy of centimeters not be frustrated by inaccurate centering or H.I. measurement. Since many systems measure the height of the antenna to the edge of the ground plane or to the exterior ARP of the receiver itself, the calibration of the tribrach affects both the centering and the H.I. measurement. The resetting of a receiver that occupies the same station in consecutive sessions is an important source of redundancy for many kinds of GNSS networks. However, integrity can only be added if the tribrach has been accurately calibrated.

The checking of the carrier-phase receivers themselves is also critical to the control of errors in a GNSS survey, especially when different receivers or different models of antennas are to be used on the same work. The *zero-baseline* test is a method that may be used to fulfill equipment calibration specifications where a three-dimensional test network of sufficient accuracy is not available. As a matter of fact, the simplicity of this test is an advantage. It is not dependent on special software or a test network. This test can also be used to separate receiver difficulties from antenna errors. Two or more receivers are connected to one antenna with a signal or antenna splitter. The antenna splitter can be purchased from specialty electronics shops and is also available online. An observation is done with the divided signal from the single antenna reaching both receivers simultaneously. Since the receivers are sharing the same antenna, satellite clock biases, ephemeris errors, atmospheric biases, and multipath are all canceled. In the absence of multipath, the only remaining errors are attributable to random noise and receiver biases. The success of this test depends on the signal from one antenna reaching both receivers, but the current from only one receiver can be allowed to power the antenna. This test checks not only the precision of the receiver measurements but also the processing software. The results of the test should show a baseline of only a few millimeters.

Auxiliary Equipment

Tools to repair the ends of connecting cables and other small implements have saved more than one GNSS observation session from failure. Experience has shown that GNSS surveying requires at least as much resourcefulness, if not more, than conventional surveying. The health of the batteries is a constant concern in GNSS. There is simply nothing to be done when a receiver's battery is drained but to resume power as soon as possible. A backup power source is essential.

Information

The information every GNSS observer carries throughout a project ought to include emergency phone numbers; the names, addresses, and phone numbers of relevant property owners; and the combinations to necessary locks. The observation schedule for static GNSS work will be revised daily based upon actual production (see Table 6.3). It should specify the start–stop times and station for all the personnel during each session of the upcoming day. In this way, the schedule will not only serve to inform every receiver operator of his or her own expected occupations but those of every other member of the project as well. This knowledge is most useful when a sudden revision requires observers to meet or replace one another.

STATION DATA SHEET

The principles of good field notes have a long tradition in land surveying, and they will continue to have validity for some time to come. In GNSS, the ensuing paper trail will not only fill subsequent archives; but it has immediate utility. For example, the station data sheet is often an important bridge between on-site reconnaissance and the actual occupation of a monument.

Though every organization develops its own unique system of handling its field records, most have some form of the station data sheet. The document illustrated in Figure 6.11 is merely one possible arrangement of the information needed to recover the station. The station data sheet can be prepared at any period of the project, but perhaps the most usual times are during the reconnaissance of existing control or immediately after the monumentation of a new project point.

Neatness and clarity, always paramount virtues of good field notes, are of particular interest when the station data sheet is to be later included in the final report to the client. The overriding principle in drafting a station data sheet is to guide succeeding visitors to the station without ambiguity. A GNSS surveyor on the way to observe the position for the first time may be the initial user of a station data sheet. A poorly written document could void an entire session if the observer is unable to locate the monument. A client, later struggling to find a particular monument with an inadequate data sheet, may ultimately question the value of more than the field notes.

TABLE 6.3

Observation Schedule

	Session 1 Start 7:10 Stop 8:10	8:10 to 8:40	Session 2 Start 8:40 Stop 9:50	9:50 to 10:15	Session 3 Start 10:15 Stop 11:15	11:15 to 11:30	Session 4 Start 11:30 Stop 12:30	12:30 to 14:00	Session 5 Start 14:00 Stop 15:00
Svs PRNs	9,12,12,16,20, 24		3,12,13,16,20 24		3,12,13,16,17, 20,24		3,16,17,20,22 23,26		1,3,17,21,23, 26,28
Receiver A *Dan H.*	**Station 1** NGS Horiz. Control	Re-Set	**Station 1** NGS Horiz. Control	Move	**Station 5** NGS Benchmark	Re-Set	**Station 5** NGS Benchmark	Re-Set	**Station 5** NGS Benchmark
Receiver B *Scott G.*	**Station 3** NGS V&H Control	Re-Set	**Station 3** NGS V&H Control	Move	**Station 6** Project Point	Re-Set	**Station 6** Project Point	Move	**Station 1** NGS Horiz. Control
Receiver C *Dewey A.*	**Station 8** NGS Horiz. Control	Move	**Station 2** Project Point	Re-Set	**Station 2** Project Point	Move	**Station 7** Project Point	Move	**Station 10** NGS Benchmark
Receiver D *Cindy E.*	**Station 13** NGS Horiz. Control	Move	**Station 4** Project Point	Re-Set	**Station 4** Project Point	Move	**Station 9** Project Point	Move	**Station 13** NGS Horiz. Control

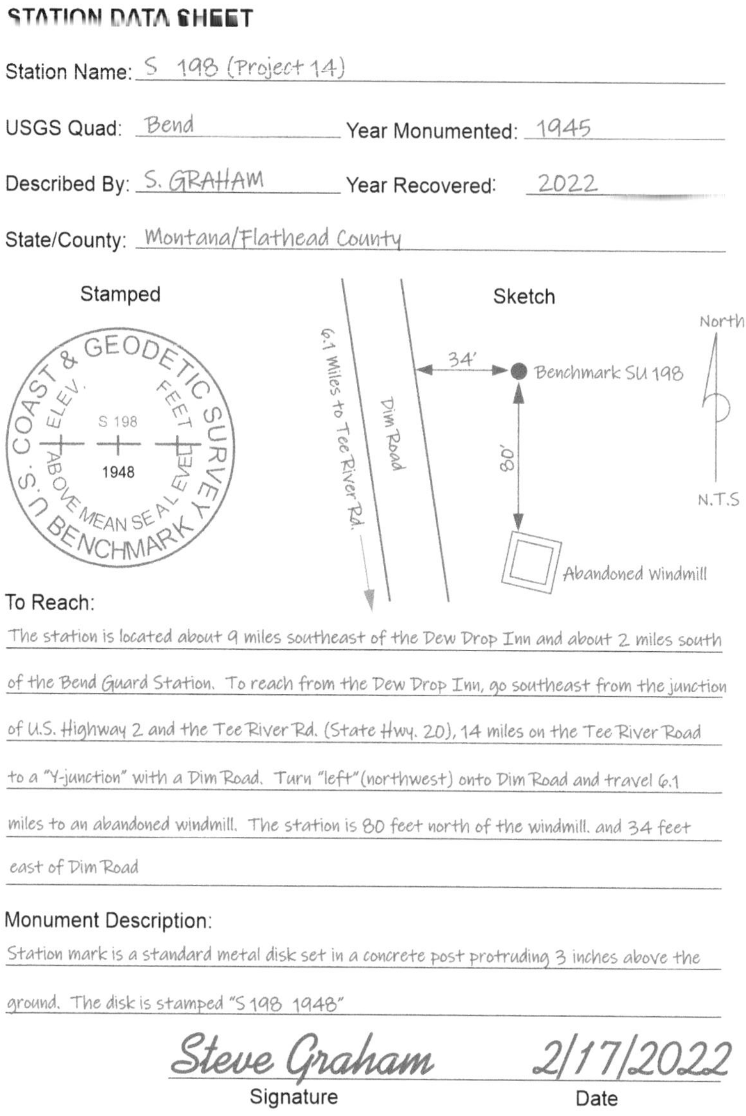

STATION DATA SHEET

Station Name: S 198 (Project 14)

USGS Quad: Bend _____ Year Monumented: 1945

Described By: S. GRAHAM _____ Year Recovered: 2022

State/County: Montana/Flathead County

Stamped	Sketch

To Reach:

The station is located about 9 miles southeast of the Dew Drop Inn and about 2 miles south

of the Bend Guard Station. To reach from the Dew Drop Inn, go southeast from the junction

of U.S. Highway 2 and the Tee River Rd. (State Hwy. 20), 14 miles on the Tee River Road

to a "Y-junction" with a Dim Road. Turn "left" (northwest) onto Dim Road and travel 6.1

miles to an abandoned windmill. The station is 80 feet north of the windmill. and 34 feet

east of Dim Road

Monument Description:

Station mark is a standard metal disk set in a concrete post protruding 3 inches above the

ground. The disk is stamped "S 198 1948"

Steve Graham   2/17/2022
Signature   Date

FIGURE 6.11   Station Data Sheet

STATION NAME

The station name fills the first blank on the illustrated data sheet. Two names for a single monument are far from unusual. In this case, the vertical control station, officially named S 198, is also serving as a project point, number 14. However, two names purporting to represent the same position can present a difficulty. For example, when a horizontal control station is remonumented, a number 2 is sometimes added to the original name of the station and it can be confusing. For example, it can be easy to mistake station, "Thornton 2," with an original station named "Thornton," that no longer exists. Both stations may still have a place in the published record but with slightly different coordinates. Another unfortunate misunderstanding can occur when inexperienced field personnel mistakes a reference mark, R.M., for the actual station itself. Taking close-up photographs is widely recommended to avoid such blunders regarding station names or authority.

PHOTOGRAPHS

The use of photographs is growing as a help for the perpetuation of monuments. It can be convenient to photograph the area around the mark as well as the monument itself. These exposures can be correlated with a sketch of the area. Such a sketch can show the spot where the photographer stood and the directions toward which the pictures were taken. The photographs can then provide valuable information in locating monuments, even if they are later obscured. Still, the traditional ties to prominent features in the area around the mark are the primary agent of their recovery.

TO-REACH DESCRIPTIONS

When driving or walking to a position can be aided by computerized mapping turn-by-turn navigation it is a great tool that may make writing to-reach descriptions unnecessary in many cases. However, GNSS control work is often done in areas where the online roadway mapping in such navigation aids is inadequate or completely absent. In those situations, the description of the route to the station is one of the most critical documents written during the reconnaissance. Even though it is difficult to prepare the information in unfamiliar territory and although every situation is somewhat different, there are some guidelines to be followed. It is best to begin with the general location of the station with respect to easily found local features. The description in Figure 6.11 relies on a road junction, a guard station, and a local business. After defining the general location of the monument, the description should recount directions for reaching the station. Starting from a prominent location, the directions should adequately describe the roads and junctions. Where the route is difficult or confusing, the reconnaissance team should not only describe the junctions and turns needed to reach a station; it is wise to also mark them with lath and flagging, when possible. It is also a good idea to note gates. Even if they are open during reconnaissance, they may be locked later. When turns are called for, it is best to describe not only the direction of the turn but the new course too. For example, in the description in Figure 6.11, the turn onto the dim road from the Tee River Road is described to the "left (north-west)." Roads and highways should carry both local names and designations found on standard highway maps. For example, in Figure 6.11, Tee River Road is also described as State Highway 20.

The "to-reach" description should certainly state the mileages as well as the travel times where they are appropriate, particularly where packing-in is required. Land ownership, especially if the owner's consent is required for access, should be mentioned. The reconnaissance party should obtain permission to enter private property and should inform the GNSS observer of any conditions of that entry. Alternate routes should be described where they may become necessary. It is also best to make special mention of any route that is likely to be difficult in inclement weather.

Where helicopter access is anticipated, information about the duration of flights from point to point, the distance of landing sites from the station, and flight time to fuel supplies should be included on the station data sheet.

FLAGGING AND DESCRIBING THE MONUMENT

Flagging the station during reconnaissance may help the observer find the mark more quickly. On the station data sheet, the detailed description of the location of the station with respect to roads, fence lines, buildings, trees, and any other conspicuous features should include measured distances and directions. A clear description of the monument itself is important. It is wise to also show and describe any nearby marks, such as R.M.s, that may be mistaken for the station or aid in its recovery. The name of the preparer, a signature, and the date round out the initial documentation of a GNSS station.

VISIBILITY DIAGRAMS

Obstructions above the mask angle of a GNSS receiver must be considered in finalizing the observation schedule. A station that is blocked to some degree is not necessarily unusable, but its inclusion in any session is probably contingent on the position of the specific satellites involved.

The diagram in Figure 6.12 is widely used to record such obstructions during reconnaissance. It is known as a *station visibility diagram*. The concentric circles are meant to indicate 10° increments along the upper half of the celestial sphere, from the observer's horizon at 0° on the perimeter to the observer's zenith at 90° in the center. The hemisphere is cut by the observer's meridian, shown as a line from 0° in the north to 180° in the south.

The prime vertical is signified as the line from 90° in the east to 270° in the west. The other numbers and solid lines radiating from the center, every 30° around the perimeter of the figure, are azimuths from north and are augmented by dashed lines every 10°.

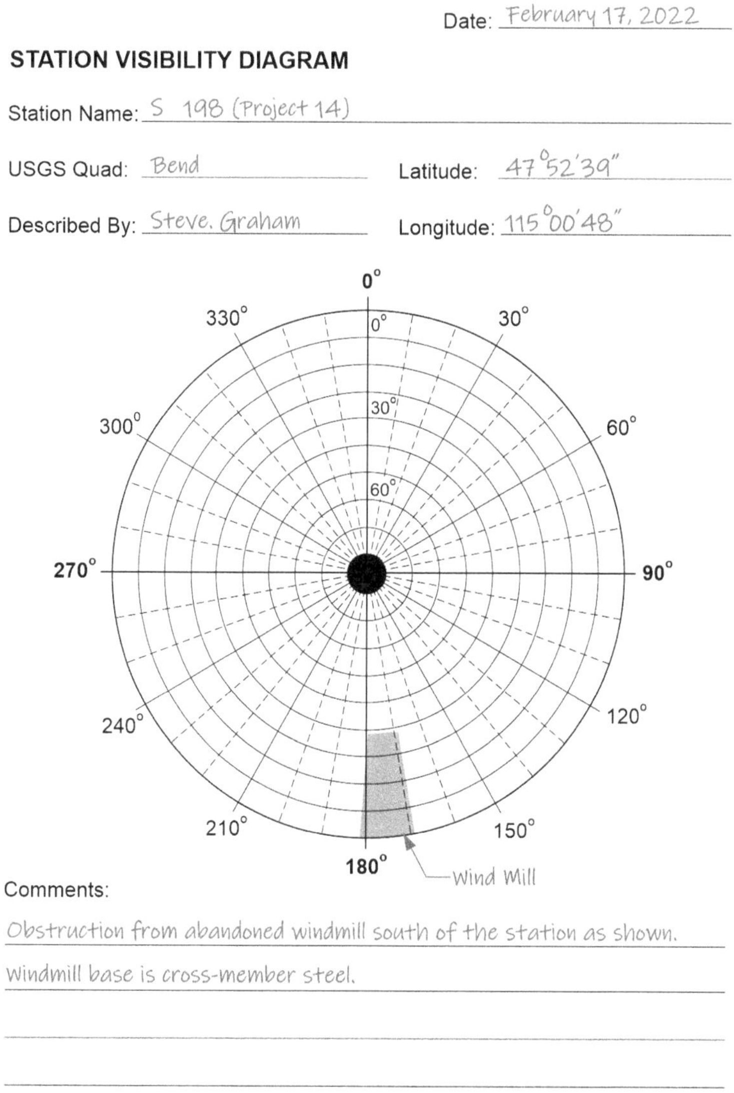

Date: February 17, 2022

STATION VISIBILITY DIAGRAM

Station Name: S 198 (Project 14)

USGS Quad: Bend   Latitude: 47°52'39"

Described By: Steve. Graham   Longitude: 115°00'48"

Comments:

Obstruction from abandoned windmill south of the station as shown.

Windmill base is cross-member steel.

FIGURE 6.12   Station Visibility Diagram

Using a compass and a clinometer, a member of the reconnaissance team can fully describe possible obstructions of the satellite's signals on a visibility diagram. By standing at the station mark and measuring the azimuth and vertical angle of points outlining the obstruction, the observer can plot the object on the visibility diagram. For example, a windmill base is shown in Figure 6.12 as a darkened figure. It has been drawn from the observer's horizon up to 37° in vertical angle from about 168° to about 182° in azimuth at its widest point. This description by approximate angular values is entirely adequate for determining when satellites may be blocked at this station.

Suppose a 1-hour session from 9:10 to 10:10, illustrated in Table 6.4, was under consideration for the observation on station S 198. The station visibility chart might motivate a careful look at SV PRN 16. Twenty minutes into the anticipated session, at 9:30 SV 16 has just risen above the 15° mask angle. Under normal circumstances, it would be available at station S 198, but it appears from the polar plot that the windmill will block its signals from reaching the receiver. In fact, the signals from SV 16 will apparently not reach station S 198 until sometime after the end of the session at 10:10.

WORKING AROUND OBSTRUCTIONS

Under the circumstances, some consideration might be given to observing station S 198 during a session when none of the satellites would be blocked. However, the 9:10 to 10:10 session may be adequate after all. Even if SV 16 is completely blocked, the remaining five satellites will be unobstructed, and the constellation still will have a relatively low PDOP. Still, the analysis must be carried to other stations that will be occupied during the same session. The success of the measurement of any baseline depends on common observations at both ends of the line. Therefore, if the signals from SV 16 are garbled or blocked from station S 198, any information collected during the same session from that satellite at the other end of a line that includes S 198 will be useless in processing the vector between those two stations.

The material of the base of the abandoned windmill has been described on the visibility diagram as cross-membered steel, so it is possible that the signal from SV 16 will not be entirely obstructed during the whole session. There may be more concern of multipath interference from the structure than that of signal availability. One strategy for handling the situation might be to program the receiver at S 198 to ignore the signal from SV 16 completely if the receiver allows it.

The visibility diagram (see Figure 6.12) and the azimuth-elevation table (Table 6.4) complement each other. They provide the field supervisor with the data needed to make informed judgments about the observation schedule. Even if the decision is taken to include station S 198 in the 9:10 to 10:10 session as originally planned, the supervisor will be forewarned that the blockage of SV 16 may introduce a bit of weakness at that station.

APPROXIMATE STATION COORDINATES

The latitude and longitude given on the station visibility diagram should be understood to be approximate. Its primary role is as input for the receiver at the beginning of its observation. The coordinate need only be close enough

TABLE 6.4
Satellite Azimuth and Elevation Table 3

Time	El	Az	El	Az	El	Az	El	Az	El	Az	El	Az	El	Az	PDOP
SV	3		12		13		20		24						
					constellation of 5 SVs										
8:50	54	235	74	274	44	28	16	308	68	169					4.8
9:00	51	229	74	255	40	32	20	310	72	163					5.7
9:10	47	224	72	238	37	35	23	311	77	153					4.9
9:20	43	219	68	226	33	38	27	313	80	134					4.0
SV	3		12		13		16		20		24				
					constellation of 6 SVs										
9:30	39	215	64	218	29	41	16	179	31	314	81	102			2.1
9:40	35	212	59	213	26	45	19	176	36	314	80	73			2.3
9:50	31	209	54	209	23	48	23	173	40	315	76	57			2.4
10:00	27	207	49	206	19	52	27	170	44	314	72	49			2.5
10:10	23	204	44	204	16	55	30	167	48	314	67	45			2.5

to the actual position to minimize the time the receiver must take to lock onto the constellation of satellites it expects to find. A receiver's ability to achieve lock-on without being given a somewhat correct beginning position and time.

MULTIPATH

Be alert to any indications on the station visibility diagram that multipath may be a concern. Before the observations are done, there is nearly always a simple solution. Discovering multipath in the signals after the observations are done is not only frustrating but often expensive. The multipath condition is by no means unique to GNSS. When a transmitted television signal reaches the receiving antenna by two or more paths, the resulting variations in amplitude and phase cause the picture to have ghosts. This kind of scattering of the signals can be caused by reflection from land, water, or man-made structures. In GNSS, the problem can be particularly troublesome when signals are received from satellites at low elevation angles; hence the general use of a 15° to 20° mask angle. The use of choke ring antennas to mediate multipath may also be considered. It is also wise, where it is possible, to avoid using stations that are near structures likely to be reflective or to scatter the signal. For example, chain-link fences that are found hard against a mark can cause multipath by forcing the satellite's signal to pass through the mesh to reach the antenna. The elevation of the antenna over the top of the fence with a survey mast is often the best way to work around this kind of obstruction. Metal structures with large flat surfaces are notorious for causing multipath problems. A long train moving near a project point could be a potential problem, but vehicles passing by on a highway or street usually are not, especially if they go by at high speed. It is important, of course, to avoid parked vehicles. It is best to remind new GNSS observers that the survey vehicle should be parked far enough from the point to avert any multipath. A good way to handle unfavorable conditions is to set an offset point.

POINT OFFSETS

An offset must, of course, stand far enough away from the source of multipath or an attenuated signal to be unaffected. However, the longer the distance from the originally desired position, the more important the accuracy of the bearing and distance between that position and the offset becomes. Recording the tie between the two correctly is crucial to avoid misunderstanding after the work is completed. Some receivers allow input of the information directly into the observations recorded in the receiver or data logger. However, during a control survey, it is best to also record the information in a field book. Offsets in GNSS control surveying are an instance where conventional surveying equipment and expertise are

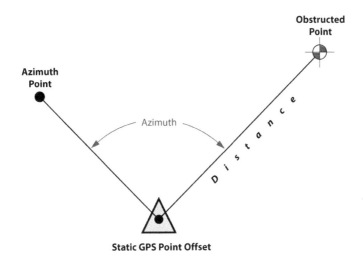

FIGURE 6.13   Static Point Offset

necessary. Clearly, the establishment of the tie requires a position for the occupation of the instrument, that is, a total station, and a position for the establishment of its orientation, that is, an azimuth (see Figure 6.13). If possible, it is good to establish three intervisible rather than two points—one to occupy and two azimuth marks. This approach makes it possible to add a redundant check to the tie. The positions on these two, or three, points may be established by setting monuments and performing static observations on them all. Alternatively, azimuthal control may be established by astronomic observations.

MONUMENTATION

The monumentation set for GNSS projects varies widely and can range from brass tablets to aerial premarks, capped rebar or even pin flags. The objective of most station markers is to adequately serve the client's subsequent use. The time, trouble, and cost in most high-accuracy GNSS work often warrant the most permanent, stable monumentation, but the suitability of a particular type of monument is an area still most often left to the professional judgment of the surveyors involved.

LOGISTICS

Scheduling

Once all the station data sheets, visibility diagrams, and other field notes have been collected, the schedule can be finalized for the first observations. There will almost certainly be changes from the original plan. Some of the anticipated control stations may be unavailable or obstructed, some project points may be blocked, too difficult to reach, or simply not serve the purpose as well as a control station at an alternate location. When the final control has been chosen, the project points have been monumented, and the reconnaissance has been completed, the information can be brought together with some degree of certainty that it represents the actual conditions in the field. Now that the access and travel time, the length of vectors, and the actual obstructions are more certainly known, the length and order of the sessions can be solidified. Despite all the care and planning that goes into preparing for a project, unexpected changes in the satellites' orbits or health can upset the best schedule at the last minute. It is always helpful to have a backup plan. The receiver operators usually have been involved in the reconnaissance and are familiar with the area and many of the stations. Even though an observer may not have visited the stations scheduled, the copies of the project map, appropriate station data sheets, and visibility diagrams will usually prove adequate to their location.

OBSERVATION

When everything goes as planned, a GNSS observation is uneventful. However, even before the arrival of the receiver operator at the control or project point, the session can get off-track. The simultaneity of the data collected at each end of a baseline is critical to the success of any measurement in static GNSS control surveys. When a receiver occupies a master station throughout a project or CORS stations are used, there may need be little concern on this subject, but many static applications depend on the sessions of many mobile receivers beginning and ending together.

The possible delays that may befall an observer on the way to a station are too numerous to mention. With proper planning and reconnaissance, the observer will likely find that there is enough time for the trip from station to station and that sufficient information is on hand for guidance to the position, but this, too, cannot be guaranteed. When the observer is late to the station, the best course is usually to set up the receiver quickly and collect as much data as possible. The baselines into the late station may or may not be saved, but they will certainly be lost if the receiver operator collects no information at all. It is at times like these that good communication between the members of the GNSS team is most useful. For example, some of the other observers in the session may be able to stay on their station a bit longer with the late arrival and make up some of the lost data. Along the same line, it is usually a good policy for those operators who are to remain on a station for two consecutive sessions to collect data as long as possible while still leaving themselves enough time to reset between the two observation periods.

SETUP

Centering an instrument over the station mark is always important. However, the centimeter-level accuracy of static GNSS gives the centering of the antenna special significance. It is ironic that such a sophisticated system of surveying can be defeated from such a commonplace procedure. A tribrach with an optical plummet or any other device used for centering should be checked and, if necessary, adjusted before the project begins. With good centering and leveling procedures, an antenna should be within a few millimeters of the station mark. Unfortunately, the centering of the antenna over the station does not ensure that its phase center is properly oriented. The contours of equal phase around the antenna's electronic center are not themselves perfectly spherical. Part of their eccentricity can be attributed to unavoidable inaccuracies in the manufacturing process and part to the changes in the frequency, azimuth, and elevation of the received signals as the satellites move overhead. To compensate for some of this offset, it is a good practice to rotate all antennas in a session to the same direction. Many manufacturers provide reference marks on their antennas so that each one may be oriented to the same azimuth, usually north. That way, they are expected to maintain the same relative position between their physical and electronic centers when observations are made.

The antenna's configuration also affects another measurement critical to successful GNSS surveying: the height of the instrument. The frequency of mistakes in this important measurement is remarkable. Several methods have been devised to focus special attention on the height of the antenna. Not only should it be measured in both feet and meters, but it should also be measured immediately after the instrument is set up and just before tearing it down to detect any settling of the tripod during the observation.

HEIGHT OF INSTRUMENT

The measurement of the height of the antenna in a GNSS survey is often not made on a plumb line. A tape is frequently stretched from the top of the station monument to the ARP on the antenna or the receiver itself. Some GNSS teams measure and record the height of the antenna to more than one reference mark on the ground plane. These measurements are usually mathematically corrected to plumb. The care ascribed to the measurement ot antenna heights is due to the same concern applied to centering. GNSS has an extraordinary capability to achieve accurate ellipsoidal heights, but those heights can be easily contaminated by incorrect H.I.s.

OBSERVATION LOGS

Most GNSS operations require its receiver operators to keep a careful log of each observation. These field notes provide a written record of the measurements, times, equipment, and other data that explains what occurred during the observation itself. It is difficult to overestimate the importance of this information. It is usually incorporated into the final report of the survey, the archives. However, the most immediate use of the observation log is in evaluation of the day's work by the on-site field supervisor. An observation log may be organized in several ways. The log illustrated in Figure 6.14 is one method that includes some of the information that might be used to document one session at one station.

Of course, the name of the observer and the station must be included, and while the date need not be expressed in both the Julian and Gregorian calendars, that information may help in quick cataloging of the data. The approximate latitude, longitude, and height of the station are usually required by the receiver as a reference position for its search for satellites. The date of the planned session will not necessarily coincide with the actual session observed. The observer's arrival at the point may have been late, or the receiver may have been allowed to collect data beyond the scheduled end of the session. There are various methods used to name observation sessions in terminology that is sensible to computers. A widely used system is noted here. The first four digits are the project point's number. In this case, it is point 14 and is designated 0014. The next three digits are the Julian day of the session, in this case, it is day 50, or 050. Finally, the session illustrated is the second of the day, or 2. Therefore, the full session name is 0014 050 2. Whether onboard or separate, the type of antenna used and the height of the antenna are critical pieces of information. The relation of the height of the station to the height of the antenna is vital to the station's later utility. The distance that the top of the station's monument is found above or below the surface of the surrounding soil is sometimes neglected. This information can not only be useful in later recovery of the monument but can also be important in the proper evaluation of photo-control panel points.

WEATHER

The meteorological data is useful in modeling the atmospheric delay. This information is best recorded at the beginning, middle, and end of each

OBSERVATION LOG

JOB NUMBER | ULY2396

OBSERVER	STATION	JULIAN DATE	DATE
Steve Graham	S 198 (Point 14)	50	2/17/2022

LATITUDE	LONGITUDE	HEIGHT
47°52'39"	115°00'48"	3,241.09 Feet

PLANNED OBSERVATION SESSION	SESSION NAME	ACTUAL OBSERVATION SESSION
START TIME: 9:10 STOP TIME: 10:10	0014 050 2	START TIME: 9:10 STOP TIME: 10:10

ANTENNA TYPE	ANTENNA HEIGHT ABOVE STATION MONUMENT	
ON-BOARD	BEFORE OBSERVATION	AFTER OBSERVATION
	METERS: 1.585	METERS: 1.585
MASK ANGLE: 15°	FEET: 5.20	FEET: 5.20

METEOROLOGICAL DATA

TIME	RELATIVE HUMIDITY	BAROMETER	THERMOMETER (D)	THERMOMETER (W)
9:30	30%	29.94	37°F	35°F

VISIBILITY DIAGRAM COMPLETED? (Y) N	TOP OF MONUMENT ABOVE THE SURFACE: 3 9N
STATION DATA SHEET COMPLETED? (Y) N	TOP OF MONUMENT BELOW THE SURFACE:

SV PRN TRACKED		COMMENTS:
3	16	Data from SV 16 appears healthy
12	20	Despite Windmill Obeservation
13	24	

FIGURE 6.14 Observation Log

session of projects. Under those circumstances, measurements of the atmospheric pressure in millibars, the relative humidity, and the temperature in degrees Centigrade are expected to be included in the observation log. However, the general use is less stringent. The conditions of the day are observed, and unusual changes in the weather are noted. A record of the satellites available during the observation and any comments about unique circumstances of the session round out the observation log.

DAILY PROGRESS EVALUATION

The planned observation schedules of a large GNSS project usually change daily. The arrangement of upcoming sessions is often altered based on the success or failure of the previous day's plan. Such a regrouping follows an evaluation of the day's data. This evaluation involves an examination of the observation logs as well as the data each receiver has collected. Unhealthy data caused by cycle slips or any other source are not always apparent to the receiver operator at the time of the observation. Therefore, a daily quality control check is a necessary preliminary step before finalizing the next day's observation schedule. Some field supervisors prefer to compute the independent baseline vectors of each day's work to ensure that the measurements are adequate. Neglecting the daily check could leave unsuccessful sessions undiscovered until the survey was thought to be completed. The consequences of such a situation could be expensive.

EXERCISES

1. Which of the following information is not available from the NGS data sheet for a passive station?
   a. The type of monument
   b. The State Plane Coordinates
   c. The latitude and longitude of the primary azimuth mark
   d. The Permanent Identifier

2. Which of the following is a quantity which, when added to an astronomic azimuth, yields a geodetic azimuth?
   a. The primary azimuth
   b. The mapping angle, also known as the convergence
   c. The Laplace correction
   d. The second term

3. How many nontrivial, or independent, and how many trivial, or dependent baselines are created in one GNSS session using four receivers?
   a. 1 independent baseline and 5 dependent baselines
   b. 3 independent baselines and 1 dependent baseline
   c. 2 independent baselines and 3 dependent baselines
   d. 3 independent baselines and 3 dependent baselines

4. Which of the following statements concerning loop closure in a GNSS network is not correct?
   a. A loop closure that uses baselines from one GNSS session will appear to have no error at all.

b. Loop closure is a procedure by which the internal consistency of a GNSS network is discovered.

c. No baselines should be excluded from a loop closure analysis.

d. At the completion of a GNSS control survey, the independent vectors should be capable of formation into closed loops that incorporate baselines from two to four different sessions.

5. How many observation sessions will be required for a GNSS control survey involving 20 stations that is to be done with 4 GNSS receivers, a planned redundancy of 2, a production factor of 1.1, and a safety factor is 0.2?

   a. 21 observation sessions
   b. 17 observation sessions
   c. 15 observation sessions
   d. 12 observation sessions

6. Which of the following options describes the best configuration for the horizontal control of a GNSS static control survey that is not for a route project?

   a. Divide the project into four quadrants and choose at least one horizontal control station near the project boundary in each quadrant.
   b. Divide the project into three parts and choose at least one horizontal control station near the center of the project in each.
   c. The primary base station should be located at the center of the project.
   d. Draw a north–south line through the center of the project, and choose at least three horizontal control stations near that line.

7. Which of the following data sheet searches is not possible on the NGS Survey Marks and Datasheets Internet site?

   a. A rectangular search based upon the range of latitudes and longitudes.
   b. A search using a particular station's geoid height.
   c. A search using a particular station's Permanent Identifier (PID).
   d. A radial search that is defining the region of the survey with one center position and a radius.

8. Which of the following is the standard symbol used in this book for project points in a GNSS control survey?

   a. A solid dot
   b. A triangle
   c. A square
   d. A circle

9 Which of the following statements about the zero baseline test is not true?

a. The zero-baseline text eliminates random noise and receiver biases from the results.

b. The zero-baseline test requires that receivers share the same antenna using an antenna splitter.

c. The zero-baseline test eliminates satellite clock biases, ephemeris errors, atmospheric biases, and multipath from the results.

d. The zero-baseline test is not dependent on special software or a test network.

10. The place of a station in a static GNSS control survey observation schedule may depend on a few factors. Which of the following would not be one of them?

a. The obstructions around the station.

b. The difficulty involved in reaching the station in bad weather.

c. The previous day's successful and unsuccessful occupations.

d. The line-of-sight with other stations in the survey.

11. What tools are necessary to adequately prepare a visibility diagram at a station that is somewhat obstructed?

a. A theodolite and EDM

b. A compass and clinometer

c. A GNSS receiver

d. A level and level rod

12. In order to ameliorate the misalignment of the antenna and its phase center:

a. The antenna height should be measured in meters and feet

b. Rotate all antennas in a session to the same direction

c. It is required to calibrate the phase center of the antenna

d. Type the antenna height in the data collector

ANSWERS AND EXPLANATIONS

1. Answer is (c)

   Explanation: NGS data sheets for passive stations provide State Plane and UTM coordinates, the latter only for horizontal control stations. They also provide mark setting information, the type of monument, and the history of mark recovery. The data sheet certainly shows the station's designation, which is its name and its Permanent Identifier (PID). Either of these may be used to search for the station in the NGS database. The PID is also found all along the left side of each data sheet record and is always two upper case letters followed by four numbers. However, it does not show the latitude and longitude of azimuth marks. That information may sometimes be found on a data sheet devoted to the azimuth mark itself.

2. Answer is (c)

   Explanation: The Laplace correction is a quantity which, when added to an astronomically observed azimuth, yields a geodetic azimuth. It is important to note that NGS uses a clockwise rotation regarding the Laplace correction. It can contribute several seconds of arc.

3. Answer is (d)

   Explanation: Where r is the number of receivers, every session yields $r - 1$ independent baselines. Four receivers used in one session would produce six baselines. Of these, three would be independent or nontrivial baselines, and three would be dependent or trivial baselines. It cannot be said that the shortest lines are always chosen to be independent lines. Sometimes there are reasons to reject one of the shorter vectors due to incomplete data, cycle slips, multipath, or some other weakness in the measurements. Before such decisions can be made, each session will require analysis after the data has been collected. In the planning stage, it is best to consider the shortest vectors as the independent lines, but once the decision is made only those three baselines are included in the network.

4. Answer is (c)

   Explanation: Any loop closures that only use baselines derived from a single common GNSS session will yield an apparent error of zero because they are derived from the same simultaneous observations. When the project is completed, the independent vectors from two to four different sessions, excluding the dependent, or trivial baselines, should be capable of formation into closed loops that incorporate baselines from two to four different sessions.

5. Answer is (c)

   Explanation: The appropriate formula is

$$s = \frac{(m \cdot n)}{r} + \frac{(m \cdot n)(p-1)}{r} + k \cdot m$$

$$s = \frac{(20)(2)}{4} + \frac{(20)(2)(1.1-1)}{4} + (0.2)(20)$$

$$s = \frac{(40)}{4} + \frac{(4)}{4} + 4$$

$$s = 10 + 1 + 4$$

$$s = 15 \text{ sessions}$$

where
  s = the number of observing sessions
  r = the number of receivers
  m = the total number of stations involved
  n = the planned redundancy
  p = the experience, or production factor
  k = the safety factor

The variable n is a representation of the level of redundancy that has been built into the network based on the number of occupations on each station. The experience of a firm is symbolized by the variable p in the formula. A typical production factor is 1.1. A safety factor of 0.1, known as k, is recommended for GNSS projects within 100 km of a company's home base. Beyond that radius, an increase to 0.2 is advised. The substitution of the appropriate quantities for the illustrated project increases the prediction of the number of observation sessions required for its completion.

6. Answer is (a)

   Explanation: For work other than route surveys, a handy rule-of-thumb is to divide the project into four quadrants and to choose at least one horizontal control station in each quadrant. The actual survey should have at least one horizontal control station in three of the four quadrants. Each of them ought to be as near as possible to the project boundary. Supplementary control in the interior of the network can then be used to add more stability to the network.

7. Answer is (b)

   Explanation: A rectangular search based upon the range of latitudes and longitudes can be performed on the NGS Internet site. It is possible to do a radial search—defining the region of the survey with one center position and a radius. You may also retrieve individual data sheets by the Permanent Identifier (PID) control point name. However, you cannot retrieve a data sheet for a station based only on its geoid height.

8. Answer is (a)

   Explanation: A solid dot is the standard symbol used to indicate the position of project points. Some variation is used when a distinction must be drawn between those points that are in place and those that must be set. When its location is appropriate, it is always a good idea to have a vertical or horizontal control station serve double duty as a project point. While the precision of their plotting may vary, it is important that project points be located as precisely as possible even at the preliminary stage. A horizontal control point is shown as a triangle, and, a vertical control point is shown as a square.

9. Answer is (a)

   Explanation: The simplicity of the zero-baseline test is an advantage. It is not dependent on special software or a test network and it can be used to separate receiver difficulties from antenna errors. Two or more receivers are connected to one antenna with a signal or antenna splitter. An observation is done with the divided signal from the single antenna reaching both receivers simultaneously. Since the receivers are sharing the same antenna, satellite clock biases, ephemeris errors, atmospheric biases, and multipath are all canceled. The only remaining errors are attributable to random noise and receiver biases.

10. Answer is (d)

    Explanation: In creating an observation, schedule consideration might be given to observing a particular station during a session when none of the satellites would be blocked by obstructions. While GNSS is not restricted by inclement weather, particular access routes may not be so immune. Despite best efforts, a planned observation may have been unsuccessful at a required station on a previous day, and it may need to be revisited. However, the line-of-sight between a particular station and another in the survey is not likely to affect the GNSS observation schedule, though such a consideration may be critical in a conventional survey.

11. Answer is (b)

    Explanation: Using a compass and a clinometer, a member of the reconnaissance team can fully describe possible obstructions of the satellite's signals on a visibility diagram. By standing at the station mark and measuring the azimuth and vertical angle of points outlining the obstruction, the observer can adequately plot the object on the visibility diagram.

12. Answer is (b)

Explanation: Unfortunately, the centering of the antenna over the station does not ensure that its phase center is properly oriented. The contours of equal phase around the antenna's electronic center are not themselves perfectly spherical. Part of their eccentricity can be attributed to unavoidable inaccuracies in the manufacturing process. To compensate for some of this offset, it is a good practice to rotate all antennas in a session to the same direction. Many manufacturers provide reference marks on their antennas so that each one may be oriented to the same azimuth. That way, they are expected to maintain the same relative position between their physical and electronic centers when observations are made.

7 Real-Time GNSS Surveying

REAL-TIME KINEMATIC (RTK) AND DIFFERENTIAL GNSS

Most, not all, GNSS surveying relies on the idea of differential positioning. Whether it is a CORS or the surveyor's own base station, the mode of a base or reference receiver at a known location logging data at the same time as a receiver at an unknown location together provides the fundamental information for the determination of accurate coordinates. This basic approach remains today. However, the majority of GNSS surveying is not done in the static post-processed mode. That is most often applied to control work. Now, the most common methods utilize receivers on reference stations that provide a correction signal to the end user via a data link, sometimes over the Internet, radio signal, or cell phone in real time. In this category of GNSS surveying work, there is sometimes a distinction made between code-based and carrier-based solutions. In fact, most systems use a combination of code and carrier measurements.

THE GENERAL IDEA

As you know, errors in satellite clocks, imperfect orbits, the trip through the layers of the atmosphere, and many other sources bias GNSS signals by the time they reach a receiver. These errors are variable, so the best way to correct them is to monitor them as they happen. A good way to do that is to set up a GNSS receiver on a station whose position is known exactly, a base station. This base station receiver's computer can calculate its position from satellite data, compare that position with its actual known position, and find the difference. The resulting error corrections can be communicated from the base to the rover. It works well, but the errors are constantly changing, so a base station must monitor them all the time, at least all the time the rover receiver or receivers are working. While this is happening, the rover or rovers move from place to place collecting data on the points. Then all you have to do is get those base station corrections and the rover's data together somehow. That combination can be done over a data link in real time or applied later in post-processing. Real-time positioning is built on the foundation of the idea that, with the important exceptions of multipath and receiver noise, GNSS error sources are largely correlated over short baselines. In other words, the closer the rover is to the base the more the errors at the ends of the baseline match. The longer the baseline, the less the errors are correlated.

RADIAL GNSS

Real-time surveying is essentially radial. There are advantages to the approach. The advantage is a large number of positions can be established in a short amount of time with little or no planning. The disadvantage is that there is typically little or no redundancy in the positions derived, especially when each of the baselines originates from the same control station. Redundancy can be incorporated, but it requires repetition of the observations, so each baseline is determined with more than one GNSS constellation. One way to do it is to occupy the project points, the unknown positions, successively with more than one rover (see Figure 7.1). It is best if these successive occupations are separated by some hours so the satellite constellation can reach a significantly different configuration. However, even very much shorter intervals yield some advantages. A more convenient but less desirable approach is to do a second occupation almost

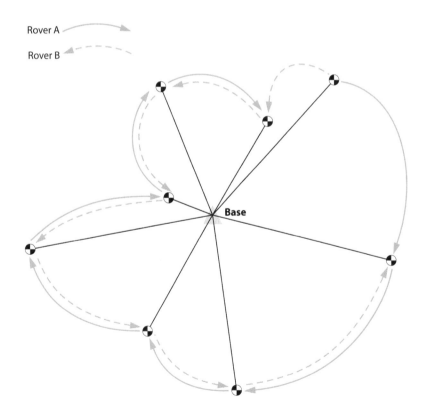

FIGURE 7.1   Radial GNSS

immediately after the first. The roving receiver's antenna is blocked or tilted until the lock on the satellites is lost. It is then reoriented on the same point; a new initialization follows and a repeat solution is achieved. This does offer a second solution, but from virtually the same constellation. It is better to have a longer time between the two occupations. Each one of them may be brief, but the longer the span between the two occupations the more the satellites of the second session reach a configuration different enough to supply a redundant position for the project point.

A third way to achieve redundancy is to occupy each point with the same rover but utilizing a different base station. This approach allows a solution to be available from two separate control stations. Obviously, this can be done with reoccupation of the project points after one base station has been moved to a new control point, or two base stations can be up and running from the very outset and throughout the work. It is best if there are two occupations on each point and each of the two utilize different base stations. More efficiency can be achieved by adding additional roving receivers. However, as the number of receivers rises, the logistics become more complicated, and a survey plan becomes necessary. Also, project points that are simultaneously near one another but far from the control station should be directly connected with a baseline between the two of them to maintain the integrity of the survey. Finally, if a base receiver loses lock and that goes unnoticed, it will completely defeat the radial survey from that station for the time it is down.

THE CORRECTION SIGNAL

The agreed-upon protocol for communication between base stations and rovers was first designed for used in marine navigation by an organization known as the *Radio Technical Commission for Maritime Services (RTCM)*. It was founded in 1947 as a U.S. State Department advisory committee. It is now an independent not-for-profit scientific, educational, and professional organization chartered in the District of Columbia. Toward those ends, it provides information on maritime radio navigation and radio communication policies and associated regulations to its members. It is supported by an international membership that includes both governmental and nongovernmental institutions. For example, RTCM standards are incorporated by reference into U.S. Federal Communications Commission (FCC) and U.S. Coast Guard regulations. It is also involved in technical standards development. In 1985, the RTCM Special Committee (SC-104) created a standard that is still more used than any proprietary formats that have come along since. RTCM is open. In other words, it is a general-purpose format and is not restricted to a particular receiver type. The message augments the information from the satellites. It was originally designed to accommodate a slow GPS data rate with a configuration like the navigation message. The data format has evolved since its inception. For example, RTCM 2.0 supported GPS code only. However, when it became clear in 1994 that including carrier-phase information in the message could improve the accuracy of the system, RTCM Special Committee 104 added four new message types to Version 2.1 to fulfill the needs of RTK. RTCM 2.1 supported both code and phase correction, but still GPS only. Version 2.2 became available in 1998. RTCM 2.2 added support for GLONASS and version 2.3 included antenna corrections, and the changes continued. Version 2.3 included antenna corrections. Next was Version 3. RTCM 3.0 utilizes a more efficient message structure than its predecessors which proves beneficial in the RTK data heavy real-time communication of code and carrier messages, antenna, and system parameters between a base and a rover. RTCM 3.1 added a network correction message and version 3.2 introduced *Multiple Signal Messages (MSM)*, which can be employed with any GNSS, such as Europe's Galileo, the Chinese BeiDou, and the Japanese QZSS systems. There are seven MSM message types. They range from MSM1 for simple pseudorange delivery to MSM7 for delivering pseudorange, carrier-phase ranges, range rate, and more. Each of them can be utilized by each GNSS, for example, the GPS MSM3 is virtually the same as the GALILEO MSM3, which is virtually the same as the BeiDou MSM3, etc. though each will have a different ID. Clearly, this allows for more consistency across all systems and the potential for additional redundancy in measurement and reliability in positioning. As of this writing, the RTCM standard stands at version 3.3, aka RTCM 10403.3, which makes it possible to communicate corrections from a base tracking all GNSS satellites to the rover.

DGNSS

The term DGNSS is sometimes used to refer to differential GNSS that is based on pseudoranges, aka the code phase (see Figure 7.2). Even though the accuracy of code phase applications was given a boost with the elimination of Selective Availability (SA) in May 2000, consistent accuracy still requires reduction of the effect of correlated ephemeris and atmospheric errors by differential corrections. Though the corrections could be applied in post-processing services that supply these corrections, most often operate in real time. In such an operation pseudorange based versions can offer meter or even submeter results. Usually, pseudorange corrections are broadcast from the base to the rover or rovers for each satellite in the visible constellation. Rovers with an appropriate input/output (I/O) port can receive the correction signal and calculate coordinates. The real-time signal comes to the receiver over a data link. It can originate at a project

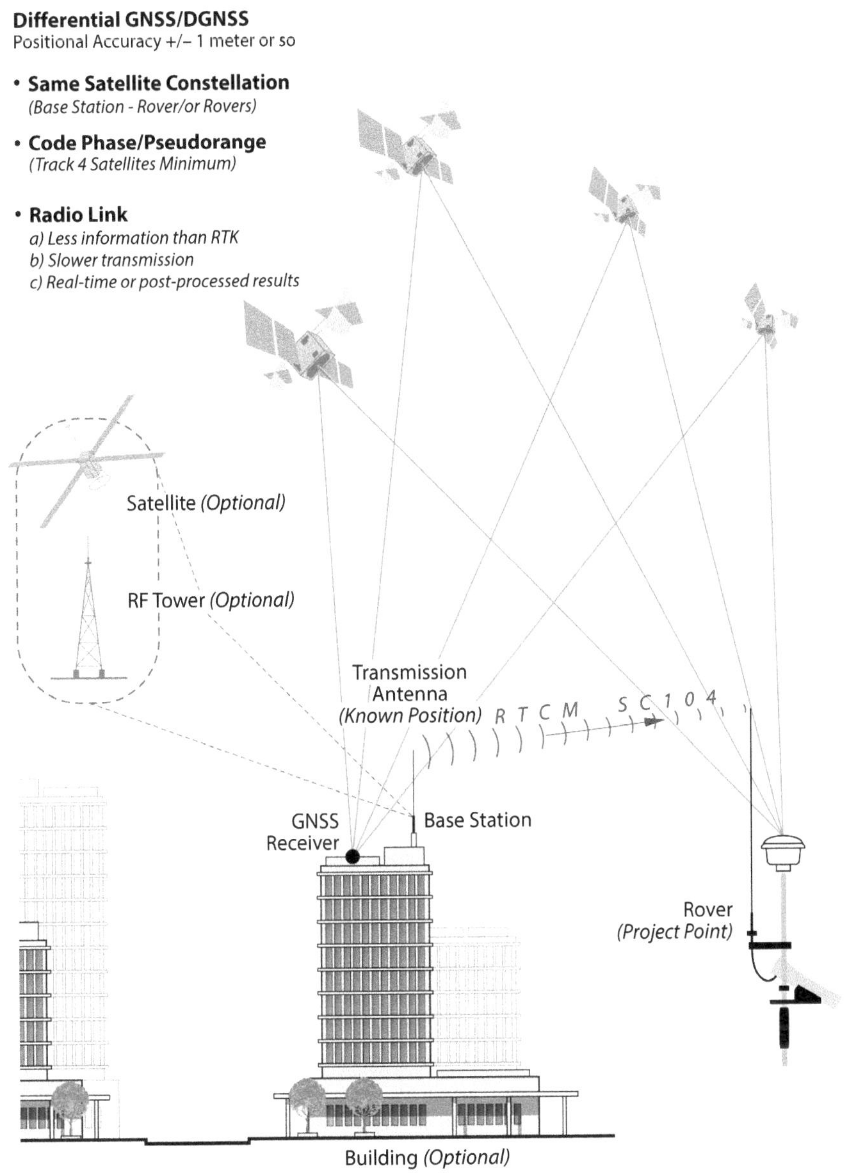

Differential GNSS/DGNSS
Positional Accuracy +/– 1 meter or so

- **Same Satellite Constellation**
  (Base Station - Rover/or Rovers)

- **Code Phase/Pseudorange**
  (Track 4 Satellites Minimum)

- **Radio Link**
  a) Less information than RTK
  b) Slower transmission
  c) Real-time or post-processed results

Satellite *(Optional)*

RF Tower *(Optional)*

Transmission
Antenna
(Known Position) R T C M   S C 1 0 4

GNSS
Receiver | Base Station

Rover
(Project Point)

Building *(Optional)*

FIGURE 7.2 DGNSS

specific base station or it can come to the user through a service of which there are various categories. Some are open to all users, and some are by subscription only. Coverage depends on the spacing of the beacons, aka transmitting base stations, their power, interference, and so forth. Some systems require two-way, some one-way, communication with the base stations. Radio systems, geostationary satellites, low-Earth-orbiting satellites, and cellular phones are some of the options available for data communication.

In any case, most of the wide variety of DGNSS services were not originally set up to augment surveying and mapping applications of GNSS; they were established to aid navigation.

LOCAL AND WIDE AREA DGPS

The correlation between most of the GNSS biases becomes weaker as the rover gets farther from the base. The term Local Area Differential GPS (LADGPS) was used when the baselines from a single base station to the roving receivers using the service were less than a couple of hundred kilometers. The term Wide Area Differential GPS (WADGPS) was used when the service used a network of base stations and distributed correction over a larger area, an area that may even be continental in scope. Many bases operating together provide a means by which the information from several of them can be combined to send a normalized or averaged correction tailored to the rover's geographical position within the system. Some use satellites to provide the data link between the service provider and the subscribers. Such a system depends on the network of base stations receiving signals from the GNSS satellites and then streaming that data to a central computer at a control center. There the corrections are calculated and uploaded to a geostationary communication satellite. Then the communication satellite broadcasts the corrections to the service's subscribers (see Figure 7.3). In all cases, the base stations are at known locations and their corrections are broadcast to all rovers that are equipped to receive their radio message carrying real-time corrections in the RTCM format. An example of such a service originated as an augmentation for marine navigation. Both the U.S. Coast Guard (USCG) and the Canadian Coast Guard (CCG) instituted DGPS services to facilitate harbor entrances, ocean mapping, and marine traffic control as well as navigation in inland waterways. Their system base stations beacons used to broadcast GPS corrections along major rivers, major lakes, the east coast, and the west coast, but the system was discontinued in June 2020. However, another U.S. DGPS pseudorange service originally initiated in the early 1990s cooperatively by the U.S. Department of Transportation and the Federal Aviation Administration (FAA) is going strong. It is known as the *Wide Area Augmentation System (WAAS)*.

WIDE AREA AUGMENTATION SYSTEM

WAAS augments GPS accuracy, availability, and integrity. The system initially designed to assist aerial navigation from takeoff, enroute through

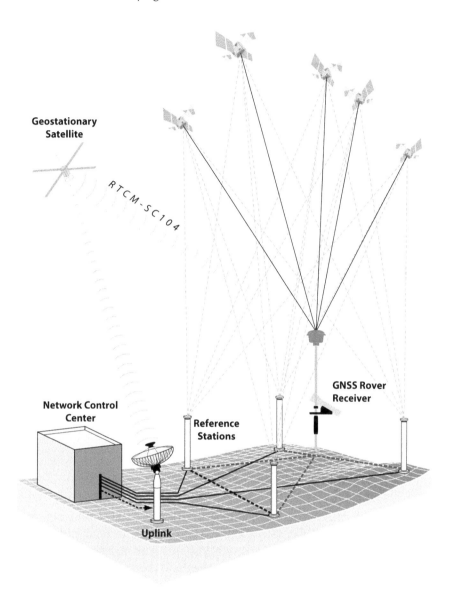

Geostationary
Satellite

RTCM-SC104

GNSS Rover
Receiver

Network Control
Center

Reference
Stations

Uplink

FIGURE 7.3 Wide Area Differential GNSS

landing, was declared fully operational and approved for safety-of-life aviation in July 2003. Three master stations collect data from the 38 reference
stations monitoring the GPS constellations signals from known locations in
the conterminous United States, Alaska, Hawaii, Puerto Rico, Canada, and
Mexico. They calculate the corrections and bounds of confidence, pack-
age it into messages and send them to six Ground Uplink Stations (GUS).
These stations upload the messages to three commercial geostationary

satellites devoted to transmission of the corrections to users on the ground. The service is freely available to users with receivers equipped to receive it and typically delivers submeter corrected horizontal accuracy.

There are many similar Satellite-Based Augmentation Systems (SBAS) around the world. The *European Geostationary Navigation Overlay Service (EGNOS)* has been operational since 2009 and available for safety-of-life service since March 2011. It has 39 monitoring stations in Europe, Africa, and North America, four master control centers, and six navigation stations from which three geostationary satellites are controlled. It typically delivers 1.2 m of differentially corrected horizontal accuracy. India's *GPS-Aided GEO Augmented Navigation (GAGAN)* system was commissioned in 2013. It includes 15 Indian reference stations, 3 Indian navigation land uplink stations, and three Indian mission control centers. It typically delivers 2.3 m differentially corrected horizontal accuracy. The Russian *System for Differential Corrections and Monitoring (SDCM)*, with 19 reference stations in the Russian Federation, provides correction to both GPS and GLONASS. It typically delivers 1.5 m differentially corrected horizontal accuracy. Japan's *Michibiki Satellite Augmentation System (MSAS)* has been operational since 2007. There are many more augmentation systems in development; South Korea's Korea Augmentation Satellite System, KASS; Africa's ASECNA SBAS for Africa and the Indian Ocean, A-SBAS; Australia and New Zealand's Southern Positioning Augmentation Network (SouthPAN), the United Kingdom's (UKSBAS) and *China's BeiDou SBAS, BDSBAS*, more about that one in Chapter 8.

GIS APPLICATION

Aerial navigation, marine navigation, agriculture, vehicle tracking, and construction utilize DGNSS. DGNSS is also useful in land and hydrographic surveying, but perhaps the fastest growing application for DGNSS is in data collection, data updating, and even in-field mapping for Geographic Information Systems (GIS). GIS data has long been captured from paper records such as digitizing and scanning paper maps. Today, DGNSS, photogrammetry, remote sensing, and conventional surveying are sources of GIS data too. GIS data collection with DGNSS requires the integration of the position of features of interest and relevant attribute information about those features. In GIS, it is frequently important to return to a particular site or feature to perform inspections or maintenance. DGNSS with real-time correction makes it convenient to load the position or positions of features into a data logger and navigate to the vicinity, but a GIS must be reasonably current for such applications to be efficient. It must be maintained. A receiver configuration including real-time DGNSS, sufficient data storage, and graphic display allows verification and updating of existing information. DGNSS allows immediate attribution and validation in the field with accurate and efficient recording of position. In the past, many GIS mapping efforts relied on ties to street centerlines, curb lines, railroads, and so forth. Such dependencies can be destroyed by demolition or new construction, but meter level positional accuracy, even in obstructed environments such as urban areas, amid high-rise buildings, is possible with DGNSS. Therefore, it can provide reliable positioning even if the landscape has changed. Its data can be integrated with other technologies, such as laser range-finders and so forth, in environments where DGNSS is not ideally suited to the situation. Finally, loading GNSS data into a GIS platform does not require manual intervention. GNSS data processing can be automated; the results are digital and can pass into a GIS format without redundant effort, reducing the chance for errors.

REAL-TIME KINEMATIC

Kinematic surveying, also known as stop-and-go kinematic surveying, is not new. The original kinematic GPS innovator, Dr. Benjamin Remondi, developed the idea in the mid-1980s. RTK is a method that provides positional accuracy nearly as good as static carrier-phase positioning, but faster. RTK accomplishes positioning in real time (see Figure 7.4). It involves the use of at least one stationary reference receiver, the base station, and at least one moving receiver, the rover. All the receivers involved observe the same satellites simultaneously. The base receivers are stationary on control points. The rovers move from project point to project point stopping momentarily at each new point, usually briefly. The collected data provides vectors between themselves and the base receivers in real time. RTK has become routine in development and engineering surveys where the distance between the base and roving receivers can most often be measured in thousands or hundreds of feet or even less. When compared with the other relative positioning methods, there is little question that the very short sessions of the RTK method can produce the largest number of positions in the least time.

RTK receivers can be single or multifrequency receivers with GNSS antennas, but multifrequency receivers are usual because RTK typically relies on carrier-phase observations corrected in real time. In other words, it depends on the fixing of the integer ambiguity and that is most efficiently accomplished with a multifrequency GNSS receiver capable of making both carrier-phase and pseudorange measurements. The process can be done in a few seconds when circumstances where the receivers are tracking a large constellation of satellites, the PDOP is small, the receivers

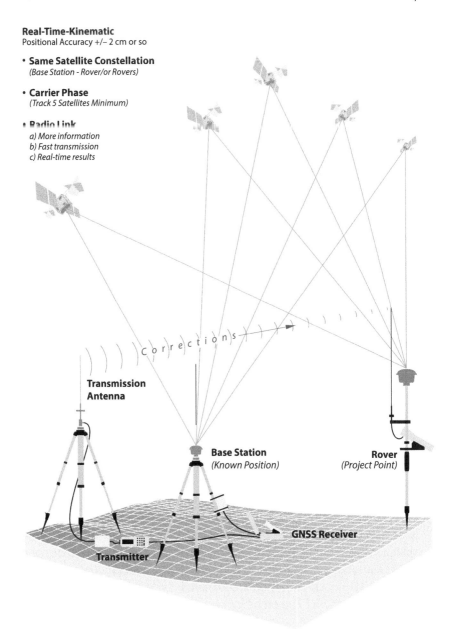

Real-Time-Kinematic
Positional Accuracy +/– 2 cm or so

- **Same Satellite Constellation**
  (Base Station - Rover/or Rovers)

- **Carrier Phase**
  (Track 5 Satellites Minimum)

- **Radio Link**
  a) *More information*
  b) *Fast transmission*
  c) *Real-time results*

Corrections

**Transmission
Antenna**

Base Station
(Known Position)

Rover
(Project Point)

GNSS Receiver

Transmitter

FIGURE 7.4   Real-Time Kinematic GNSS

are multifrequency, there is no multipath, and the receiver noise is low. Therefore, it is a good idea to restrict RTK to situations where there is a good correlation of atmospheric biases at both ends of the baseline. In other words, RTK is best used when the distance between the base and the rover is relatively short.

WIRELESS LINK

RTK requires a real-time wireless connection be maintained between the base station and the rover. When a radio link is used, the radio receiving antennas for the rovers will either be built into the GNSS antenna or be present as separate units. It is usual that the radio antennas for the data transmitter and the rover are omnidirectional whip antennas; however, at the base, it is usually on a separate mast and has a higher gain than those at the rovers. The position of the transmitting antenna affects the performance of the system significantly. It is usually best to place the transmitter antenna as high as is practical for maximum coverage, and the longer the antenna, the better its transmission characteristics. However, be very careful to look up and be sure there are no dangers above, such as overhead power lines. In fact, it is best if the base station receiver occupies a control station that has no overhead obstructions, is unlikely to be affected by multipath, is somewhat away from the action if the work is on a construction site and is within line of sight of the rovers when possible. If line of sight is not practical, as little obstruction as can be had along the radio link is best.

The data radio transmitter consists of an antenna, a radio modulator, and an amplifier. The modulator converts the correction data into a radio signal. The amplifier increases the signal's power, which determines how far the information can travel. Well, not entirely; the terrain and the height of the antenna have something to do with it too. RTK work requires a great deal of information to be successfully communicated from the base station to the receivers. To have sufficient capacity to handle the load the base station transmitter ought to be VHF, UHF, spread spectrum-frequency hopping, or direct. UHF spread spectrum radio modems are the most popular for DGNSS and RTK applications. The typical gain on the antenna at the base is 6 dB. However, while DGNSS operations may need no more than 200 bits-per-second (bps), updated every 10 seconds or so, RTK requires at least 2400 bps updated about every 1/2 second or less. Like the power of the transmission, the speed of the link between the base and rover, the data rate can be a limiting factor in RTK performance. It is also important to note that it takes some time for the base station to calculate corrections, and it takes some time for it to put the data into packets in the correct format and transmit them. Then the data makes its way from the base station to the rover over the data link. It is decoded and must go through the rover's software. The time this takes is called the *latency* of the communication between the base station and the rover. It can be as little as a fraction of a second or as long as a couple of seconds, and since the base station corrections are only accurate for the moment they were created; the base station must send a range rate correction along with them. Using this rate correction, the rover can back date the correction to match the moment it made that same observation. When a platform is moving quickly, like a Mobile Mapping System (MMS) or an Unmanned Aerial Vehicle (UAV), latency can be troublesome.

As mentioned earlier, RTK is at its best when the distance between the base station and the rovers is short. Baseline length may be limited by the effective range of the radio data link. In areas with high radio traffic, it can be difficult to find an open channel. It is remarkable how often the interference emanates from other surveyors in the area doing RTK as well. Most radios connected to RTK GNSS surveying equipment operate between UHF 400–475 MHz or VHF 170–220 MHz, and emergency voice communications also tend to operate in this same range, which can present problems from time to time. That is why most radio data transmitters used in RTK allow the user several frequency options within the legal range. The usual data link configuration operates at 4800 baud or faster. The transmitter at the base station is usually the larger and more powerful of the two radios. However, the highest wattage radios, 35 Watts or so, cannot be legally operated in some countries. Lower-power radios, from 1/2 W to 2 W, are sometimes used in such circumstances. The radio at the rover has usually lower power and is smaller. The FCC is concerned with some RTK GNSS operations interfering with other radio signals, particularly voice communications. It is important for GNSS surveyors to know that voice communications have priority over data communications.

The FCC requires cooperation among licensees that share frequencies. Interference should be minimized. For example, it is wise to avoid the most typical community voice repeater frequencies. They usually occur between 455–460 MHz and 465–470 MHz. Part 90 of the Code of Federal Regulations, 47 CFR 90, contains the complete text of the FCC Rules including the requirements for licensure of radio spectrum for private land mobile use. The FCC does require an application be made for licensing a radio transmitter. Fortunately, when the transmitter and rover receivers required for RTK operations are bought simultaneously, radio licensing and frequency selection are often arranged by the GNSS selling agent. Nevertheless, it is important that surveyors do not operate a transmitter without a proper license. Please remember that the FCC can levy fines for several thousand dollars for each day of illegal operation. More can be learned by consulting the FCC Wireless Fee Filing Guide. There are also other international and national bodies that govern frequencies and authorize the use of signals elsewhere in the world. In some areas, certain bands are designated for public use, and no special permission is required. For example, in Europe, it is possible to use the 2.4 GHz band for spread spectrum communication without special authorization with certain power limitations. Here in the United States, the band for spread spectrum communication is 900 MHz.

It is vital, of course, that the rover and the base station are tuned to the same frequency for successful communication. The receiver also has an antenna and a demodulator. The demodulator converts the signal back to an intelligible form for the rover's receiver. The data signal from the base station can be weakened or lost at the rover from reflection, refraction, atmospheric anomalies, or even being too close. A rover that is too close to the transmitter may be overloaded and not receive the signal properly, and, of course, even under the best circumstances, the signal will fade as the distance between the transmitter and the rover grows too large.

THE VERTICAL COMPONENT IN RTK

The output of RTK can appear to be similar to that of optical surveying with an electronic distance measuring (EDM) and a level. Nevertheless, it is not a good idea to consider the methods equivalent. RTK offers some advantages and some disadvantages when compared with more conventional methods. For example, RTK can be much more productive since it is available 24 hours a day and is not really affected by weather conditions. However, when it comes to the vertical component of surveying RTK and the level are certainly not equal.

GNSS can be used to measure the differences in ellipsoidal height between points with good accuracy. However, unlike a level—unaided GNSS cannot deliver orthometric height without calculation from a geoidal model. In other words, orthometric elevations are not directly available from the GNSS measurements. The accuracy of orthometric heights in GNSS is dependent on the veracity of the geoidal model and/or the *Digital Elevation Model* (*DEM*) used and the care with which they are applied. Fortunately, ever improving geoid models have been, and still are, available from NGS, however, please see Chapter 5 for the coming changes in this regard, i.e., NAPGD. Since geoidal heights can be derived from these models, and ellipsoidal heights are available from GNSS, it is certainly feasible to calculate orthometric heights, especially when a geoid model is onboard the RTK systems. However, it is important to remember that without a geoid model, RTK will only provide differences in ellipsoid heights between the base station and the rovers. It is not a good idea to presume that the surface of the ellipsoid is sufficiently parallel to the surface of the geoid and ignore the deviation between the two.

SOME PRACTICAL RTK SUGGESTIONS

A multifrequency receiver is a benefit in doing RTK as it is almost as if there were one and a half more satellites available to the observer. Most often, the more satellites that are available, the faster the integer ambiguities are resolved. When using your own base station, it is best to set it up over a known position first, before configuring the rover. After the tripod and tribrach are level and over the point, attach the GNSS antenna to the tribrach and, if possible, check the centering again. Set up the base station transmitter in a sheltered location close to the radio transmitter's antenna but at least 10 feet from the base station's GNSS receiver's antenna. It is best if the airflow of the base station transmitter's cooling fan is not restricted. The radio transmitting antenna is often mounted on a range pole attached to a tripod. Set the radio transmitting antenna as far as possible from obstructions and as high as stability will allow. Before setting up the radio transmitting antenna, to eliminate the danger of electrocution, be certain there are no power lines with which the antenna could make contact. The base station transmitter's power is usually provided by a deep-cycle battery. Even though the attendant power cable is usually equipped with a fuse, it is best to be careful not to reverse the polarity when connecting it to the battery. It is also best to avoid bending or kinking any cables and have the base station transmitter properly grounded. After connecting the base station receiver to the GNSS antenna, to the battery, and to the data collector, if necessary, carefully measure the GNSS antenna height. This measurement is often the source of avoidable error, both at the base station and the rovers. Many surveyors measure the height of the GNSS antenna to more than one place on the antenna, and it is often measured in both meters and feet for additional assurance. Select a channel on the base station transmitter that is not in use, and be sure to note the channel used so that it may be set correctly on the rovers as well.

When the RTK work is done, it is best to review the collected data from the data logger. Whether or not fixed height rods have been used, it is a good idea to check their antenna heights. Incorrect antenna heights are a very common mistake. Another bulwark against blunders is the comparison of different observations of the same stations. If large discrepancies arise, there is an obvious difficulty. Along the same line, it is worthwhile to check for discrepancies in the base station coordinates. Clearly, if the base coordinates are wrong, the work created from that base is also wrong. Finally, look at the residuals of the final coordinates to be sure they are within reasonable limits. Remember that multipath and signal attenuation can pass by the observer without notice during the observations but will likely affect the residuals of the positions where they occur. A base station on a coordinated control position must be available. Its observations must be simultaneous with those at the roving receivers and it must observe the same constellation of satellites. It is certainly possible to perform a differential survey in which the position of the base station is either unknown or based on an assumed coordinate at the time of the survey. However, unless only relative coordinates are desired, the position of the base station must be known or determined in the end. In other words, the base station must occupy a control position, even if that control is established later. Utilization of the DGNSS techniques requires a minimum of four satellites for three-dimensional positioning. RTK ought to have at least five satellites for initialization. Tracking five satellites provides insurance against losing one abruptly; also, it adds considerable strength to the results. While cycle slips are always a problem, it is imperative in RTK that every epoch contains a minimum of the necessary satellite data without cycle slips. This is another reason to always track at least five satellites when doing RTK.

There is an alternative to the radio link method of RTK; the corrections can be carried to the rover using a cell phone. The cell phone connection does tend to ameliorate the signal interruptions that can occur over the radio link, and it offers a somewhat wider effective range in some circumstances. However, cell phone coverage may be unavailable in remote areas. Nevertheless, the use of cell phones in this regard is also a characteristic of Real-Time Network (RTN) solutions.

REAL-TIME NETWORK SERVICES

There is no question that real time dominates the GNSS surveying applications. It is applicable to much of engineering, surveying, air navigation, mineral exploration, machine control, hydrography, and a myriad of other areas that require centimeter-level accuracy and immediate positioning. However, the requirements for RTK, setting up a GNSS reference station on a known position, the establishment of a radio frequency transmitter and all attendant components before a single measurement can be made are both awkward and expensive. This, along with the limitation on the length of baselines for centimeter level work, has made RTK a bit more cumbersome and less flexible than most surveyors prefer. To alleviate these difficulties, services have arisen around the world to provide RTCM real-time corrections to surveyors by a different means. RTN also known as Network RTK (see Figure 7.5) has been implemented by both governments and commercial interests. The services are sometimes free and sometimes require a subscription or the payment of a fee before the surveyor can access the broadcast corrections over a data link via a modem such as a cell phone or some other device. Nevertheless, there are definite advantages including the elimination of individual base station preparation and the measurement of longer baselines without rapid degradation of the results.

These benefits are accomplished by the services gleaning corrections from a whole network of Continuously Operating Reference Stations (CORS) rather than just a single base. In this way, quality control is facilitated by the ability to check corrections from one CORS with those

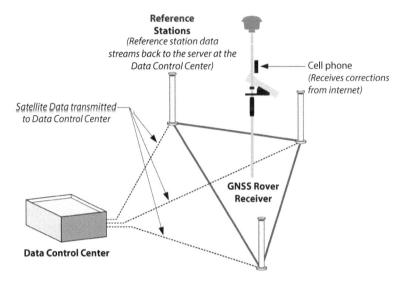

FIGURE 7.5   Real-Time Network GNSS

generated from another, and should a CORS go off-line or give incorrect values, other CORS in the network can take up the slack with little accuracy loss. Another benefit of this approach and perhaps the central idea underlying RTN differential corrections is the combination of observations from several CORS at known positions used to derive a model of an entire region. So rather than being considered as isolated beacons with each covering its own segregated area, the CORS are united into a network. The data from the network can then be used to produce a virtual model of the area of interest. From this model, distance-dependent biases such as ionospheric, tropospheric, and orbit errors can be calculated. Once the roving receiver's place within that network is established, it is possible to predict the errors at that position with a high degree of certainty. Not only can the CORS network be used to model errors in a region more correctly, but the multibase solution also can improve redundancy. Solving several baselines that converge on a project point simultaneously rather than relying on just one from typical RTK adds more certainty to the resulting coordinate.

Implementing an RTN requires data management and communication. The information from the CORS must be communicated to the central master control station, where all the calculations are done. Their raw measurement data, orbits, and so forth, must be managed as they are received in real time from each of the CORS that make up the network. Along with the modeling of the distance-dependent errors, all the integer ambiguities must be fixed for each CORS in real time. This is probably the most significant data processing difficulty required of an RTN, especially considering that there are usually large distances between the CORS. To accomplish it, post-computed ephemerides, antenna phase center corrections, and all other available information are brought to bear on the solution such as tropospheric modeling and ionospheric modeling.

Modeling is subject to variation in both space and time. For example, ionospheric and orbit biases are satellite specific, whereas tropospheric corrections can be estimated station by station. However, the ionospheric, dispersive, biases change more rapidly than the tropospheric and orbit biases, which are nondispersive. Therefore, ionospheric corrections must be updated more frequently than orbit and tropospheric corrections. While it is best to keep the modeling for ionosphere within the limited area around three or so CORS, when it comes to tropospheric and orbital modeling the more stations used the better. Finally, the pseudorange and/or carrier-phase residuals must be determined for all the carriers by using one of many techniques to interpolate the actual distance-dependent corrections for the surveyor's particular position within the network. Then the subsequent corrections must be communicated to the surveyor in the field, which typically requires the transmission of a large amount of data. There is more than one way the correction can be determined for a particular position within an RTN. So far, there is no clear best method. One approach is the creation of a position sometimes known as a Virtual Reference Station, VRS, and the attendant corrections. This approach requires a two-way communication link. Users must send their approximate positions to the master control center, usually as a string in the standard format that was defined by the *National Marine Electronics Association* (*NMEA*). The master control center returns corrections for an individual VRS, usually via RTCM and then the baseline processing software inside the rover calculates its position using the VRS, which seems to the receiver to be a single nearby reference station.

REAL-TIME GPS TECHNIQUES

Station diagrams, observation logs, and to-reach descriptions are usually unnecessary in real-time GNSS surveying. However, some components of static GNSS control methods are useful. One such technique is offsetting points to avoid multipath and signal attenuation. The need to offset points is prevalent in real-time GNSS (see Figure 7.6). For example, an offset point must often be established far enough from the original position to avoid an obstructed signal but close enough to prevent unacceptable positioning errors. While the calculation of the allowable vertical and horizontal measurement errors can be done trigonometrically, the measurements themselves will be different than those for an offset point in a static survey. For example, rather than the total station point and an azimuth point used in static work, a magnetic fluxgate digital compass and laser may be used

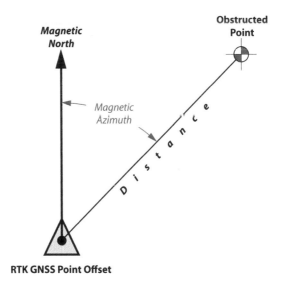

FIGURE 7.6   DGNSS or RTK Point Offset

to measure the tie from the offset point to the original point. It is worth noting that magnetic declination must be accommodated and metal objects avoided when using a magnetically determined direction. Such internal compasses should be carefully checked before they are relied upon. The length of the tie may be measured by an external laser, a laser cabled directly into the GNSS receiver, or even a tape and clinometer. Lasers are much more convenient since they can be used to measure longer distances more reliably and taping requires extra field crew members.

Rather than recording the bearing and distance in a field book for post-processing, the tie is usually stored directly in the data collector. In fact, often, the receiver's real-time processor can combine the measured distance and the direction of the *sideshot* with the receiver's position and calculate the coordinate of the originally desired position.

Dynamic Lines

A technique that is used especially in mobile GNSS application is the creation of dynamic lines. The GNSS receiver typically moves along a route to be mapped logging positions at predetermined intervals of time or distance. These points can then be joined together to create a continuous line. Obstructions along the route present a clear difficulty for this procedure. Points may be in error or lost completely due to multipath or signal attenuation. Also, in choosing the epoch interval, the capacity of the receiver's memory must be considered, especially when long lines are collected. If the interval chosen is too short, the receiver's storage capacity may be

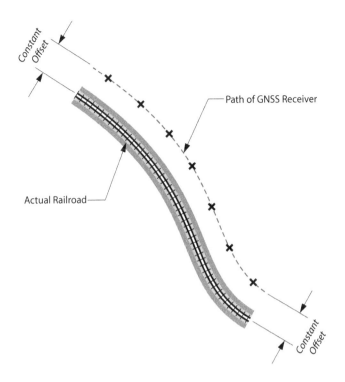

Constant Offset

Path of GNSS Receiver

Actual Railroad

Constant Offset

FIGURE 7.7   Line Offset

overwhelmed. If the interval is too long, important deflections along the way may be missed.

Where it is impossible or unsafe to travel along the line to be collected in the field, the dynamic line may be collected with a consistent offset (see Figure 7.7). This technique is especially useful in the collection roads and railroads, where it is possible to estimate the offset with some certainty due to the constant width of the feature. It is also possible, of course, to collect routes with individual discrete points with short occupations where that approach recommends itself.

Planning

Multipath and signal attenuation are particularly troublesome for the real-time GNSS. While the visibility diagrams mentioned earlier are not directly applicable, it is nevertheless prudent to plan the work so that at least five GNSS satellites are available above the mask angle in the area where data is to be collected. It is also useful to lower the mask angle when conditions warrant it. There are, of course, trade-offs to such strategies that may involve a reduction in positional integrity. The balance between accuracy and productivity is always a consideration.

A FEW RTK PROCEDURES

As mentioned earlier, redundancy in RTK work can be achieved by occupying each newly established position twice, and it is best if the second occupation is done using a different base station than was used to control the first. If this technique is used, the control points occupied by base stations should not be too close to one another. Each time the base is set up and before it is taken down, it is best to do a check shot on at least one known control point to verify the work. Please note that the point on which the check shot is taken should have been originally established from a base other than the base being checked. Clearly, an incorrect base position would not be revealed by checking into a point it has controlled. To ensure that the GNSS constellation during the second occupation differs substantially from that of the first, it is best if the second occupation takes place a significant amount of time after the first.

It is important to ensure that the centering is correct during the short occupations of real-time work. If a fixed height rod is used, it is best to eliminate the possibility of incorrect height of instrument measurement by measuring the height of the fixed height rod initially to confirm the value used is correct.

Concerning heights, if orthometric heights in real time are desired, a geoidal model is required and it is best if it is the most recent. However, please note that work retraced with a different geoidal model than was used initially will likely show vertical differences at the reoccupied points.

Some rover configurations facilitate *in-fill* surveys. It is an RTK survey style used when there is a risk of radio link interruption. When the correction signals from the base station fail to reach the rover, the collected raw data is stored in the memory of the receiver, perhaps at a different rate than the real-time data, for post-processing after the work is completed.

SITE CALIBRATION

The area of interest that is the project area, covered by an RTK survey, is usually relatively small and defined. Typically, a *site calibration*, aka localization, is performed to prepare such a GNSS project to be done using plane coordinates. A site calibration establishes the relationship between geographical coordinates—latitude, longitude, and ellipsoidal height—with plane coordinates—northing, easting, and orthometric heights across the area. In the final analysis, the relationship is expressed in three dimensions: translation, rotation, and scale. Because of the inevitable distortion that a site calibration must model, one of the prerequisites for such localization is the enclosure of the area by the control stations that will be utilized during the work. In the horizontal plane, the method of using plane coordinates on an imaginary flat reference surface with northings and eastings, or y- and x-coordinates, assumes a flat Earth. That is incorrect, of course, but a viable simplification if the area is small enough and the distortion is negligible. Such local tangent planes fixed at discrete control points by GNSS site calibration have been long used by land surveyors. Such systems demand little, if any, manipulation of the field observations, and once the coordinates are derived, they can be manipulated by straightforward plane trigonometry. In short, Cartesian systems are simple and convenient. However, there are difficulties as the area grows as was mentioned in Chapter 5. For example, each of these planes often has a unique local coordinate system derived from its own unique site calibration. The axes, the scale, and the rotation of each one of these individual local systems will not be the same as those elements of its neighbor's coordinate system. Therefore, when a site calibration is done and a local flat plane coordinate system is created, it is important to keep all the work in that system inside the limits created by the control points used in its creation. In the simplest case, a single point calibration, a flat plane is brought tangent to the Earth at one point, but a more typical approach is the utilization of three or four points enclosing the area of interest to be covered by the independent local coordinate system. Working outside of the limits created by those points should be avoided as it involves working where the distortion has not been modeled.

It might be said that a site calibration is the best fit of a plane onto a curved surface in which the inevitable distortion is distributed in both the horizontal and vertical planes. The vertical aspect is particularly important. It is called upon to adjust the measured GNSS ellipsoid heights to a desired local vertical datum. Therefore, it must account for undulations in the geoid because the separation between the ellipsoid and geoidal models may not be consistent over the project area. It can usually be modeled approximately as a trend across the area of interest so that the site calibration typically produces an inclined plane in the vertical aspect. Toward that end, the set of control points used to establish the site calibration must have both geographical coordinates—latitude, longitude, and ellipsoidal height-and plane coordinates—northing, easting, and orthometric heights in the desired local system. It is best if these control points are from the National Spatial Reference System (NSRS), when possible, that enclose the project and are distributed evenly around its boundary.

Real-time or static differential GNSS have long been the preferred methods of data processing for surveyors and geodesists. Typically, the differential processing technique depends on one of at least two receivers standing at a control station whose position is known, the base. It follows that the size of the positional error of the base receiver is knowable. By finding the difference between the biases at the base and the biases at the rover, the positional error at the other end of the baseline can be estimated.

Through the process of differencing, corrections are generated, which reduce the three-dimensional positional error at the unknown point by reducing the level of the biases there. The approach can generally provide up to sub-meter position from single-frequency pseudorange observations. Differentially processed carrier-phase observations can typically reach accuracies of a few centimeters in static work and decimeter-level or better accuracies on a moving platform. These facts have led to the construction of networks of CORS on control stations and around the world to support differential processing. There are many networks and associated services the NOAA ORS Network, The International GNSS Network, Europe's EUREF Permanent Network (EPN), EGNOS, the Japanese Multi-functional Satellite Augmentation System (MSAS), India's GPS And Geo-Augmented Navigation (GAGAN) system, the Australian Fiducial Network (AFN), Russia's SDCM, as well as many commercial networks and services. The list is constantly growing. Some of these services stream real-time differential corrections to users. These RTNs support the now well-known RTK methods. The convenience of these RTNs has contributed substantially to the extraordinary expansion of relatively high-accuracy GNSS applications that rely on differential processing. However, there is an alternative in both real-time and post-processed work. It is known as Precise Point Positioning (PPP).

PRECISE POINT POSITIONING

It was mentioned earlier that single point positioning can be a real-time solution using a single receiver measuring to a minimum of four satellites simultaneously (see Figure 2.8). There is no question that this is the most common GNSS solution outside of the geodesy and surveying disciplines and it is, in a sense, the fulfillment of the original idea of GPS. However, its weakness is that a receiver making such a pseudorange measurement must rely on the information it collects from the satellite's navigation message to learn the positions of the satellites, the satellite clock offset, the ionospheric correction, etc. This data contains errors. Under such circumstances, a single point position cannot be highly accurate, but what if the positions of the satellites, the satellite clock offset, the orbital information, etc., were not derived from the broadcast navigation message?

Consider PPP a global, primarily carrier-phase, GNSS technique with some unique advantages. The user need not establish control stations or have access to corrections from close-by reference stations. It is not restricted by the limits on the baseline lengths imposed by differential processing. Its corrections are expressed in a global reference frame, typically the International Terrestrial Reference Frame, which offers better overall consistency in absolute positioning than does a local or regional solution.

PPP is particularly useful where CORS networks are not available. While it requires precise satellite orbit and clock error information, the data is freely available from the International GNSS Service (IGS). There are other noncommercial sources but many commercial service providers too. The IGS collects and archives GNSS data from its worldwide network of more than 506 CORS. Then up to eight IGS analysis centers are involved in the processing and formulation of precise satellite ephemerides and clock solutions. These data, being more accurate than the broadcast orbits and the clock corrections, enable higher accuracy single point positioning.

Best PPP results are obtained with a multifrequency receiver. Multifrequency receivers can mitigate the ionospheric delay. A single-frequency receiver can be used for PPP but requires ionospheric correction be available along with the orbit and clock information. The single-frequency solution will likely be in the decimeter range at best, while the multifrequency position might achieve centimeters. Both post-processed and real-time positioning are available with PPP and both can be supported by IGS products. Concerning post-processed work, IGS and the National Oceanic and Atmospheric Administration (NOAA) have worked together to provide accurate ephemerides and clock data in three forms: Ultra-Rapid, Rapid, and Final. The Ultra-Rapid includes and is available online 3 to 9 hours after an observation is completed. It is posted four times daily at 03, 09, 15, and 21 hours UTC. The predictions are 6 hours old on average. Each posting contains 48 hours of both observed and predicted ephemerides. The first 24 hours are based on the most recent data from the IGS tracking network. These data allow users to process their observations using the positions of the satellites and the state of the clocks derived at the time the satellites were being tracked by the user. The following 24 hours are predicted orbits. It takes longer for the Rapid to come online. It is posted at 17hr UTC the following day. The Final product takes the longest of all, 12–18 days. As would be expected, the accuracy of the ephemeris and the clock data of each increment increases. The Final is the best. There are also PPP post-processing services whose websites accept users' RINEX file uploads, automatically compute GNSS receiver positions and return them.

Implementation of PPP in real time requires access to streaming data. IGS provides such an open Real-Time Service (RTS) to the public by subscription. The service has been made possible through partnership with other agencies, specifically Natural Resources Canada (NRCan), the German Federal Agency for Cartography and Geodesy (BKG), and the European Space Agency's Space Operations Centre in Darmstadt, Germany (ESA/ESOC). However, the IGS network data, through their multiple analysis and data centers, are the foundation of the service. The RTS's precise ephemeris and clock data is available on the Internet every

25 seconds via an open-source protocol known as the Network Transport of RTCM (NTRIP), which has been an RTCM standard for the real-time collection and distribution of GNSS information since 2004. These precise satellite orbits and clocks, along with code and phase observations from a dual-frequency receiver, provide the data from which the PPP algorithm derives accurate positions. This is currently a GPS-only service, but there are plans for it to be expanded to the Russian GLONASS system and other GNSS constellations.

PPP DISADVANTAGE

The real downside of real-time PPP is its typically long convergence time, the time it needs to stabilize at an accuracy level. While a decimeter position from a single-frequency receiver might converge in a few minutes, a much more accurate dual-frequency solution may take half an hour or more due to the time necessary for phase ambiguity resolution. The time is extended because PPP cannot build double differences in the same way as is done in relative positioning. The undifferenced ambiguities are not necessarily integers and are therefore estimated as real numbers, the float solution. Nevertheless, methods are being developed by organizations such as the IGS PPP Ambiguity Resolution (PPP-AR) working group by which the undifferenced phase ambiguities can be fixed to integer values. The goal is for PPP-AR with real-time data providing precise satellite orbits and the clocks to deliver accuracies commensurate with those available from differenced measurements.

EXERCISES

1. What is the meaning of latency as applied to DGNSS and RTK?
   a. The baud rate of a radio modem in real-time GNSS.
   b. The time taken for a base station to compute corrections, transmit them, and the rover to implement them in real-time GNSS.
   c. The frequency of the RTCM SC104 correction signal in real-time GNSS.
   d. The range rate broadcast with the corrections from the base station.

2. Which of the following errors are not reduced by using DGNSS or RTK methods?
   a. Atmospheric errors
   b. Satellite clock bias
   c. Ephemeris bias
   d. Multipath

3. Who was the original kinematic GNSS innovator?
   a. Ann Bailey
   b. Benjamin Remondi
   c. George F. Syme
   d. R.E. Kalman

4. Which of the following statements about rules governing RTK radio communication is correct?
   a. According to the FCC regulations, voice communications have priority over data communications.
   b. Code of Federal Regulations, 48 CFR 91, contains the complete text of the FCC Rules.
   c. Most typical community voice repeater frequencies are about 900 MHz.
   d. In Europe, it is possible to use the 4.2 GHz band for spread spectrum communication without special authorization.

5. Which of the following data rates is closest to that required for RTK?
   a. 200 bps updated every 10 seconds
   b. 1400 bps updated about every 5 seconds
   c. 2400 bps updated about every 1/2 of a second
   d. 3500 bps updated about every 0.05 of a second

6. How many satellites are the minimum for the best initialization with RTK?
   a. Four satellites
   b. Five satellites
   c. Six satellites
   d. Eight satellites

7. Which of the following is not part of a typical data radio transmitter used in RTK?
   a. A radio modulator
   b. An amplifier
   c. A demodulator
   d. An antenna

8. Why is it advisable that some time elapse between successive point occupations in a GNSS survey?
   a. To eliminate multipath when it corrupted the first occupation of the station.

b. To allow the satellite constellation to reach a significantly different configuration than it had during the first occupation of the station.
c. To overcome an overhead obstruction during the first occupation of the station.
d. To eliminate receiver clock errors.

9. Which of the following is a weakness of the real-time PPP?
   a. The length of time necessary to resolve the integer ambiguity
   b. PPP does not use base stations
   c. RTCM message is not received by the receiver
   d. The PPP uses only one receiver

ANSWERS AND EXPLANATIONS

1. Answer is (b)

   Explanation: In DGNSS and RTK, it takes some time for the base station to calculate corrections, and it takes some time for it to transmit them and for the rover to implement them. The base station's data is put into packets in the correct format. The data makes its way from the base station to the rover over the data link. It must then be decoded and go through the rover's software. The time it takes for all of this to happen is called the latency of the communication between the base station and the rover. It can be as little as a quarter of a second or as long as a couple of seconds.

2. Answer is (d)

   Explanation: In autonomous point positioning, all the biases discussed earlier affect the GNSS position. However, when errors are well correlated between the base and the rover, as in DGNSS and RTK, these biases are mitigated. The atmospheric errors, too, are reduced. In fact, most errors in GNSS are mediated by the relatively short distance between receivers, but the same cannot be said for multipath. When it is present, it is just as troublesome in real-time work as it is in other GNSS surveying.

3. Answer is (b)

   Explanation: Kinematic surveying, also known as stop-and-go kinematic surveying, is not new. The original kinematic GNSS innovator, Dr. Benjamin Remondi, developed the idea in the mid-1980s.

4. Answer is (a)

   Explanation: In the United States, most transmitters connected to RTK GNSS surveying equipment operate between 450 and 470 MHz, and voice communications also operate in this same range. It is important for GNSS surveyors to know that voice communications have priority over data communications. The FCC requires cooperation among licensees that share frequencies. Interference should be minimized. For example, it is wise to avoid the most typical community voice repeater frequencies. They usually occur between 455 and 460 MHz and 465 and 470 MHz. Part 90 of the Code of Federal Regulations, 47 CFR 90, contains the complete text of the FCC Rules. There are also other international and national bodies that govern frequencies and authorize the use of signals elsewhere in the world. In some areas, certain bands are designated for public use, and no special permission is required. For example, in Europe, it is possible to use the 2.4 GHz band for spread spectrum communication without special authorization with certain power limitations. Here in the United States, the band for spread spectrum communication is 900 MHz.

5. Answer is (c)

   Explanation: DGNSS operations may need no more than 200 bps, bits-per-second, updated every 10 seconds or so. RTK requires at least 2400 bps updated about every 1/2 second or less.

6. Answer is (b)

   Explanation: Utilization of the DGNSS techniques requires a minimum of four satellites for three-dimensional positioning. RTK ought to have at least five satellites for initialization. Tracking five satellites is a bit of insurance against losing one abruptly; also, it adds considerable strength to the results. While cycle slips are always a problem it is imperative in RTK that every epoch contains a minimum of four satellite data without cycle slips. This is another reason to always track at least five satellites when doing RTK.

7. Answer is (c)

   Explanation: A data radio transmitter consists of a radio modulator, an amplifier, and an antenna. The modulator converts the correction data into a radio signal. The amplifier increases the signal's power and then the information is transmitted via the antenna. The receiver has a demodulator. The demodulator converts the signal back to an intelligible form for the rover's receiver.

8. Answer is (b)

Explanation: Some time should elapse between successive occupations on a point so the satellite constellation can reach a significantly different configuration than that which it had during the first occupation. Please recall that GNSS measurements are not actually made between the occupied stations but directly to the satellites themselves.

9. Answer is (a)

Explanation: Real-time PPP currently has disadvantages. One of the most persistent is the time necessary to resolve the integer ambiguity because the ambiguity cannot be assumed to be an integer as it is in a differenced solution. As things stand, the convergence can take 20 minutes or more.

8 GPS Modernization and Global Navigation Satellite System (GNSS)

GPS MODERNIZATION

The configuration of the GPS Space Segment is well-known (see Figure 8.1). The satellites are on orbit at a nominal height of about 20,000 km above the Earth. There are three carriers: L1 (1575.42 MHz), L2 (1227.60 MHz), and L5 (1176.42 MHz). A minimum of 24 GPS satellites ensures 24-hour worldwide coverage, but there are more than that minimum on orbit. There are spares on hand and in space too. The redundancy is prudent because GPS is critical to positioning, navigation, and timing, of course. It is also critical to the smooth functioning of financial transactions, air traffic, ATMs, cell phones, and modern life in general around the world. This very criticality requires continuous modernization.

GPS was put in place with amazing speed considering the technological hurdles and reached its fully operational capability (FOC) on July 17, 1995. The oldest satellites in the current constellation were launched in the late 1990s. If you imagine using a personal computer of that vintage today, it is not surprising that there are plans in place to update the system. In 2000, U.S. Congress authorized the GPS III effort. The introduction of new ground stations, the Next Generation Operational Control System (OCX) roll-out, new satellites, additional civilian and military navigation signals, and improved availability is underway. This chapter is about some of the changes in the modernized GPS, its inclusion in the Global Navigation Satellite System (GNSS), and more.

SATELLITE BLOCKS

BLOCK I, BLOCK II/IIA, BLOCK IIR, AND BLOCK III SATELLITES

Since the beginning of the system, a total of 77 GPS satellites have been launched. As of this writing, 31 of them are operational. Here is an illustration that summarizes the improvements made in the satellite blocks that have made up the GPS constellation over the years (see Table 8.1).

DOI: 10.1201/9781003405238-8

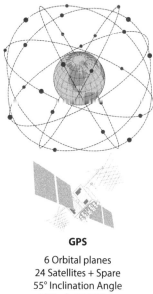

GPS

6 Orbital planes
24 Satellites + Spare
55° Inclination Angle
Altitude 20,200km

FIGURE 8.1   GPS Constellation

As these changes have been made, there has been a corresponding drop in the annual root mean square (RMS) of the GPS signal-in-space range error. It decreased from 1.6 m in 2001 to 0.7 m in 2014 and continues to improve as new generations of satellites come online.

BLOCK I

The first of the 11 successful Block I satellites was launched in 1978 from Vandenberg Air Force Base. The last of them was launched in 1985. One launch failed, NAVSTAR 7. They were all retired by late 1995. These satellites needed frequent help from the Control Segment. They could operate independently for only 3 1/2 days. The Control Segment handled the necessary momentum dumping for the satellites and maintained their attitudes using hydrazine thrusters. The inclination of these satellites relative to the equator was 63° instead of the inclination of 55° used for subsequent GPS blocks. They had a design life of 4.5 years, though some operated for double that. They were powered by 7.25 square meters of solar panels, and they also had three rechargeable nickel-cadmium batteries. In subsequent blocks of satellites, design lives increased and dependence on the Control Segment decreased. However, some of the features of Block I were carried forward into the subsequent blocks of GPS satellites. They had nuclear detonation detection sensors onboard; a feature which continued in future GPS satellite blocks. It was clear from the beginning that atomic frequency

TABLE 8.1
GPS Modernization

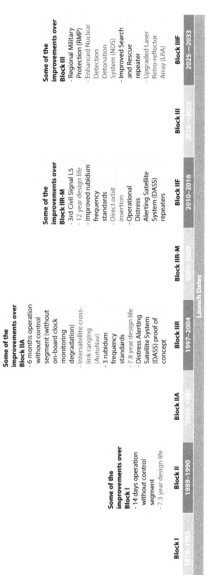

Block I	Block II	Block IIA	Block IIR	Block IIR-M	Block IIF	Block III	Block IIIF
1978–1985	1989–1990	1990–1997	1997–2004	2005–2009	2010–2016	2018–2023	2025–2033
				Launch Dates			
Demonstrated GPS - L1 (CA) navigation signal - L1 & L2 (P Code) navigation signal - 4.5 year design life	**Some of the improvements over Block I** - 14 days operation without control segment - 7.3 year design life	**Some of the improvements over Block II** - 6 months operation without control segment (with degradation) - 7.5 year design life - Radiation hardened	**Some of the improvements over Block IIA** - 6 months operation without control segment (without on-board clock monitoring degradation) - Intersatellite cross-link ranging (AutoNav) - 3 rubidium frequency standards - 7–8 year design life - Distress Alerting Satellite System (DASS) proof of concept	**Some of the improvements over Block IIR** - 2nd Civil Signal L2 (L2C) - M-Code on L1/L2 - L5 Demo - Flexible power levels for military signals - Anti-Jam Flex Power	**Some of the improvements over Block IIR-M** - 3rd Civil Signal L5 - 12 year design life - Improved rubidium frequency standards - Direct orbit insertion - Operational Distress Alerting Satellite System (DASS) repeaters	**Some of the improvements over Block IIF** - Increased earth coverage power - 15 year design life - 4th Civil Signal (L1C) - On-board Laser Retro-reflector Arrays (LRA) - 2–4 crosslink antennas for more efficient uploads - Spot Beam for Anti-Jam (AJ) - Navigation Integrity	**Some of the improvements over Block III** - Regional Military Protection (RMP) - Enhanced Nuclear Detection System (NDS) - Improved Search and Rescue repeater - Upgraded Laser Retro-reflector Array (LRA)

standards, clocks, were necessary for the proper functioning of the system. Therefore, the Block I satellites had cesium and rubidium frequency standards onboard, a feature that GPS satellites continue to share. The first three Block I satellite carried three rubidium clocks. Unfortunately, they stopped working after about a year in space. The rubidium standards were improved, equipment was added to keep the frequency standards at a constant temperature during flight, and 1 cesium frequency standard was added to subsequent satellites in this block.

BLOCK II

The Block II satellites were about twice as heavy as the Block I satellites. The first of them was launched in 1989. The Block II satellites often exceeded their 7.3 design life. The last was decommissioned in 2007 after 17 years of operation. They could be autonomous, without contact with the Control Segment, for up to 14 days. They used reaction wheels to stabilize themselves, orbits were corrected with a hydrazine propulsion system and each carried 2 rubidium and 2 cesium clocks. The uploads from the Control Segment to the Block II satellites were encrypted unlike uploads during Block I. The signals from the Block II satellites were periodically and purposely disrupted. Specifically, the onboard clocks were intentionally dithered in a procedure known as Selective Availability (SA). None of the Block II satellites are functioning today. The last was decommissioned in March 2007.

BLOCK IIA

Block IIA satellites were an improved version of Block II. The first of 19 Block IIA satellites was launched in 1990. They were radiation hardened against cosmic rays, built to provide SA, anti-spoofing (AS) capability, and onboard momentum dumping. This SA continued until May 2, 2000, when it was discontinued. Block IIA satellites could store more of the Legacy Navigation message than the Block II satellites and could therefore operate without contact with the Control Segment for 6 months. However, if that had been done their broadcast ephemeris and clock correction would have degraded. Two Block IIA satellites, SVN 35 (PRN 05) and SVN 36 (PRN 06), were equipped with Laser Retro-reflector Arrays (LRA). The second of these was launched in 1994. The retro-reflectors facilitated satellite laser ranging (SLR), which provided data to separate clock and ephemeris biases from one another and, thereby, independent orbit validation. Like the Block II satellites, the Block IIA satellites were equipped with two rubidium and two cesium frequency standards and were expected to have a design life of 7.5 years, which they exceeded. The last of them, SVN-34, was retired in April 2020.

BLOCK IIR

The first launch of the next block, Block IIR satellites in January 1997 was unsuccessful. The following launch in July of 1997 succeeded. As of this writing, there are 7 Block IIR satellites in orbit, which are operational. There are differences between the Block IIA and the Block IIR satellites. The Block IIR satellites have a design life of 7.8 years and can determine their own position using inter-satellite crosslink ranging called AutoNav. This involves their use of reprogrammable processors onboard to do their own fixes in flight. They can operate in that mode for up to 6 months and still maintain full accuracy. The Control Segment can also change their software while the satellites are in flight and, with a 60-day notice, move them into a new orbit. Unlike some of their direct predecessors, these satellites are equipped with three rubidium frequency standards. Some of the Block IIR satellites also have an improved antenna panel that provides more signal power. They are more radiation hardened than their predecessors and they cost about a third less than the Block II satellites did.

The Block IIR satellites broadcast the same fundamental GPS signals that have been in place for a long time. Their frequencies are centered on L1 and L2. As mentioned before, the Coarse/Acquisition code or C/A code is carried on L1 and has a chipping rate of 1.023 million chips per second. It has a code length of 1023 chips over the course of a millisecond before it repeats itself. There are 32 different code sequences that can be used in the C/A code. The Precise code or P(Y) code on L1 and L2 has a chipping rate that is ten times faster than the C/A code at 10.23 million chips per second. The P(Y) code has a code length of about a week, approximately 6 trillion chips before it repeats. If this code is encrypted it is known as the P(Y) code or simply the Y-code. Some Block IIR satellites carry Distress Alerting Satellite System (DASS) repeaters. These DASS repeaters are used to relay distress signals from emergency beacons and were part of a proof of the concept of satellite-supported search and rescue that was completed in 2009.

BLOCK IIR-M

As of this writing, there are 7 Block IIR-M satellites on orbit and operational in the current constellation. These are IIR satellites that were modified before they were launched. The modifications upgraded these satellites so that they broadcast two new codes; a new military code, the M-code, a new civilian code, the L2C code, and demonstrate a new carrier, L5. The L2C code is broadcast on L2 only, and the M-code is on both L1 and L2. The L2C code helps in the correction of the ionospheric delay and the M code improves the military anti-jamming efforts through flexible power capability. One of the Block IIR-M satellites, SVN 49, transmitted on L1, L2, and L5, unfortunately, it has been unhealthy since its launch. The first of these Block IIR-M satellites was launched in the fall of 2005, and the last in the summer of 2009.

BLOCK IIF

The first Block IIF satellite was launched in the summer of 2010. There are 12 Block IIF satellites on orbit and operational as of this writing. The first was launched in May 2010, and the last in February 2016. Their design life is 12 years. They have 240 W of transmission power. Block IIF satellites have faster processors and more memory onboard than did previous blocks. They broadcast all the previously mentioned signals, and one more, a new carrier known as L5. This is a signal that was demonstrated on Block IIR-M. It is now available from all the Block IIF satellites. At 1176.45 MHz, the L5 signal is within the *Aeronautical Radio Navigation Services* (*ARNS*) frequency band 1164–1215 MHz and can service aeronautical applications. The improved rubidium frequency standards on Block IIF satellites have a reduced white noise level. The Block IIF satellite's launch vehicles can place the satellites directly into their intended orbits, so they do not need the *apogee kick motors* (AKM) their predecessors required. All GPS Blocks I and II, except Block IIF, needed boosts from AKM to reach their medium Earth orbit (MEO). GPS III satellites, more massive than Block IIF satellites, will rely on liquid apogee engines to reach their operational orbit. All the Block IIF satellites will carry DASS repeaters. Their onboard *navigation data units* (*NDU*) support the creation of new navigation messages with an improved broadcast ephemeris and clock corrections. Like the Block IIR satellites, the Block IIF can be reprogrammed on orbit.

BLOCK III

Block III satellites are on orbit. The first of ten, formerly called Block IIIA and now simply as Block III, was launched in December 2018. The next was launched in August 2019. It was followed by the third GPS III satellite in June 2020, the fourth in November 2020, and the fifth in June 2021. These are the first GPS satellites to be given nicknames, i.e., Vespucci, Magellan, Matthew Henson, Sacagawea, Neil Armstrong, and Amelia Earhart. These will be followed by an updated version called GPS IIIF. Their 300 W of transmission power broadcasts include new signals. It will be resistant to hostile jamming. They will also carry DASS repeaters. When the whole GPS constellation has DASS repeaters on board, there will be global coverage for satellite-supported search and rescue, and at least four DASS-equipped satellites will always be visible from anywhere on Earth. This system will enhance the international *COSPAS-SARSAT* satellite-aided search and rescue (SAR) system and will be interoperable with similar Russian (SAR/GLONASS), European (SAR/Galileo), and other GNSS systems. Block III satellites have cross-link capability to support inter-satellite ranging and transfer; telemetry, tracking, and control (TT&C) capability, and from two to four directional crosslink antennas. This means they can be updated from a single ground station instead of requiring each satellite to be in the range of a ground antenna to be updated. This and their high-speed upload and download antennas help increase the upload frequency from once every 12 hours to once every 15 minutes. Each Block III satellites have three enhanced rubidium atomic frequency standards (RAFS clocks), and a fourth slot is available for a new clock, i.e., a hydrogen maser. All have onboard LRA (aka retro-reflectors). The satellite laser tracking available with this payload provides data from which it is possible to distinguish between clock error and ephemeris error.

These satellites broadcast the Legacy Navigation (LNAV) message on C/A, CNAV on L2C and L5, and CNAV-2 on L1C and MNAV. The L1C signal was designed with international cooperation to maximize *interoperability* with Galileo's Open Service (OS) Signal, China's BeiDou, Japan's Quazi-Zenith Satellite System (QZSS), and India's NavIC. Codes available from earlier blocks, i.e., the M-code, L5, the P(Y) code, and the C/A code, are broadcast with increased power from the Block III satellites. The broadcast of the M-code is changed in an interesting way. It continues to be radiated with a wide angle to cover the full Earth just as in the Block IIR-M satellites, but the Block III M-code can be broadcast through a large deployable high-gain antenna to produce a directional spot beam. The spot beam has much more power (−138 dBW) compared with (−158 dBW) the wide-angle M-code broadcast (see Figure 8.2). It has the anti-jam (AJ) capability to be aimed to a region several hundreds of kilometers in diameter. Two antennas mean that, to those inside the spot beam, the satellite appears to be two satellites occupying the same position.

Approximately 22 Block IIIF satellites will follow Block III. The launches will begin around 2025 or 2026 and will continue through 2034 or so. They will have improved nuclear detonation detection capability, a search and rescue repeater, and broadcast the M-code more powerfully.

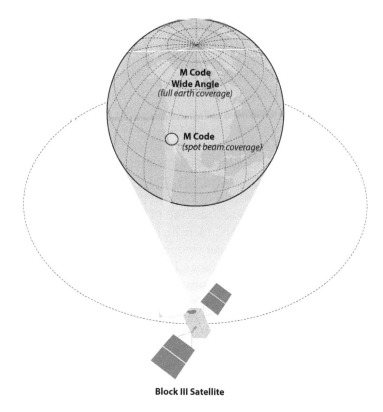

Block III Satellite

FIGURE 8.2   Spot Beam

POWER SPECTRAL DENSITY DIAGRAMS

Many of the improvements in GPS are centered on the broadcast of new signals. Power spectral density (PSD) function diagrams are a convenient way to visualize the structure and improved capabilities available with the new signals. They illustrate the signals power per bandwidth in Watts per Hertz as a function of frequency.

In GPS and GNSS literature, the PSD diagram is often represented with the frequency in MHz on the horizontal axis and the density, the power, represented on the perpendicular axes in decibels relative to 1 Hertz per Watt or dBW/Hz (see Figure 8.3).

dBW/Hz

Perhaps a bit of background is in order to explain those units. A bel unit originated at Bell Labs to quantify power loss on telephone lines. A decibel is a tenth of a bel. A decibel (dB) is a dimensionless number, a ratio that can acquire dimension by being associated with measured units. Here are some of the quantities with which it is sometimes associated: seconds of

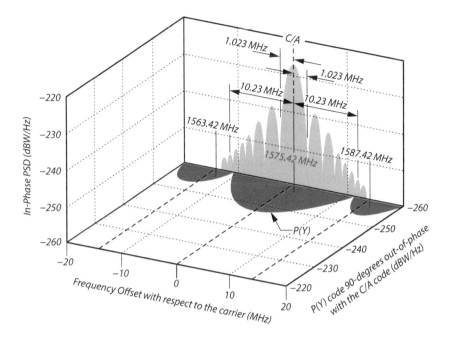

FIGURE 8.3 Legacy L1 Signal

time, symbolized dBs, bandwidth measured in Hertz, symbolized dBHz, and temperature measured in Kelvins, symbolized dBK. Since signal power is of interest here dB will be described with respect to 1Watt, the symbol used is dBW.

dBW is a short concise number that can conveniently express the wide variation in signal power levels. dBW can represent quite large and quite small amounts of power more handily than other notations. For example, consider a value of interest in GPS signals. The value is very small, one-tenth of a millionth billionth of a watt. Expressing it as 0.0000000000000001 W is a bit exhausting. It would be more convenient expressed in dBW a value that can be derived using the formula

$$P_{\mathrm{dBW}} = 10\log_{10}\frac{P_W}{1\ \mathrm{W}}$$

where P_W is the power of the signal.

$$P_{\mathrm{dBW}} = 10\log_{10}\frac{10^{-16}\ \mathrm{W}}{1\ \mathrm{W}}$$

$$P_{\mathrm{dBW}} = 10\log_{10}10^{-16}\ \mathrm{W}$$

$$-160\ \mathrm{dBW} = 10\log_{10}10^{-16}\ \mathrm{W}$$

The expression 160 dBW is immediately useful. Here is an example. A change in 3 decibels is always an increase or a decrease of 100% in power level. Stated another way, a 3-decibel increase indicates a doubling of signal strength, and a 3-decibel decrease indicates a halving of signal strength. Therefore, it is easy to see that a signal of −163 dBW has half the power of a signal of −160 dBW. Considering the broadcasts from the current constellation of satellites, the minimum power received from the P(Y) code on L1 by a GPS receiver on the Earth's surface is about −161.5 dBW as broadcast by Block IIR-M satellites, and the minimum power received from the C/A code on L1 is approximately −158.5 dBW. This difference between the two received signals is not surprising since, at the start of their trip to Earth, they are transmitted by the satellite at power levels that are also 3 decibels apart. One might wonder why there are such differences between the power of the transmitted GPS signal, called the *Effective Isotropic Radiated Power* (*EIRP*), and the power of the received signal. The large difference is mostly because of the 20,000 km distance from the satellite to a GPS receiver on Earth. There is also an atmospheric loss and a polarization mismatch loss, but the biggest loss by far is along the path in free space.

Much of his loss is a function of the spreading of the GPS signal in space as described by the inverse square law. The intensity of the GPS signal varies inversely to the square of the distance from the satellite. In other words, by the time the signal makes that trip and reaches the GPS receiver, it is quite weak (see Figure 8.4). It follows that GPS signals are easily degraded by vegetation canopy, urban canyons, and other interference.

A GPS signal has power, of course, but it also has bandwidth. PSD is a measure of how much power a modulated carrier contains within a specified bandwidth. That value can be calculated using the following formula and allowing that there is an even distribution of 10^{-16} W over the 2.046 MHz C/A bandwidth of the C/A code:

$$\text{Power density}\left(\frac{\text{dBW}}{\text{Hz}}\right) = 10\log_{10}\left[\frac{\text{Power (W)}}{\text{Bandwidth (Hz)}}\right]$$

$$\text{Power density}\left(\frac{\text{dBW}}{\text{Hz}}\right) = 10\log_{10}\left[\frac{10^{-16} \text{ W}}{2.046 \times 10^6 \text{ Hz}}\right]$$

$$\text{Power density}\left(\frac{\text{dBW}}{\text{Hz}}\right) = 10\log_{10}\left[4.888^{-23}\right]$$

$$\text{Power density}\left(\frac{\text{dBW}}{\text{Hz}}\right) = 10(-22.3)$$

$$\text{Power density} = -223 \text{ dBW/Hz}$$

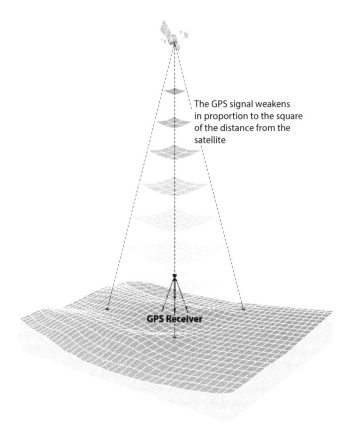

The GPS signal weakens in proportion to the square of the distance from the satellite

GPS Receiver

FIGURE 8.4   Inverse Square Law

The calculation is also frequently normalized and done presuming an even distribution of 1 W over the 2.046 MHz C/A bandwidth of the C/A code. In other words, the following calculation presumes an even distribution of the power over 1 W instead of the 10^{-16} W used in the previous calculation:

$$\text{Power density}\left(\frac{\text{dBW}}{\text{Hz}}\right) = 10\log_{10}\left[\frac{\text{Power (W)}}{\text{Bandwidth (Hz)}}\right]$$

$$\text{Power density}\left(\frac{\text{dBW}}{\text{Hz}}\right) = 10\log_{10}\left[\frac{1\text{ W}}{2.046\times10^6\text{ Hz}}\right]$$

$$\text{Power density}\left(\frac{\text{dBW}}{\text{Hz}}\right) = 10\log_{10}\left[4.888^{-7}\right]$$

$$\text{Power density}\left(\frac{\text{dBW}}{\text{Hz}}\right) = 10(-6.31)$$

$$\text{Power density} = -63\text{ dBW/Hz}$$

L1 LEGACY SIGNALS

The PSD diagrams show the increase or decrease, in decibels, of power, in Watts with respect to frequency in Hertz. Figure 8.3 illustrates the PSD diagrams of the well-known codes on L1. In the diagram, the C/A code on the L1 signal is centered on the frequency 1575.42 MHz, and a portion of the bandwidth over which it is spread, approximately 20.46 MHz, is shown evenly split, 10.23 MHz, on each side of the center frequency. The horizontal scale shows the offset in MHz from 1575.42 center frequency. Other scales show the decibels relative to 1 Watt per Hertz (dBW/Hz). The P(Y) code is in quadrature that is 90 degrees from the C/A code. In both cases, most of the power is close to the center frequency. The C/A code has many lobes, but the P(Y) code with the same bandwidth but ten times the clock rate has just one main lobe.

NEW SIGNALS

An important aspect of GPS modernization is the advent of some new and different signals that are augmenting the old reliable codes. In GPS, a dramatic step was taken in this direction on September 21, 2005, when the first Block IIR-M satellite was launched. One of the significant improvements coming with the Block IIR-M satellites was increased power by virtue of the new antenna panel. The Block IIR-M satellites also broadcast new signals, such as the M-code.

THE M-CODE

Eight to twelve of these replenishment satellites were modified to broadcast a new military code, the M-code. This code is carried on both L1 and L2 and will replace the P(Y) code in a few years. The M-code has the advantage of allowing the Department of Defense (DoD) to increase the power of the code to prevent jamming. There was consideration given to raising the power of the P(Y) code to accomplish the same end, but that strategy was discarded when it was shown to interfere with the C/A code.

The M-code was designed to share the same bands with existing signals, on both L1 and L2, and still be separate from them. See those two peaks in the M-code in Figure 8.5. They represent a split-spectrum signal about the carrier. Among other things, this allows minimum overlap with the maximum power densities of the P(Y) code and the C/A code, which occur near the center frequency. That is because the actual modulation of the M-code is done differently. It is accomplished with binary offset carrier (BOC) modulation, which differs from the binary phase shift key (BPSK) used with the legacy C/A and P(Y) signals. BOC modulation is known as a

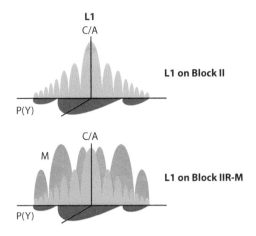

FIGURE 8.5   L1 with M-Code

split-spectrum modulation. The signal's spectrum is divided into two parts and can be shaped to accommodate inter-system-compatibility. The BOC modulated M-code has its greatest power density at the edges, at the nulls, and away from P(Y) and C/A. It is characteristic for such modulation to be written BOC (α, β) where α indicates the frequency of the square wave modulation of the carrier, also known as the subcarrier frequency factor, and β describes the frequency of the pseudorandom noise (PRN) modulation, also known as the spreading code factor. In the case of the M-code, the notation BOC (10, 5). Both are multiples of 1.023 MHz. In other words, their actual values are:

$$\alpha = 10 \times 1.023 \text{ MHz} = 10.23 \text{ MHz and}$$
$$\beta = 5 \times 1.023 \text{ MHz} = 5.115 \text{ MHz}$$

This architecture mitigates interference with the existing codes and simplifies implementation at the satellites and receivers. Suffice it to say that this aspect and others of the BOC modulation strategy offer good spectral separation between the M-code and the older legacy signals. The M-code is spread across 24 MHz of the bandwidth. It carries a packetized MNAV navigational message permitting more flexible data payloads and it utilizes *forward error correction (FEC)*. FEC is an advanced technique used to detect and correct errors in transmitted data without the need for retransmission. It adds redundancy to the navigation data bits, which doubles the binary data rate but allows the receiver to decode them with a much smaller probability of error. However, the biggest benefit of FEC may be that the receiver can read the navigation data bits in poorer signal-to-noise conditions with greater certainty.

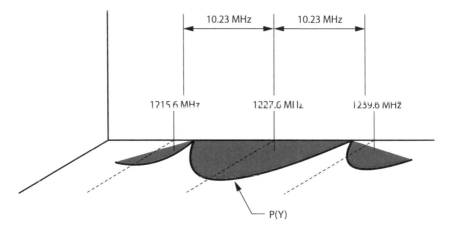

FIGURE 8.6   Legacy L2 Signal

The M-code is designed to be autonomous. It is tracked by direct acquisition. This means that, as mentioned in Chapter 1, the receiver correlates the signal coming in from the satellite with a replica of the code that it has generated itself. In other words, a military receiver can determine its position with the M-code alone, whereas with the P(Y) it must first acquire the C/A code to do so.

L2C

Figure 8.6 shows an L2 signal diagram centered on 1227.60 MHz as it was when it carried only one military signal, the P(Y) code. That has changed, of course. L2 now carries the military M-code, just discussed, and where there had been no civilian code at all there is now L2C (see Figure 8.7).

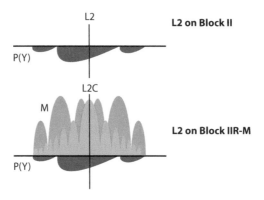

FIGURE 8.7   L2 with M-Code and L2C

A new military code on L1 and L2 may not be terribly exciting to civilian users, but the IIR-M satellites and those that followed have something else going for them. They broadcast a new civilian code, L2C. The "C" is for civil. This is a code that was first announced back in March of 1998. Even though its 2.046 MHz from null-to-null gives it a very similar power spectrum to the C/A code, it is important to note that L2C is not a copy of the C/A code even though that was the original idea. The original plan was that L2C would be a replica of the venerable C/A code but carried on L2 instead of L1. This concept changed when Colonel Douglas L. Loverro, Program Director for the GPS Joint Program Office (JPO), was asked if perhaps it was time for some improvement of C/A. The answer was yes. The C/A code is somewhat susceptible to both waveform distortion and narrow-band interference, and its cross-correlation properties are marginal at best. So, the new code on L2, known as L2 Civil, or L2C was announced. It is more sophisticated than C/A.

L2C is composed of two PRN signals the Civil-Moderate length code, CM, and the Civil Long code, CL. They both utilize the same modulation scheme, BPSK, as the legacy signals and both signals are broadcast at 511.5 kilobits per second (Kbps). This means that CM repeats its 10230 chips every 20 milliseconds and CL repeats its 767250 chips every 1.5 seconds. But wait a minute, how can you do that? How can you have two codes in one? L2C achieves this by time multiplexing. Since the two codes have different lengths, L2C alternates between chips of the CM code and chips of the CL code (see Figure 8.8). It is called chip-by-chip time multiplexing. So even though the actual chipping rate is 511.5 KHz, half the chipping rate of the C/A code, with the time multiplexing, it still works out that taken together, L2C ends up having the same overall chip rate as L1 C/A code, 1.023 MHz. This provides separation from the M-code. L2C has better autocorrelation and cross-correlation protection than the C/A

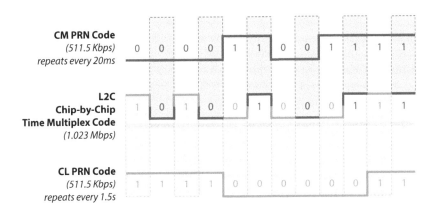

FIGURE 8.8   L2C Signal Structure

code because both of the CM and CL codes are longer than the C/A code. Longer codes are easier to keep separate from the background noise. In practice, this means these signals can be acquired with more certainty by a receiver and that receiver can then maintain lock-on them more surely. This is especially useful in marginal situations where the sky is obstructed. There is another characteristic of L2C that pays dividends when the signal is weak. CM carries the newly formatted 25 bits per second navigation data and is known as the *data channel*, but CL does not. It is the *pilot signal*. This is an idea that harks all the way back Project 621B at the very beginnings of GPS. It was known in the 1960s and 1970s that a pilot signal would support longer integration, especially when reception was weak, but it was not implemented until recently.

Even though the L2C signals' transmission power is 2.3 dB weaker than C/A on L1 and even though it is subject to more ionospheric delay than the L1 signal, L2C is still much more user friendly. The long data-less CL pilot signal has 250 times (24 dB) better correlation protection than C/A. This is due in large part to the fact that the receiver can track the long data-less CL with a phase-locked loop instead of a squaring *Costas loop* as is necessary to maintain lock-on CM, C/A, and P(Y). This allows for improved tracking from what is, in fact, a weaker signal and a subsequent improvement in protection against continuous wave interference. To illustrate how this works in practice, here is one normal sequence by which a receiver would lock onto L2C. First, there would be acquisition of the CM code with a frequency locked or Costas loop, next, there would be testing of the 75 possible phases of CL, and finally acquisition of CL. The CL, as mentioned, can be then tracked with a basic phase-locked loop. Using this strategy, even though L2C's transmission power is weaker than C/A, there is an improvement in the threshold of nearly 6 dB by tracking the CL with the phase-locked loop. Compared to the C/A code, L2C has 2.7 dB greater data recovery and 0.7 dB greater carrier tracking.

L5

L5, the new carrier, was first broadcast on Block IIF and subsequent satellites. It is centered on 1176.45 MHz, 115 times the fundamental clock rate. As you see from Figure 8.9, the basic structure of L5 looks like that of L1. There are two PRN codes on this 20 MHz carrier. The two codes are modulated using quad phase shift key (QPSK), and they are broadcast in quadrature to each other. However, borrowing a few pages from the developments just mentioned, the in-phase (L5I) signal carries data while the quadraphase signal (L5Q) is data-less. L5 also utilizes chip-by-chip time multiplexing in broadcasting its two codes, as does L2C in broadcasting CM and CL. Both L5 codes have a 10.23 MHz chipping rate, the same

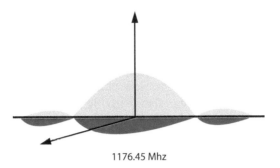

1176.45 Mhz

FIGURE 8.9   L5

as the fundamental clock rate. This is the same rate that has been available on the P(Y) code from the beginning of the system. However, this is the fastest chipping rate available in any civilian code. L5 has the only civilian codes that are both ten times longer and ten times faster than the C/A code. Since the maximum resolution available in a pseudorange is typically about 1% of the chipping rate of the code used, the faster the chipping rate the better the resolution. L5 has about more power than the signals on L1. It also has a wider bandwidth than legacy civil signals. It does not carry military signals and achieves an equal power split between its two signals. In this way, L5 lowers the risk of interference and improves multipath protection. It also makes the data-less signal easier to acquire in unfavorable and obstructed conditions and L5 users benefit from its place at 1176.45 MHz within the ARNS frequency band 1164–1215 MHz. Therefore, it is not prone to interference with ground-based navigation aids and is available for aviation applications. While no other GPS signal occupies this band, L5 does share space with one of the Galileo signals, E5. L5 also incorporates FEC.

PRACTICAL ADVANTAGES

Great, so what does all that mean in English? Having two civilian signals, one on L1 (C/A) and one on L2 (L2C), makes it possible to model the ionosphere using code phase. In the past ionospheric modeling was only available to multi-carrier observations or by reliance on the atmospheric correction in the LNAV. The L2C signal also ameliorates the effect of local interference. This increased stability means improved tracking in obstructed areas like woods, near buildings, and urban canyons. It also means fewer cycle slips. So, even if it is the carrier-phase that ultimately delivers the wonderful high-positional accuracy on which we depend, the codes get us in the game and keep us out of trouble every time we turn on the receiver. In other words, our receivers have been combining pseudorange and carrier-phase observables in innovative ways for some time now to measure the ionospheric delay, detect multipath, do wide-laning, and so forth, but those techniques can be improved. While the old methods work, the results can be noisy and not quite as stable as they might be, especially over long baselines. It is cleaner to get the signal directly from the two clear civilian codes. The slow chipping rate, short code length, and low power of L1 C/A mean that it has the low correlation protection When a receiver is in an environment where it collects some satellite signals that are quite strong and others that are weak, such as inside buildings or places where the sky is obstructed, correlation protection is vital because the strong signal can block the collection of the weak signals. To avoid tracking the strong signal it does not want instead of a weak signal that it does the receiver is forced to test every single signal. This problem is much reduced with the more robust tracking available from L2Cs CL and L5s longer code lengths. It is also reduced with the L5 higher power and faster chipping rate. In short, the civilian codes on L2 and L5 have better cross-correlation protection and narrowband interference protection than L1 C/A. Nevertheless, there are a few obstacles to full utilization of the signals as they are still considered pre-operational as is the new L1C signal.

L1C ANOTHER CIVIL SIGNAL

Another civil signal broadcast by the Block III satellites is known as L1C (see Figure 8.10). As a result of a June 2004 agreement between the United States and the European Union (EU), this signal is broadcast by GPS and Galileo's L1 OS signals. It is also broadcast by the Japanese

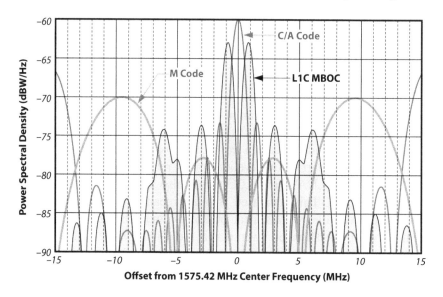

FIGURE 8.10   L1 with M-Code and L1C

Quasi-Zenith Satellite System (QZSS), the Indian NavIC, and China's BeiDou system. These developments open extraordinary possibilities of improved accuracy and efficiency when one considers there may be a combined GNSS constellation of 50 or more satellites all broadcasting the same civilian signal. This is possible because each of these different satellite systems utilizes carrier frequencies centered on the L1, 1575.42 MHz. Perhaps that has something to do with the fact that L1, having the highest frequency, experiences the least ionospheric delay of the carrier frequencies. There are many signals on L1. Looking at GPS alone, there is the C/A code, the P(Y) code, the M-code, and now L1C. To meet the challenge of introducing yet another code on the crowded L1 frequency and still maintain separability, the L1C design shares an M-code characteristic, i.e., BOC. It also has some similarities with L2C. For example, it has a data-less pilot signal ($L1C_P$) and a signal with a data message ($L1C_D$) whose codes are of the same length as CM on L2C 10,230 chips and are broadcast at 1.023 Mbps. The data signal uses BOC (1, 1) modulation, and the pilot uses time multiplexed binary offset carrier (TMBOC). The TMBOC is BOC (1, 1) for 29 of its 33 cycles but switches to BOC (6, 1) for 4 of them. The pilot component has 75% of the signal power and the data portion of the signal has 25%. This means that L1C has good separation from the other signals on L1 and a good tracking threshold. A receiver can reach its first fix to the satellite broadcasting L1C faster.

THE NEW NAVIGATION MESSAGES

Remember how L2C's CM data channel carries a new navigation message? That message is called L2-CNAV. The navigation message data carried on L5I is known as L5-CNAV. Though their content may vary somewhat, the structure of both of these data message is the same (see Figure 8.11).

However, L2-CNAV is sent at 25 bits-per-second and L5-CNAV is broadcast at 50 bits-per-second like the LNAV. In any case, they are both improvements over LNAV. For example, you may recall that a complete transmission of the LNAV almanacs required 12.5 minutes. L2-CNAV can send almanacs for the full constellation twice in the same amount of time. L5-CNAV can send all that in about 6 minutes. These capacities allow a receiver to get a faster first fix. LNAV can support 32 satellites. L2-CNAV and L5-CNAV can both support of 32 satellites with 75% or less of their bandwidth. They have the capability to support 63 satellites. Earth orientation parameters, more precise orbits and more compact almanac data are available to the receivers faster and are repeated more often by L2-CNAV and L5-CNAV than the LNAV. Both have a packetized architecture comprised of 12-second 300-bit message packets. One of every four of these packets includes clock data; two of every four contain

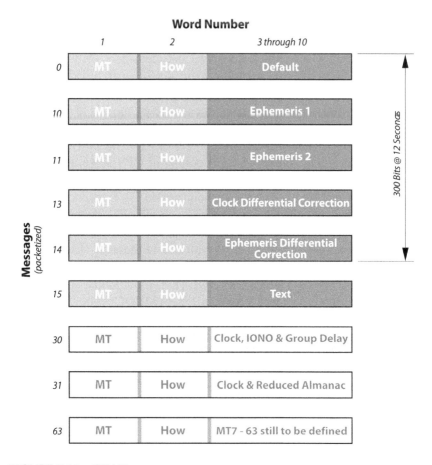

FIGURE 8.11   CNAV

ephemeris data, and so on. This flexibility facilitates faster data collection and delivers more timely data. They include information about the time offset between GPS and other GNSS, such as GLONASS and Galileo, allowing for interoperability with other global time-transfer systems. They include corrections that separate clock and ephemeris errors from each other. There is a flag in every packet that can alert a user within 6 seconds if the satellite data is unreliable. This will be a substantial benefit to safety-of-life applications when L2C and L5 are declared fully operational. Only a fraction of the available packet types is in use at this point. There could be packets containing differential corrections like that available from satellite-based augmentation systems (SBAS) and other useful things in the future.

There is another new navigation message called CNAV-2 message. It is carried by L1C on the Block III satellites. As described earlier, the frames and subframes of the LNAV message are repeated in fixed patterns. CNAV-2 is also organized in frames, but each of its frames is divided into three subframes of different lengths. The first contains a time interval identifier, the second subframe includes clock and ephemerides data, and the third subframe changes from frame to frame. Sending a complete CNAV-2 data message requires multiple frames. Similar navigation message models have been adopted by Galileo, BeiDou, and QZSS with adaptations to accommodate their time and reference systems and some variation in the representation of the individual quantities. CNAV-2 is conveyed at 50 bits-per-second like the LNAV message, but a satellite will most often achieve the first fix faster using CNAV-2.

In order to match GPS as much as possible, the legacy and new GPS navigation messages are broadcast by the Japanese QZSS. The Legacy Navigation (LNAV) messages L2-CNAV, L5-CNAV, and CNAV2 on their signals have necessary differences in orbital parameters, almanacs, ephemerides, etc. Like MNAV, L2-CNAV, L5-CNAV, and CNAV2 can achieve better error rates with FEC.

GPS EVOLUTION

New spacecraft with better electronics, better navigation messages, and newer and better clocks are just part of the story. Beginning with the launch of the first IIR-M satellite, new civil signals began to appear, starting with L2C followed by L5 on the Block IIF satellites and L1C on Block III (see Figure 8.12). These signals tend to have longer codes,

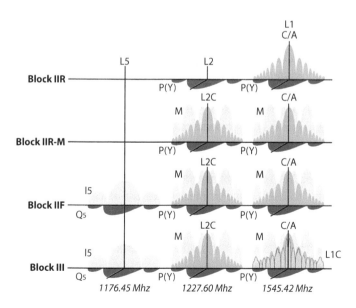

FIGURE 8.12 GPS Signal Evolution. (Adapted from Lazar, Steve, Crosslink [Summer 2002], Aerospace Corporation.)

faster chipping rates, and more power than the C/A and P(Y) codes have. In practical terms, these developments lead to faster first acquisition, better separation between codes, reduced multipath, and better cross-correlation properties. These improvements have made it possible to announce the planned retirement of the venerable legacy P(Y) code. An announcement in the U.S. government's 2019 Federal Radionavigation Plan included a caution concerning GPS receivers that achieve dual-frequency capability by exploiting aspects of the P(Y) signal on L2 today. They will no longer have that option in a few years. The P(Y) signal will be retired. The good news is that the same announcement commits the government to maintaining the existing GPS L1 C/A, L1 P(Y), L2C, and L2 P(Y) signal characteristics that enable codeless and semi-codeless GPS access until at least two years after there are 24 operational satellites broadcasting L5 with fully functional navigation messages. This is estimated to occur in 2027. Nevertheless, it is a good idea to plan for the retirement of P(Y).

GLOBAL NAVIGATION SATELLITE SYSTEM

The GPS system is one component of the worldwide effort known as the GNSS. Another component of GNSS is the GLONASS system of the Russian Federation, a third is the Galileo system administered by the EU, and the fourth is the Chinese BeiDou Satellite Navigation and Positioning System. Regional constellations are sometimes included in GNSS definition, such as the Japanese QZSS and the Navigation with Indian Constellation or NavIC system. All are positioning, navigation, and timing systems.

Their scope and effectiveness are increased by both *ground-based augmentation systems* (*GBAS*) and *space-based augmentation systems* (*SBAS*). GBAS include infrastructure that collect, calculate, and then broadcast GNSS differential corrections via very high-frequency (VHF) data links and VHF Data Broadcast (VDB), to assist aircraft in approach and landing at airports. SBAS are differential GNSS services that rely on a network of satellite monitoring stations to track constellations, collect data, and upload to geostationary satellites that, in turn, broadcast navigation messages to users. One immediate effect of GNSS is the substantial growth of the available constellation of satellites, the more signals that are available the better. The objectives are interoperability and compatibility. Interoperability is the idea that properly equipped receivers will be able to obtain useful signals from all available satellites in all the constellations and have their solutions improved rather than impeded by the various configurations of the different satellite broadcasts. Compatibility refers to the ability of the various GNSS Positioning, Navigation, and Timing (PNT) services to be used separately or together without interfering with each other.

GLONASS

SATELLITES

Russia's *Globalnaya Navigationnaya Sputnikovaya Sistema* (*Global Orbiting Navigation Satellite System*), known as GLONASS, did not reach full operational status before the collapse of the Soviet Union. Its first Uragan (Hurricane in Russian) satellites reached orbit in October 1982, a bit more than 4 years after the GPS constellation was begun. There were 87 Uragan satellites launched and a nearly full constellation of 24 made up of 21 satellites in 3 orbital planes, with 3 on-orbit spares, in 1996. However, only about 7 healthy satellites remained on orbit about 1000 km lower than the GPS satellites in 2001. They were only expected to have a design life of 3 years. The situation was not helped by the independence of Kazakhstan, subsequent difficulties over the Baikonur Cosmodrome launch facility, and lack of funds. The system was in poor health when a decision was taken in August 2001 outlining a program to rebuild and modernize GLONASS. Improvements followed. Today, Russia's GLONASS is operational and has worldwide coverage. A complete GLONASS constellation of improved satellites is in place at an altitude of 19,100 km inclined 64.8 degrees toward the Equator (see Figure 8.13).

Compared with the original Uragan satellites, the Uragan-M, which comprise most of the current GLONASS constellation, have several improvements (see Figure 8.14). They have a pressurized container. They have an increased lifetime of 7 years, improved solar array orientation,

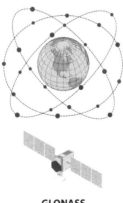

GLONASS

3 Orbital planes
24 Satellites + 3 spares
64.8° Inclination angle
Altitude: 19,100 km

FIGURE 8.13   GLONASS Constellation

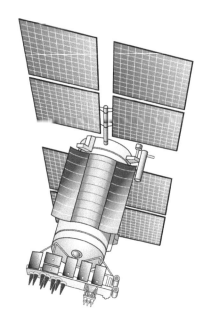

FIGURE 8.14   GLONASS Uragan-M

better clock stability, and better maneuverability. They are three-axis stabi-
lized and have three onboard cesium clocks. Launched between 2003 and
2014, they were augmented with an L2 frequency for civilian users in 2004.

Introduced in 2011 the GLONASS K1 satellite is lighter than the
Uragan-M. It has an unpressurized bus, a 10-year service life and costs
less to produce (see Figure 8.15). They have two cesium and two rubidium

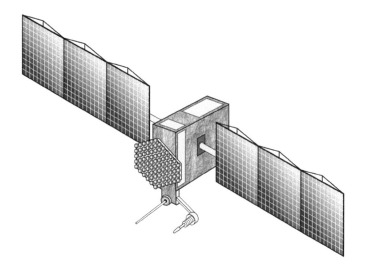

FIGURE 8.15   GLONASS K1

clocks onboard a transmitter for a third L-band civilian signal. They carry the international search and rescue COSPAS-SARSAT distress alert detection and routing system. They have a capability for calibration and remote clock synchronization via laser ranging, a radio cross link for data exchange, and an optical cross link. As the Uragan-M satellites age, they will be replaced by the smaller GLONASS-K1 satellites, which will be followed by further improved versions known as GLONASS-K2 and later GLONASS-KM.

Code Division Multiple Access

A receiver collecting signals from GPS, or from most other GNSS constellations for that matter, collects a unique segment of the PRN codes from each satellite. For example, a particular segment of the 37-week-long P(Y) code is assigned to each GPS satellite; i.e., SV14 is so named because it broadcasts the 14th week of the P(Y) code. Also, each GPS satellite broadcasts its own completely unique segment of the C/A code. Even though the segments of the P(Y) code and the C/A code coming into a receiver on L1 are unique to their satellite or origin, they all arrive at the receiver on the same frequency, 1575.42 MHz. The same is true of the codes on GPS L2. They all arrive at the same frequency, 1227.60 MHz. This approach is known as *Code Division Multiple Access (CDMA)*.

CDMA technology was originally developed by the U.S. military during World War II to secure communications in the presence of jamming. CDMA allows all the satellites in a GNSS constellation to broadcast on the same frequency (or frequencies) and still ensures the receivers can uniquely pick out each satellite's signal from the others. While the satellites share the common carrier frequency (or frequencies), each has its own distinct PRN code. A satellite's PRN code is modulated onto the common frequency (or frequencies). It is the exclusiveness of that code that allows the receiver to differentiate a particular satellite's signal from all the others. In short, the technique serves many simultaneous users over the same frequency (or frequencies). CDMA does not use frequency channels or time slots. CDMA usually involves a narrow band message multiplied by a wider bandwidth PRN signal. The increased bandwidth is wider than is necessary to broadcast the data information and is called a spread spectrum signal. To make this work, it is important that each of the PRN codes have high autocorrelation and low cross-correlation properties. High autocorrelation promotes efficient de-spreading and recovery of the unique code coming from a particular satellite which includes matching it with the PRN code available for that satellite inside the receiver. Low cross-correlation means that the autocorrelation process for a particular satellite's signal will not be interfered with by any of the other satellite's signals that are coming in from the rest of the constellation at the same time.

Frequency Division Multiple Access

GLONASS uses a different strategy. All the satellites in the constellation transmit the same codes, but each transmits them on its assigned frequency. The satellites all transmit L-band signals, but unlike GPS, each code a receiver collects from any one of the GLONASS satellites is the same. Also, unlike GPS, each GLONASS satellite broadcasts its codes at its own unique frequency. This is known as *Frequency Division Multiple Access* (*FDMA*) (see Figure 8.16). It ensures signal separation known as an improved Spectral Separation Coefficient (SSC). However, the system does require more complex hardware and software than CDMA.

The three GLONASS L-bands have a range of frequencies to assign to satellites. There are three intervals. The first is between ~1598.0625 and ~1607.0625 MHz in which the separation between each individual carrier is 0.5625 MHz. The second is between ~1242.9375 and ~1249.9375 MHz with separation between the individual carriers of 0.4375 MHz. The third is L3. It is close to the Galileo E5b frequency. It is from 1201.743 to 1208.511 MHz and is available on the K satellites. There is a separation between these individual carriers of 0.4375 kHz. However, within those ranges, there can be up to 25 channels of L-band signals; currently, there are 16 channels on each to accommodate the available satellites.

Like most GNSS constellations, GLONASS has two services; the standard service is open and unencrypted, analogous to the GPS C/A code. The high-precision, aka pinpoint, service like the GPS P(Y) code is for authorized, primarily military, users only. The OS is available on three carriers, L1, L2, and L3, and the high precision on two, L1 and L2. All GLONASS satellites broadcast existing FDMA signals as has been done

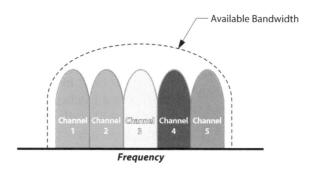

FIGURE 8.16   Frequency Division Multiple Access (FDMA)

since the launch of the first GLONASS satellite in order to ensure backward compatibility with legacy receivers still in use, but there is a CDMA OS signal emanating from GLONASS too. It is centered at 1202.025 MHz. GLONASS-K1 began transmitting this additional CDMA signal on L3 in 2011. Today, the 3 K-satellites in the constellation and the M satellites launched since 2014 can transmit a CDMA signal on L3, which is called L3OC. It has a familiar structure. There is a data component and a pilot component which have equal power and are in phase quadrature with each other. There are plans for the GLONASS K2 satellites to feature a full suite of modernized CDMA signals in the existing L1 and L2 bands.

There may be some changes to the FDMA itself in the future. Because it is difficult to accommodate 24 satellites with 16 channels, Russia has altered the architecture a bit. GLONASS will now assign the same frequency to antipodal satellites. Those are satellites that are in the same orbital plane but are always on opposite sides of the Earth from each other, and so will never be above an observer's horizon at the same time. This not only reduces the amount of the radio spectrum used by GLONASS, but it may improve its broadcast ephemeris information. Utilizing so many frequencies makes it difficult to accommodate the wide variety of propagation rates and keep the ephemeris information sent to the receivers within good limits. In other words, using the same codes for each satellite means that FDMA is better than CDMA at limiting narrow-band interference, but FDMAs use of multiple frequencies can increase the variations in the group delay within GLONASS receivers. Therefore, it is good that the GLONASS-K1 satellites transmit not only both the open and authorized FDMA on L1 and L2 but also the open CDMA signals on L3. In the future, CDMA may be broadcast by the GLONASS-K2 and GLONASS-KM satellites on L1 and L2 too.

As mentioned in Chapter 4, Russia has established an SBAS for real-time and precise point positioning (PPP) service over the whole of the Russian Federation. It is known as SDCM, which stands for System for Differential Correction and Monitoring. It is of note that the SDCM reference stations are equipped with multifrequency receivers that track, monitor the integrity of, and provide augmentation for both GPS and GLONASS satellites. To achieve worldwide coverage, ROSCOSMOS, Russia's State Corporation for Space Activities, claims there are 53 SDCM ground stations 46 in Russia, 3 in Antarctica, and 1 each in Minsk, Belarus; Erevan, Armenia; Nur-Sultan and Kizilorda, both in Kazakhstan, as of this writing. These stations have hydrogen maser atomic clocks and links for real-time data transfer. Integrity and correction data are currently relayed over the L1 band through the transponders on two geostationary satellites, Luch-5B and Luch 5V. Users access it via the Internet.

CONTROL SEGMENT

The GLONASS Ground Control Center and Time Standards are near Moscow. The System Control Center (SSC) is in Krasnoznamensk from which the ground segment is coordinated. The five TT&C stations are well distributed across the country from St Petersburg and Schelkovo in the west, Yeniseysk in the center, to Ussuriysk, and Komsomolsk in the east. They collect the status of the GLONASS satellites, send control commands, and do two-way ranging for orbit determination. Three of them, Schelkovo, Yeniseysk, and Komsomolsk, are collocated with uplink stations and there are two more at Vorkuta and Petropavlovsk. Three times a day orbit and clock data are uploaded from these five dual antenna stations to the GLONASS satellites. The many monitoring stations distributed over the Russian territory, the former Soviet Union territories, and elsewhere collect the pseudorange and carrier-phase measurements from which orbit integrity, clock, and satellite performance are evaluated. Other distance measurements are contributed by the nine SLR stations. GLONASS satellites have carried retroreflectors from the outset, so these laser measurements have informed orbit determination, validated accuracy estimates and added to the realization of the GLONASS reference frame from the very beginning of the constellation. In the past, this capability has alleviated the disadvantage of the limited geographical distribution of the GLONASS ground stations somewhat. However, their distribution is expanding. One station was installed in Brasilia in 2013, and more are planned in Brazil. There is a GLONASS station in South Africa, and plans to install them in other countries such as India, Angola, Indonesia, Central, and South America.

Since January 13, 2014, Parametry Zemli 1990 (PZ-90.11) has been the model of the Earth from which the GLONASS ephemerides are calculated. It includes the reference ellipsoid, gravity field factors, transformation parameters to other reference systems and a right-handed three-dimensional geocentric coordinate system. In other words, it conforms to the recommendations of the International Earth Rotation and Reference Systems (IERS) and Bureau International De l'Heure (BIH), as they were described in Chapter 5. It is a realization of the International Terrestrial Reference System (ITRS) at epoch 2010.00 with a zero-rotation rate with respect to the ITRF2008.

The GLONASS FDMA OS navigation message is comprised of five frames and strings. There are 15 strings in each frame. The first four strings deliver the transmitting satellite's ephemeris, data the receiver needs for positioning. That information is repeated every 30 seconds. The rest of the strings in the frame provide almanac data for up to five satellites. The five frames taken together are known as a superframe. Superframe transmission takes 2.5 minutes, and it is continuously repeated. The whole ephemeris is updated every 30 minutes. However, unlike the Keplerian orbital information in the GPS navigation messages, the GLONASS orbits, position, and velocity are given in ECEF three-dimensional Cartesian coordinates with acceleration corrections for lunar and solar perturbation. The GLONASS FDMA Navigation Message also differs from the GPS navigation messages in that it does not include corrections for ionospheric delay, so single-frequency receivers cannot rely on it for help with that bias.

TIME MANAGEMENT

GLONASS System Time (GLST) began on January 1, 1996, and it has been 3 hours ahead of UTC since. This makes sense because the realization of UTC on which it is based, UTC (SU), SU for the Soviet Union, is monitored and corrected by the Main Metrological Center of Russian Time and Frequency Service (VNIIFTRI) in Mendeleevo, near Moscow, which is 3 hours ahead of Greenwich. The four hydrogen frequency standards of the GLONASS Central Synchronizer (CS) time are the foundation of GLST. The GLONASS-M satellite clocks are kept within 8 nanoseconds of GLST and 30 nanoseconds of GPS Time (GPST) less the integer leap seconds. In fact, improving the stability of the GLONASS satellites' onboard clocks with precision thermal stabilization is among the efforts underway to improve the whole system's accuracy. It has improved their stability from 5×10^{-13} to 1×10^{-13} over 24 hours.

However, please recall there are no leap seconds introduced to GPST. The same may be said of Galileo and BeiDou. However, things are different in GLONASS. GLST is aligned directly to UTC (SU), leap seconds and all. It is the only GNSS time system that includes them. That means that leap seconds are incorporated into the standard. So, there is no constant integer-second difference between GLST and UTC as there is with the other GNSS. Nevertheless, conversion from GLST to UTC (SU) and GLST to GPST is provided in a string of each frame of the GLONASS navigation message.

GALILEO

The Galileo Joint Undertaking (GJU) was established in 2002 by the European Commission (EC) and the European Space Agency (ESA) to oversee Galileo's development phase. Its duties were transferred to the European GNSS Agency (GSA) in 2007. Today, the Galileo system, with governance divided between the GSA, the EC, and the ESA, is approaching full operational status and is intended to have worldwide coverage.

It will be compatible with the other GNSS constellations. It is certainly compatible with GPS. Galileo and GPS share two carriers, GPS L1 is aligned with Galileo E1 at 1575.42 MHz, and GPS L5 is aligned with Galileo E5a at 1176.45 MHz. They both use CDMA modulation. BPSK is used on Galileo E5a just as it is on GPS L5. GPS System Time (GPST) and Galileo System Time (GST) are very similar. Both of their navigation messages are based on reference frames that are tightly aligned to ITRF, GPS WGS84 (G2139); Galileo Terrestrial Reference Frame (GTRF). In short, they are interoperable. They work together efficiently without degrading the performance of either, which is a very practical benefit for users.

SATELLITES

Just over a dozen years after the idea was first proposed, the work on Galileo culminated in the ESAs establishment of the early ground segment and the launch of Galileo In-Orbit Validation Experiment-A (GIOVE-A) on December 28, 2005. The name GIOVE, Italian for Jupiter, was a tribute to Galileo Galilei, discoverer of Jupiter's moons. The first follow-on satellite GIOVE-B was launched in 2008. GIOVE-B was more like the satellites that eventually came to comprise today's Galileo constellation than was GIOVE-A. One of the motivations for launching GIOVE-A and GIOVE-B was to prevent the right to the frequencies that had provisionally been set aside for Galileo by the International Telecommunications Union from expiring. They did their job and bought time for Europe to build additional satellites without facing a confiscation of its frequency reservations. The experimental satellites also facilitated investigation of the transmitted Galileo navigation signals, provided measurements of the radiation environment, and demonstrated the hydrogen maser frequency standard and other technologies. After that, both satellites were moved to higher altitudes and retired in 2012.

Two In-Orbit Validation (IOV) satellites were launched from French Guiana on October 21, 2011, and two more on October 12, 2012. They are Galileo-IOV PFM (GSAT0101, Thijs), Galileo-IOV FM2 (GSAT0102, Natalia), Galileo-IOV FM3 (GSAT0103, David), and Galileo-IOV FM4 (GSAT0104, Sif). The first ground location using the four satellites, the required minimum, was accomplished at the ESA's Navigation Laboratory in Noordwijk, the Netherlands, on March 12, 2013. Three of the four original IOV satellites are still in service as of this writing. The fifth and sixth Galileo satellites were also launched from French Guiana in 2014. The satellites initially entered elliptical orbits instead of the required circular orbits. These IOV Galileo satellites have been launched aboard Russian Soyuz rockets and it is possible that the orbital difficulty was caused by a malfunction of the third stage of the Soyuz launch vehicle. There were

Galileo
3 Orbital planes
24 Satellites + 6 spares
56° Inclination angle
Altitude: 23,222 km

FIGURE 8.17   Galileo Constellation

11 launches of Galileo satellites in the decade from 2011 to 2021. The 11th launch on December 5, 2021, carried the first pair of Galileo's Batch 3 satellites. They completed their in-orbit testing (IOT) at the end of April 2022. The plan is for there to be 30 Galileo satellites on orbit at a nominal height of about 23,222 km above the Earth in 3 planes, 10 in each plane (Figure 8.17). As of this writing, there are 23 Galileo satellites on orbit and useable, and each of them has two clocks, one rubidium and one passive hydrogen maser (PHM), which are operated in parallel. During maintenance or failure, the ground segment can switch between them.

Signals and Services

The Galileo signals are E1, which is aligned with GPS L1 at 1575.42 MHz; E6 at 1278.75 MHz; and E5 at 1191.795 MHz. Each of these has a pilot component, which is data-less, and a data component (see Figure 8.18). The pilot part enhances correlation and allows longer integration just as it does on L2C and L5 in GPS. E5 has two sidebands which are 15.345 MHz above and below the E5 carrier frequency carrier itself. These sidebands are called E5b on 1207.14 and E5a on 1176.45 MHz. In other words, there are two carrier frequencies within the E5 band. However, for excellent multipath protection, it can also be tracked as one quite large band. It is worth noting that E5a is aligned with GPS L5 and like that signal is intended to provide higher precision and higher availability than others. Any discussion of interoperability between GPS and Galileo must consider these overlapping signals because it is helpful that the signals center on

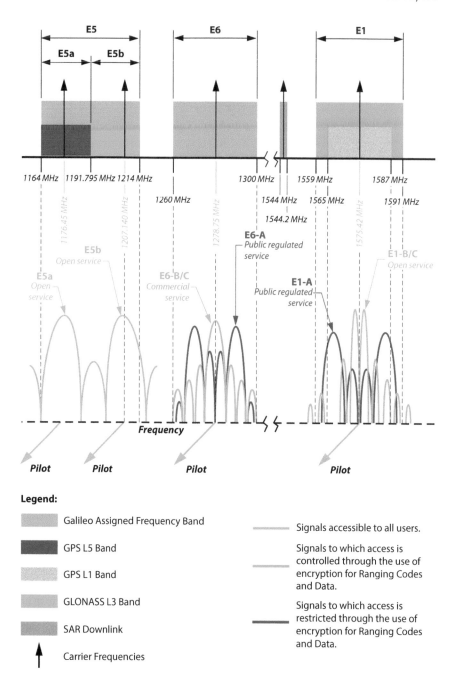

FIGURE 8.18 GALILEO–GPS–GLONASS Signals (Courtesy of Hein, Guenter W., Director of the Institute of Geodesy and Navigation, University FAF Munich.)

the same frequency if they are to be used in a combined fashion and there is more signal sharing. The third GLONASS civil reference signal on L3, available from their K-satellites, will be within a new frequency band that includes 1201.743–1208.511 MHz and will overlap Galileo's E5b signal. Also, Galileo's E6 signal shares the 1278.75 MHz band with QZSS's L6 signal and transmits a free augmentation message for a high-accuracy service (HAS) used to provide PPP.

The Galileo satellites broadcast signals in several frequency ranges including 1176–1207 MHz, near GPS L5 (see Figure 8.18). As mentioned earlier, Galileo's E5a signal is centered exactly at 1176.45 MHz as is GPS L5 and both Galileo's L1 and the GPS Ll frequencies are centered at 1575.42 MHz. There the GPS signal is based on the BPSK and the Galileo signal is accomplished with the BOC method.

An important characteristic of BOC modulation is that the code's greatest power density is at the edges, at the nulls, which mitigates interference with the existing codes just as it did with the M-code on GPS. Since the spectral separation of GPS L1 and Galileo E1 signals is accomplished by the use of these different modulation schemes, interference between the codes on Galileo and GPS is avoided. Even though they overlap, they can be used together. This strategy also allows jamming of civil signals, should that prove necessary, without affecting either GPS or Galileo services. The fortuitous coincidences of frequencies between GPS and Galileo did not happen without discussion. As negotiations proceeded between the United States and the EU, one of the most contentious issues arose just as the EU was moving to get Galileo off the ground. They announced their intention to overlay Galileo's Public Regulated Service (PRS) code on the U.S. military's M-code. The possibility that this would make it difficult for the U.S. Department of Defense to jam the Galileo signal in wartime without also jamming the U.S. signal was considered. It became known as the M-code overlay issue. In June 2004, the United States and the EU reached an agreement that ensured Galileo's signals would not harm the navigation warfare capabilities of the United States and the North Atlantic Treaty Organization (NATO).

Galileo has defined levels of service that will be provided by the system. They include the OS, which like the OSs available from GPS and GLONASS is free for timing and positioning applications. It is available on the data-pilot pairs on E5a, E5b, and E1-B/C, and since it is on separate frequencies, it presents the ability to reduce the ionospheric bias. The PRS is under government control. It is encrypted and restricted to public security and civil authorities. The PRS provides significant jamming protection because it has a wider bandwidth than the OS signals. PRS is provided on E6-A and E1 for emergency services, law enforcement, intelligence services, and customs. The Commercial Service (CS) will be an encrypted

added value service available on E6 B/C when Galileo reaches full opera
tional capability (FOC). The search and rescue (SAR) service adds Galileo
to the COSPAS-SARSAT effort already mentioned in the discussions of
GPS and GLONASS. Transponders on the satellites transfer the user's distress signals to the dedicated ground segment broadcast centers, i.e., SAR
control center in Toulouse, France, which initiate rescue operations. There
have been several exercises and real-life emergencies during which the
system has demonstrated success in determining locations, sending alerts,
and notifying users that help has been dispatched.

CONTROL SEGMENT

The ground command and control systems of the Galileo constellation are
divided in two. The Ground Control Segment (GCS) manages the functioning of the satellites and maintains the constellation. The Ground Mission
Segment (GMS), through its global networks of stations, measures and
administers the signal generation, calculates the navigation message and
distributes data back up to the satellites. There are ten Up-Link Stations
(ULS), two each at five individual sites. They communicate the calculated integrity, time, and other mission information to the satellites via the
C-band. The worldwide network of Ground Sensor Stations (GSS) collects
L-band range measurements and provides that data to the two centers, the
Fucino Control Centre in Italy and the Oberpfaffenhofen Control Centre
in Germany, for orbit determination, time synchronization, and signal
monitoring. There are six TT&C stations; Kourou, French Guiana; Kiruna,
Sweden; Nouméa, New Caledonia; Sainte-Marie, Réunion; Redu, Belgium
and Papeete, Tahiti that collect and communicate the satellite generated
telemetry. They are also responsible for sending the control commands to
the satellites. The continuity of these services and operations is ensured
by the redundancy built into the control center and the data dissemination
systems. The early operation control centers in Toulouse (French Space
Agency, CNES) and Darmstadt (ESA Operations Centre, ESOC) control
the Galileo satellites from the time they separate from the launch vehicle
until they reach their orbital slots. The satellite's payload health is then
verified by IOT, which is administered from a facility in Redu, Belgium.

TIME MANAGEMENT

There are similarities between GPST and GST. They are both linked to
UTC, but neither is subject to leap seconds, so they both had an initial
offset between themselves and UTC. That number changed, of course, the
moment more leap seconds were introduced into UTC. They are both continuous time systems like International Atomic Time (TAI) and have a

constant offset from TAI. There are also some differences between them. GPST began on UTC midnight (0hr) on January 6, 1980. GST started at UTC midnight (0hr) on August 22, 1999. GPS depends on UTC as delivered by UTC (USNO) and is always within 1 microsecond, one-millionth of its rate, and usually less than 10–20 nanoseconds. The Galileo Time Service Provider (GTSP) steers GST to within 50 nanoseconds, 50 billionths of a second, of UTC as averaged from five European timing laboratories. Galileo depends on the collaboration of UTC (PTB) Istituto Nazionale Di Ricerca Metrologica (INRiM), l'Observatoire, National Physical Laboratory, SP, INRIM (Istituto Nazionale di Ricerca Metrologica), OP (LNE-SYRTE, Observatoire de Paris), PTB (Physikalisch-Technische Bundesanstalt), ROA (Real Instituto y Observatorio de la Armada). GST is maintained by the hydrogen masers and cesium atomic clocks of the two redundant Galileo Precise Time Facilities (PTFs), one in Fucino, Italy, and one in Oberpfaffenhofen. Interoperability between GPS and Galileo is facilitated by the Galileo navigation message carrying parameters to shift between GPST and GST. It also provides GST to UTC information.

BeiDou

In the early 2000s, China considered collaborating in the EU led GNSS project that became Galileo but ultimately decided to build their own. So now the fourth global GNSS system, joining those undertaken by the United States—GPS, Russia—GLONASS, and Europe—Galileo, is the Chinese BeiDou system, named after the Big Dipper (see Figure 8.19).

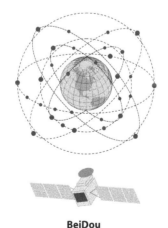

BeiDou
7 GEO + 27 MEO + 10 IGSO = 44 Satellites
55° Inclination angle
Altitude: 36,000 km, 21,500 km

FIGURE 8.19   BeiDou Constellation

It provides position, navigation, and timing (PNT) and Search and Rescue (SAR) services worldwide. BeiDou's development has been through three steps. Each step increased the coverage of their Satellite Navigation system.

BDS-1

The first generation was a territorial system known as BeiDou-1 (BDS-1). It was a regional satellite navigation system that served the portion of the Earth from 70°E longitude to 140°E longitude and from latitude 5°N to 55°N. It relied on three satellites with one backup. The first satellites were launched into geostationary (GEO) orbits in late 2000; BeiDou-1A (aka BeiDou-West) at 140°E longitude and BeiDou-1B (aka BeiDou-East) at 80°E longitude. Three years later, a third satellite, BeiDou-1C, joined them at 110.5°E longitude. With the launch of the fourth, BeiDou-1D in 2007, the first BeiDou-1 system was operational, but the territorial system was decommissioned at the end of 2012.

BDS-2

The second step was the regional BeiDou-2 (BDS-2) generation of the BeiDou Navigation Satellite System. It was developed between 2004 and 2012, became operational in 2011, and served the whole Asian-Pacific region. The first BeiDou-2 satellite, named Compass-M1, was placed in a MEO like those in GPS, Galileo, and GLONASS constellations at an inclination of 55.5°, very like the GPS constellation. More MEO satellites followed between 2007 and 2012, with sequential names from Compass-M1 through M6 (without an M2) launched. During the period from 2009 to 2012, 6 Geosynchronous Equatorial Orbiting (GEO) BeiDou-2 satellites from Compass-GEO-1 to GEO-6 were launched. Their positions are at 58.75°E longitude, 80.0°E longitude, 110.5°E longitude, 140.0°E longitude, and 160.0°E longitude. From 2010 to 2011, the 5 High Earth Orbit (HEO) BeiDou-2 satellites from Compass-IGS01 to IGS05 were launched and arranged so that one of them is always over the Chinese region. They achieved an altitude of approximately 36,000 km. The acronym IGSO means inclined geosynchronous orbit satellites. The IGS01, IGS02, and IGS03 satellites are at ~120°E. IGS04 and IGS05 satellites are at ~95°E. All the IGSO satellites have an inclination of 55°. BDS-2 began trial operations in late 2011 and followed with service to a region bounded by 55°E longitude to 180°E longitude and latitude 55°S to 55°N.

BDS-2 Signals and Services

The BDS-2 carriers are centered on three frequencies B1 at 1561.098 MHz, B2 at 1207.14 MHz, and B3 at 1268.52 MHz (see Figure 8.20). Like GPS L5,

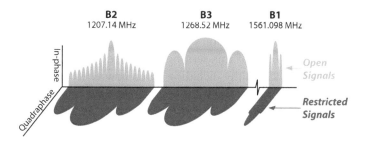

FIGURE 8.20   BDS-2 Bands

each carrier is modulated with two signals an in-phase (I) component and a quadraphase (Q) component. That means there are six signals generated at a 10.23 MHz chipping rate on these three carriers. The in-phase parts of the B1 and B2 signals, B1I and B2I, are open to civilians and include both a ranging code and a navigation message. These signals have some similarities with GPS. Their PRN signals are generated using BPSK just like the GPS C/A code. They utilize CDMA, and they carry a navigation message.

There are two service types. As mentioned, the OS signals are available to the public. They are shown in blue in Figure 8.20. They offer an autonomous, not differentially corrected, positional accuracy of 10 m, 0.2 m/s velocity accuracy, and timing accuracy within 20 nanoseconds. The Restricted, aka Authorized Service signals, shown in red in Figure 8.20, are not available to the public. They are for the military, specifically the militaries of China and Pakistan.

Only the B1I and B2I signals are official OS signals. Tracking of the B3I signal as an open signal is not officially endorsed, but its ranging code properties became known, and triple-frequency BeiDou receivers were built, which facilitated valuable surveying processing techniques and PPP with it. However, the B3I ranging code could be encrypted at some point. These signals have a minimum received power of −163 dBW. They also support a free near real-time regional SBAS in and around China.

The navigation message broadcast by the MEO/IGSO satellites is called D1. It is much like the GPS LNAV message. It is transmitted at 50 bps and includes the broadcasting satellite's information, almanacs for all satellites, and time offsets to other GNSS system clocks and UTC. It has 30-bit words and 10-word subframes, i.e., 300 bits, and capacity for almanac data for 30 satellites. The navigation message broadcast by the GEO satellites is called D2. It also has 30-bit words and 10-word subframes, but it is broadcast at 500 bps. The D2 message includes pseudorange corrections in subframes 2 and 3 with enough capacity to accommodate corrections for 18 satellites. D2 also has ionospheric corrections and clock corrections to other GNSS systems in subframe 5.

BDS-3

The third step, BDS-3 in BeiDou's evolution began in 2013 and was completed in June 2020. As of this writing, there are 44 operational BeiDou satellites on orbit and healthy including seven new GEO satellites and five held over from the regional BDS-2. The holdovers have the same positions they had before. The BDS-3 MEO and IGSO satellites also have the same orbits they had in BDS-2, an inclination of 55° and at altitudes of 21,500 and 36,000 km, respectively. There are 27 MEO in three orbital planes which supply the global *Radio Determination Satellite Services* (*RDSS*), Radio Navigation Satellite Service (RNSS), and Search and Rescue (SAR) services. The coverage of China and the Asia-Pacific region is augmented by ten IGSO satellites.

Like Galileo, BDS-3 satellites utilize two types of frequency standards, aka clocks. The MEO satellites are equipped with one to four PHMs as primary clocks and two RAFs secondary clocks. The Inclined Geosynchronous Satellite Orbiting (IGSO) and GEO satellites carry two PHMs and two RAFSs. Each BeiDou satellite carries a LRA, the GEO and IGSO satellites have 90 prisms, and the MEO satellites have 42. Each satellite also carries a C-band horn antenna, an S/L-band dish antenna, and phased array antennas for transmission of the signals in B1, B2, and B3 frequency bands.

BDS-3 Signals and Services

BDS-3 introduced five new global signals. All of them have a data in-phase component and a pilot component in quadrature, but just three of these latest signals are open; B1C, B2a, and B2b. They are shown in blue in Figure 8.21. The other BDS-3 signals shown in Figure 8.21, in red, are restricted. Perhaps the most important aspect of BDS-3s open signals are their compatibility and interoperability with GPS, GLONASS, and Galileo. For example, the in-phase data portion of B1C is on the carrier shared with GPS L1 and Galileo E1, 1575.52 MHz. The in-phase data portion B2a is on the carrier shared with GPS L5 and Galileo E5a, 1176.45 MHz.

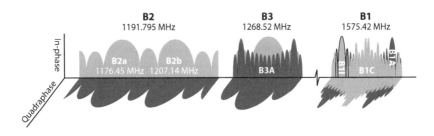

FIGURE 8.21   BDS-3 Bands

B2a is one side lobe of B2. The other side lobe is B2b. These three BDS-3 signals are transmitted by the MEO satellites and the IGSO satellites and are thereby available globally.

The data component of B2a is modulated with BPSK and includes a B-CNAV2 navigation data stream that is transmitted at 25 bps. The data component of B2b is modulated with BPSK and includes a B-CNAV3 navigation data stream with basic navigation information and integrity information, but it is broadcast at twice the data rate of B2a. The data component of B1C is modulated with a BOC and includes a B-CNAV1 navigation data stream.

The received power from B2a MEO transmissions is −156 dBW from MEO and −158 dBW from IGSO satellites. The received power from B2b MEO transmissions is −160 dBW from MEO and −162 dBW from IGSO satellites. The received power from B1C transmissions is −159 dBW from MEO and −161 dBW from IGSO satellites. In short, the BDS-3 MEO satellites have comparable signal strength code and phase accuracy as GPS, Galileo, and QZSS satellites and are appropriate for interoperability. However, the legacy open signals B1I and B3I, still being transmitted to ensure a smooth transition to BDS-3, are recommended for single-frequency services. For dual-frequency work note that B1C and B2a have better signal strength and higher code and carrier-phase accuracy than the legacy signals. For example, pairing the new BDS-3 signals, B1C/B2a, makes sense for that work.

BeiDou Message Services

BeiDou supports two different general services: RDSS and RNSS. They are broadcast simultaneously. The RDSS includes a satellite-based 2-way short message service through 3 GEO satellites equipped with L-band/C-band inbound transponders and C-band/S-band outbound transponders in China and the Asia-Pacific area (75° to 135°E and 10° to 55°N). The service allows 1000 Chinese characters within China and the Asia-Pacific region with an overall capacity of 10 million messages/hr. The global RDSS message service allows 40 characters and uses inter-satellite crosslinks for an overall capacity of 200,000 messages/hr through 14 MEO satellites.

BeiDou Augmentation Services

A wide area differential correction is available directly from the constellation, the BeiDou Ground-Based Enhancement System (BGBES). Information from a network of 150 reference stations is processed in the BeiDou Operation and Control Center Segment from which the calculated corrections are sent from the uplink stations to the BeiDou GEO satellites. Those corrections are broadcast via the previously mentioned D2 navigation message to the BeiDou user's receivers. In other words, the GEO satellites provide a satellite-based augmentation service (SBAS) called BDSBAS via B1C and B2a signals for users in and around China, i.e., between 70°E longitude to 145°E longitude and 5°N latitude to 55°N latitude. There are two service modes, single frequency with dual constellations and dual frequency with multiple constellations.

BeiDou's CONTROL/GROUND SEGMENT

The BeiDou Control/Ground Segment Master Control Station (MCS) is responsible for the operational control of the system, including orbit determination, navigation messages, and ephemerides. It is comprised of two-time synchronization and upload stations and a network of 7 Class-A and 22 Class-B widely distributed monitoring stations. There are BeiDou ground stations in Argentina, and plans to build them in Iran and Thailand. However, BeiDou is less dependent on a global network of stations than other GNSS because it has the capability to communicate telemetry and control between satellites with inter-satellite links. Like the control of other GNSS constellations, the MCS receives data from the monitoring stations' continuous constellation tracking, calculates the necessary data, and depends on the upload stations to send the data generated by the MCS to the satellites.

The BeiDou satellite's ephemerides are based on the China Geodetic Coordinate System 2000 (CGCS2000), which is the framework for the BeiDou Coordinate System, BDCS. BDCS shares the same reference ellipsoid with and is within a few centimeters of ITRF. It is a right-handed system which is in accord with the IERS specifications; origin at the Earth's center of mass, z-axis through the IERS Reference Pole (IRP), x-axis intersecting the IERS Reference Meridian (IRM), y-axis perpendicular.

BeiDou TIME MANAGEMENT

BeiDou's MCS also coordinates time synchronization with BeiDou Time (BDT). Like GPST and Galileo Time, BeiDou system time (BDT) accumulates continuously without leap seconds, though leap second information along with the offset between BDT and GPST/GST are available in the navigation message broadcasts. BDT began at 00:00:00 UTC on January 1, 2006. It is synchronized within 100 ns of UTC (NTSC) as maintained by the National Time Service Center, China Academy of Science.

The LRAs all BeiDou satellites carry play a significant role in time comparisons, laser time transfer (LTT), and precise satellite orbit determination. Using them, the system can measure the signal travel time between the ground station and the satellite within a few centimeters (sub-nanosecond) for very accurate monitoring and synchronization of the atomic frequency standards.

THE QUASI-ZENITH SATELLITE SYSTEM

SATELLITES

The Japanese QZSS is a regional GNSS that was authorized in 2002. The first demonstration satellite QZSS satellite, QSZ-1, was launched in 2010 by the Japan Aerospace Exploration Agency (JAXA) from the Tanegashima Space Center. It was nicknamed Michibiki-1, meaning guide, and was expected to have a design life of 10 years, which it exceeded. It was replaced by Michibiki-1R in 2021, which is now part of a 4-satellite constellation. There is 1 GEO satellite, QZ-4, at the equator. The three Inclined Geosynchronous Satellites Orbiting (IGSO) satellites. Each of the three IGSO satellites is on a highly inclined, slightly elliptical geosynchronous, but not geostationary, orbit.

A geosynchronous satellite can be inclined with respect to the equator, a geostationary orbit is on the same plane as the equator. In other words, the IGSO satellites do not stay in the same place in the sky from the point-of-view of those observing on the ground. They all follow an asymmetrical figure-8 pattern ground trace (see Figure 8.22), but along that route, each of them stays 120° apart from the other two. They orbit at ~32,000 to ~40,000 km, much higher than typical GNSS medium orbit satellites, and therefore, must have higher RF transmission power.

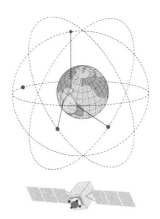

Quasi-Zenith Satellite System (QZSS)

1 GEO Satellite and 3 IGSO Satellites
~43° Inclination Angle

FIGURE 8.22   QZSS Constellation

QZSS is designed to augment the accuracy and reliability of GPS throughout the Asia-Oceania region, but particularly in Japan's mountains and urban canyons. Therefore, one of the QZSS satellites is always nearly directly overhead in Japan and dwells there for 8 hours, the origin of the name quasi-zenith name. JAXA plans to increase the 4-satellite constellation seven satellites in the future.

SIGNALS AND SERVICES

The system shares the same radio frequency, message structure, and format of GPS but has different PRN sequences. It transmits six signals. The first four are the familiar GPS positioning, navigation, and timing (PNT) signals, L1-C/A and L1C; at 1575.42 MHz; L2C at 1227.60 MHz; and L5 at 1176.45 MHz. These QZSS broadcasts were declared healthy in the summer of 2011 and have been so ever since. It is like there are more GPS satellites in the Asia-Oceania region.

The other two less familiar QZSS signals are for augmentation, one for submeter level code-phase positioning and the other for carrier-phase centimeter-level positioning. The L1-SAIF (Submeter-class Augmentation with Integrity Function) is broadcast on the L1 frequency, 1575.42 MHz. It is interoperable with GPS and provides a real-time Submeter Level Augmentation Service (SLAS) signal to code-phase single-frequency GPS L1 C/A users. The L1-SAIF error corrections are generated by the L1-SAIF Master Station (L1SMS) in real time based on the Japanese GPS Earth Observation Network System (GEONET) of more than 1200 locations. It is very like the Wide Area Augmentation System (WAAS) and is interoperable with other GPS-SBAS systems. The other augmentation signal, LEX (L-band Experiment), aka L6, has a high data rate, 2 kbps, and shares the 1278.75 MHz frequency with Galileo E6. It provides a regional open real-time Centimeter Level Augmentation Service (CLAS) to multifrequency carrier-phase users. LEX is used for surveying, machine control, precision farming, etc. It also supports a real-time GNSS orbit and clock estimation system named MultiGNSS Advanced Demonstration tool for Orbit and Clock Analysis—Precise Point Positioning (MADOCA-PPP) that began trial service in September of 2022. QZSS L6 can transmit that service to remote areas where there is no other access to corrections.

These QZSS SBAS services are somewhat unique in that they are not broadcast from geostationary satellites. Due to their highly inclined orbits, a QZSS satellite is always overhead in the region, which makes them a better means of distributing error-correction signals in the urban canyons and other difficult areas in the area than a geostationary satellite could be. QZSS offers several other services too. It has an encrypted PRS that provides users with heightened signal security. It has an early warning service that transmits short alerts of natural disasters, and a related messaging service relays user's safety confirmations. The messages travel via a GEO satellite to the control center, which forwards them to their ultimate destination by e-mail.

CONTROL SEGMENT

The QZSS MCS at the Tsukuba Space Center (TKSC) northeast of Tokyo develops the ephemerides, time, and navigation messages which are uploaded to the satellite constellation by the TT&C ground station which is collocated with the monitoring station on the main island in the Okinawa prefecture. The station is operated remotely from Tsukuba. The other QZSS monitoring stations on Japanese territory are at Ogasawara, Koganei, and Sarobetsu. There are five more in areas governed by other nations. They are in Hawaii, Guam, Bangkok, Bangalore, and Canberra. These stations receive the L-band navigation signals from both the QZSS and GPS constellation, and an independent receiver is in place for each. The collected data is subsequently transmitted to the MCS. The QZSS ground segment also includes laser ranging and tracking control stations (TCS). The main TCS station is collocated at JAXA's TKSC. These stations monitor the TT&C station, control reaction wheel unloading, orbit keeping, and issue commands for other routine operations. The QZSS orbital reference is the Japan Geodetic System (JGS). which is aligned with ITRF.

TIME MANAGEMENT

There are two independent RAFS on board the QZS-1 satellite which are identical to those in the GPS IIF satellites and are accompanied by a voltage-and oven-controlled crystal oscillator (VCOCXO). They are typically steered to follow the primary RAFS frequency but can also be controlled from the ground through the NICT's time transfer subsystem (TTS). That system relies on two-way satellite time and frequency transfer (TWSTFT). This process involves two NICT facilities, one at Koganei and another on Okinawa. Each station has an atomic clock, an antenna, and the equipment by which to track and compare the QZSS satellites' onboard clocks to their ground-based frequency standard time reference known as Quasi-Zenith Satellite System Time (QZSST). That standard is defined by the receiver clock of the Koganei station, which is connected to UTC (NICT) and thereby measures the offset between QZSST and UTC (NICT). It also compares it with the United States Naval Observatory (USNO) so that the offset between QZSST and GPST can be made available from the QZSS navigation message. The redundant TMS station at NICT's Subtropical Environment Remote-Sensing Center in Okinawa is not directly connected to UTC (NICT). Like GPS and Galileo time standards QZSST does not incorporate leap seconds. It is a continuous time scale that has a fixed −19 seconds offset from TAI.

NavIC

The building of India's GNSS, formerly IRNSS, is now known by the operational name Navigation with Indian Constellation, NavIC, a Sanskrit word for sailor or navigator (see Figure 8.23). It was authorized by the Indian government in 2006 and developed by the Indian Space Research Organization (ISRO). The system provides PNT service over India and the region approximately 1500 km around it, specifically the area from latitude 30°S latitude to 50°N latitude and from 30°E longitude to 130°E longitude.

SATELLITES

The space segment of NavIC is now comprised of eight satellites which are configured to be continuously overhead in the region. All the satellites are named with the prefix IRNSS-1 and were launched from the Satish Dhawan Space Centre at Sriharikota, India. Three of the eight satellites in the constellation are in geostationary orbits; IRNSS-1F is at 32.5°E longitude, IRNSS-1C at 83°E longitude, and IRNSS-1G is at 129.42°E longitude.

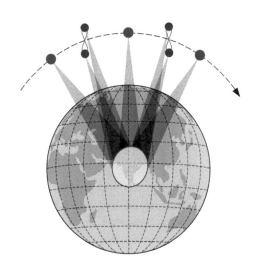

3 GEO satellites - Inclination angle < 5°
5 IGSO satellites - Inclination angle ~29°

FIGURE 8.23   NavIC Constellation

Five of the eight satellites are in geosynchronous orbits with an apogee of 24,000 km and an inclination with respect to the equator of 29°. The small inclination is appropriate to the coverage of the low latitudes of the region. The equator crossing of two of the geosynchronous satellites (IRNSS-1D and IRNSS-1E) is at 111.75°E longitude. The other two (IRNSS-1B and IRNSS-1I) cross it at 55°E longitude. IRNSS-1A launched back in July 2013 also crossed the equator at 55°E longitude, but its rubidium clocks failed in 2016. The following year, IRNSS-1H was launched to replace IRNSS-1A, but when it did not separate from its launch vehicle, it became unusable. Finally, a working replacement, IRNSS-1I reached orbit in 2018. IRNSS-1A remains somewhat useful but just supplies a short message service with which the command center can send urgent dispatches, such as weather warnings, to specific areas.

All the satellites carry rubidium clocks, corner cube retro-reflectors for laser ranging and C-band transponders. They broadcast in the L5 band (1176.45 MHz) with a bandwidth of 24 MHz. L5 was chosen because of the nearly complete consumption of L1 and L2 bands by other GNSS and to expedite interoperability with other constellations. The S-band (2492.028 MHz) with a width of 16.5 MHz was chosen to be NavIC's second frequency. It is less affected by the ionosphere than the L-band and it is not much populated since it was only recently designated by the ITU for navigation services. The ISRO plans to augment the current NavIC constellation with the launch of a new series of satellites named NVS-01 through NVS-05. These satellites will broadcast on both L5 and the S-band. They will also provide a new civil signal on the L1 band. The first of those launches will be to replace IRNSS-1G.

Signals and Services

Currently, NavIC provides two levels of service, a public Standard Positioning Service (SPS), which is available to all the users unencrypted and an encrypted Restricted Service (RS) only provided to the authorized user. Both are available on L5 and on the S-band. The SPS signal is modulated by BPSK at 1 MHz and the RS uses BOC (5, 2). The navigation signals are broadcast on the S-band.

Control Segment

The system's control segment's ranging, tracking, and commanding are facilitated by the continuous visibility of all the NavIC satellites in the region. The ISRO Navigation Center (INC) at Byalalu near Bangalore gathers the necessary data from a variety of distributed stations and then generates the navigation parameters to ensure proper navigation operations.

For example, the receivers in the IRNSS Range and Integrity Monitoring Stations (IRIMS) constantly track the constellation and send their one-way pseudorange and carrier-phase measurement data to the INC, where the atmospheric delays and other biases are determined. The IRNSS CDMA Ranging Stations (IRCDR) send the results of their two-way ranging measurements, which the INC use to estimate the orbit and validate the one-way range measurements. SLR Stations also send the measurements they made to the retroreflector arrays on the satellites for orbit validation. Given all these data streams, the INC processes range data, evaluates, validates, and predicts orbital parameters and clock parameters 24 hours in advance of the necessary uplink to the satellites. AutoNav data are also uplinked. The satellites can store 7 days of AutoNav, ephemeris, and clock parameters.

INC calculates the primary and secondary navigation parameters. A substantial part of the secondary navigation data is the parameters of grid-based ionospheric modeling. It also includes time offsets to other GNSS. All are uplinked to the satellites from the IRNSS Spacecraft Control Facility (IRSCF), which is comprised of two Spacecraft Control Centers (SCC) stations and the Spacecraft Control Earth Stations (SCES). The main SCC station at Hassan is backed up by a station at Bhopal. The IRSCF receives the navigation parameters from the INC, uploads them to the satellites, and performs Tracking, Telemetry, and Control (TT&C) operations. It also monitors telemetry, generates and encodes telecommands, and sends them to the SCES, which track the spacecraft, acquires their telemetry, and upload the telecommands.

TIME MANAGEMENT

IRNSS System Time is realized by an ensemble of atomic clocks, steered to, and maintained within 50 ns (2 sigma) of International Atomic Time (TAI). It is linked with UTC labs. IRNSS Network Time is generated, maintained, and distributed at the Byalalu IRNSS Network Timing Centre (IRNWT). The IRNSS Data Communication Network (IRDCN) provides reliable digital communication between the IRIMS, IRCDR, IRNWT, IRSCF, and INC through a network of terrestrial and satellite links.

PROGRESS

Much has changed since the early days of GPS. There are more than 120 GNSS satellites on orbit now from GPS, GLONASS, Galileo, BeiDou, QZSS, NavIC, and an ever-larger number of GNSS satellites on their way. Just speaking of global systems, the BeiDou constellation is complete, Galileo will shortly reach FOC and the legacy systems, GPS and GLONASS, continue to modernize. The regional systems, BDS-2, QZSS, and NavIC, keep on growing and developing. The availability of so many satellites and signals enables new applications in areas where their scarcity has been a hindrance in the past. For civil users, new signals provide more protection from interference, ability to compensate for ionospheric delays with pseudoranges, and wide-laning or even tri-laning capabilities. For military users, there is greater anti-jam capability and security. For everybody, there are improvements in accuracy, availability, integrity, and reliability. In a sense, the more satellites, the better the performance, particularly among trees and in urban canyons, those places where signal bounce, scatter, and multipath abound.

More augmentation systems are coming online, too, and those already in existence are upgrading to support multiple frequencies and multiple constellations. Today, GNSS receivers of all kinds are multifrequency and multi-constellation capable. Even millions of cell phones now include some with dual-frequency receivers. Free services offering real-time centimeter level accuracy are now available directly from some GNSS satellites themselves. More and more GNSS is at the heart of surveying and mapping systems made of sensor combinations. Mobile mapping, Unmanned Aerial Systems (UAS), aka drones, and others in which GNSS is integrated with many devices that provide relative positioning but depend on GNSS for the systems' absolute positioning. The increase in the number of base stations and related services is remarkable making real-time solutions, RTN, PPP, etc., available nearly everywhere. Surveyors, long at the forefront of GNSS technology, continue to rely on receivers that exploit all frequencies and constellations. Receivers that can utilize any type of service (static, RTK, RTN, post-processed and real-time PPP). However, as less sophisticated devices improve surveyors are using them in their work too.

All this progress means more measurements in a shorter time without degradation in accuracy, in part because interference can be ameliorated more easily. The time to first fix for carrier-phase receivers, the period when the receiver is solving for the integers, downloading the almanac, and so forth, initialization, is shortened significantly. Fixed solution accuracy is achieved more quickly. In the past, multifrequency carrier-phase solutions were accurate but noisy, but with the new signals available multi-frequency solutions are directly enhanced. While a GNSS capable receiver may offer a user improved availability and reliability, it may not necessarily offer higher accuracy than is achieved with GPS alone. However, the achievement of high accuracy more conveniently and in more places is certainly available with GNSS.

The algorithms currently necessary for the achievement of high accuracy with carrier-phase ranging may be simplified since many of the new GNSS signals carry a civilian code. Code correlation is a more straightforward problem than carrier differencing. Also, the more diverse the maintenance of the components of GNSS, the less chance of over-all system failure; the United States, Russia, the EU, and China all have infrastructure in place to support their contribution to GNSS. Under such circumstances, simultaneous outages across the entire GNSS constella-tion are extremely unlikely. Despite similarities, there are some issues in the consistency issues with GNSS. For example, the rollout of the new systems is not coordinated. In other words, GPS and GLONASS mod-ernizations, Galileo deployment, and BeiDou enhancements are certainly not synchronized. Despite much cooperation, there is no clear agreement among nations that launches and operational capabilities happen in the same time frame. Also, there are some differences in multiple access schemes, time standards, and other GNSS system design parameters. For the full potential of the systems to be realized, the components of GNSS need to be interoperable. In other words, as many satellites as possible delivering signals that can be used in conjunction with one another at any time, any place without interfering with one another. Fortunately, there is signal compatibility and sharing such as is apparent with L1C. Interoperability is achieved by partial frequency overlap using different signal structures and/or different code sequences for spectral separa-tion. In fact, high accuracy and interoperability are not only a matter of convenience—robust, reliable solutions are a necessity in many spheres. Consider safety-of-life uses for things such COSPAS-SARSAT, routing of emergency vehicles, or the GNSS-based automated machine control sys-tem now in wide use. Mining, agriculture, aircraft control, and so forth, are depending more and more on satellite PNT systems. These industries have high costs and high risks. They not only require high accuracy but reliability as well. If GNSS can deliver inexpensive receivers tracking the maximum number of satellites broadcasting the maximum number of signals, it will live up to the fondest hopes of not only many individuals but also many industries as well.

Today, a single receiver can track all the old and new GNSS satellite signals with a significant performance improvement. Nevertheless, the main attraction of interoperability between these systems is the greatly increased number of satellites and signals, better satellite availability, bet-ter dilution of precision, immediate ambiguity resolution on long baselines with multifrequency data, better accuracy in urban settings, and fewer multipath worries. Today, with a mask angle of 10°, there are some periods when 35 or more GNSS satellites are above a user's horizon. GNSS receiv-ers have grown from only a handful to the huge variety available today. Some push the envelope to achieve ever-higher accuracy; others offer less sophistication and lower cost. The civilian user's options are broader with GNSS than any previous positioning system.

EXERCISES

1. Which of the following statements is true of the current GPS constellation?
   a. There are four orbital planes.
   b. There are eight satellites in each orbital plane.
   c. The satellites orbit approximately 20,000 nautical miles above the Earth.
   d. Each satellite completes an orbit in 12 sidereal hours.

2. Which of the following statements about L2C is not correct?
   a. It is broadcast by the Block IIF satellites.
   b. It was originally announced in 1998.
   c. It is carried on L1 and L2.
   d. It carries L2-CNAV.

3. The L5 is a bit stronger than L1. How much stronger?
   a. L5 has about two times more power than L1
   b. L5 has about three times more power than L1
   c. L5 has about four times more power than L1
   d. L5 has about five times more power than L1

4. Which GPS satellite block first broadcast L5?
   a. Block I
   b. Block II
   c. Block IIR
   d. Block IIF

5. Which of the following GNSS constellations does not currently supply global coverage?
   a. BeiDou
   b. GPS
   c. QZSS
   d. GLONASS

6. Which of the following statements about CDMA (Code Division Multiple Access) and FDMA (Frequency Division Multiple Access) is correct?
   a. CDMA: Satellites in a GNSS constellation transmit on the same frequency (or frequencies), but each transmits its own distinct code. FDMA: Satellites in the GNSS constellation transmit the same codes, but each transmits them on its assigned frequency.

b. FDMA technology was originally developed by the U.S. military during World War II to secure communications in the presence of jamming.
c. GLONASS satellites only broadcast FDMA signals.
d. CDMA: Satellites in the GNSS constellation transmit the same codes, but each transmits them on its assigned frequency. FDMA: Satellites in a GNSS constellation transmit on the same frequency (or frequencies), but each transmits its own distinct code.

7. Which of the following statements about the GPS Block IIR-M satellites is correct?
a. The first Block IIR-M satellites were launched in the summer of 2010.
b. Block IIR-M satellites broadcast L2C.
c. Block IIR-M can broadcast the M-code through a large deployable high-gain antenna to produce a directional spot beam.
d. Block IIR-M satellites broadcast L1C.

8. Which of the following received GNSS signal values is the most powerful?
a. GPS Block IIR P(Y) received @ −164.5 dBW
b. Galileo E5b Data received @ −158.0 dBW
c. BeiDou B2b Open Service received @ −163.0 dBW
d. GPS Block IIF L5 in-phase received @ −157.0 dBW

9. Which of the following GNSS standard time systems incorporates leap seconds?
a. GPS
b. GLONASS
c. Galileo
d. BeiDou

10. What is COSPAS-SARSAT?
a. A ground-based augmentation service
b. A satellite laser ranging system
c. An international satellite-aided search and rescue service
d. A Very Long Baseline Interferometry station

11. Which of the following is not a central carrier frequency broadcast by more than one GNSS constellation?
a. 1176.45 MHz
b. 2492.028 MHz

    c. 1278.75 MHz
    d. 1575.42 MHz

12. Which of these signals broadcast by QZSS includes a centimeter level augmentation service (CLAS)
    a. L1C
    b. LEX
    c. L1-SAIF
    d. L2C

ANSWERS AND EXPLANATIONS

1. Answer is (d)
    Explanation: The satellites are in six orbital planes, four or more satellites per plane in orbit about 11,000 nautical miles above the Earth. Each completes an orbit every 12 hours in sidereal time.

2. Answer is (c)
    Explanation: L2C is broadcast by the Block IIF satellites. This is a code that was announced way back in March 1998. It is on L2, not on L1, and is known as L2C. This new code is not merely a copy of the C/A code. L2C is a bit more sophisticated. It carries L2-CNAV.

3. Answer is (a)
    Explanation: L5 has about 3 dB more power than L1, about double. It also has a wider bandwidth than legacy civil signals. L5 does not carry military signals; it achieves an equal power split between its two signals. In this way, L5 lowers the risk of interference and improves multipath protection. It also makes the data-less signal easier to acquire in unfavorable and obstructed conditions.

4. Answer is (d)
    Explanation: L5 was the first broadcast by the fourth-generation GPS satellites, Block IIF.

5. Answer is (c)
    Explanation: BeiDou, GPS, and GLONASS are all operational and supply worldwide coverage. The Japanese Quasi-Zenith Satellite System (QZSS) is a regional GNSS. QZSS is designed to augment the accuracy and reliability of GPS in Japan's urban canyons and the Asia-Oceania region. It is intended to ensure that one of its satellites is always nearly directly overhead in Japan. This is the origin of the term quasi-zenith.

6. Answer is (a)

Explanation: CDMA technology was originally developed by the U.S. military during World War II to secure communications in the presence of jamming. CDMA allows all the satellites in a GNSS constellation to broadcast on the same frequency (or frequencies) and still ensures the receivers can uniquely pick out each satellite's signal from the others. While the satellites share the common carrier frequency (or frequencies), each has its own distinct PRN code. A satellite's PRN code is modulated onto the common frequency (or frequencies). It is the exclusiveness of that code that allows the receiver to differentiate a particular satellite's signal from all the others. In short, the technique serves many simultaneous users over the same frequency (or frequencies). CDMA does not use frequency channels or time slots. GLONASS uses a different strategy. All the satellites in the constellation transmit the same codes, but each transmits them on its assigned frequency. The satellites transmit L-band signals, and unlike GPS, each code a GLONASS receiver collects from any one of the GLONASS satellites is the same. Also, unlike GPS, each GLONASS satellite broadcasts its codes at its own unique frequency. This is known as Frequency Division Multiple Access (FDMA). However, the GLONASS-K1 satellites transmit not only both the open and authorized FDMA on L1 and L2 but also the open CDMA signals on L3. FDMA signals as has been done since the launch of the first GLONASS satellite in order to ensure backward compatibility with legacy receivers still in use, but the CDMA open service signal emanating from GLONASS is centered at 1202.025. GLONASS-K1 began transmitting these additional CDMA signals on L3 in 2011. Today, the 3 K-satellites in the constellation and the M satellites launched since 2014 can transmit a CDMA signal on L3, which is called L3OC. In the future, CDMA may be broadcast on L1 and L2 by the GLONASS-K2 and GLONASS-KM satellites.

7. Answer is (b)

Explanation: The IIR-M satellites are IIR satellites that were modified before they were launched. The modifications upgraded these satellites so that they radiate two new codes; a new military code, the M-code, a new civilian code, the L2C code, and demonstrate a new carrier, L5. The first Block IIR-M satellite was launched in the fall of 2005, and the last in the summer of 2009. The first Block IIF satellite was launched in the summer of 2010. The Block III satellites broadcast L1C, and their M-code can be broadcast through a large deployable high-gain antenna to produce a directional spot beam. The spot beam has approximately 100 times more power (–138 dBW) compared with (–158 dBW) the wide-angle M-code broadcast.

8. Answer is (d)

Explanation: A change in 3 decibels is an increase or a decrease of 100% in power level. Stated another way, a 3-decibel increase indicates a doubling of signal strength, and a 3-decibel decrease indicates a halving of signal strength. The Galileo E5b Data signal of –158.0 dBW is more than twice as powerful as the GPS Block IIR P(Y) signal of –164.5 dBW. The BeiDou B2b Open Service of –163.0 dBW is less powerful than the GPS Block IIF L5 in-phase signal of –157.0 dBW. The received power of the GPS Block IIF L5 in-phase signal of –157.0 dBW is the most powerful of the signals listed.

9. Answer is (b)

Explanation: There are no leap seconds introduced to GPS Time. The same may be said of Galileo and BeiDou. However, things are different in GLONASS. GLONASS System Time (GLST) is aligned directly to UTC, leap seconds, and all. It is the only GNSS time system that is. That means that leap seconds are incorporated into the standard. In other words, there is no constant integer-second difference between GLST and UTC as there is with the other GNSS constellation's time scales.

10. Answer is (c)

Explanation: COSPAS-SARSAT is an international satellite-aided distress alert detection and routing system for search and rescue. Transponders on the satellites transfer the user's distress signals to the dedicated ground segment broadcast centers, which initiate rescue operations.

11. Answer is (b)

Explanation: All the frequencies listed are broadcast as central carrier frequencies by more than one GNSS constellation except the S-band 2492.028 MHz with a bandwidth of 16.5 MHz was chosen by NavIC to be their second frequency because it is less affected by the ionosphere than the L-band, because it was recently designated by the ITU for navigation services and because it is not much populated.

12. Answer is (b)

Explanation: QZSS transmits six signals. The first 4 are the familiar GPS Positioning, Navigation, and Timing (PNT) signals, L1-C/A, L1C; at 1575.42 MHz; L2C at 1227.60 MHz; L5 at 1176.45 MHz. The other two are augmentation signals. The L1-SAIF (Submeter-class Augmentation with Integrity Function) is broadcast on the L1 frequency, 1575.42 MHz. It is interoperable with GPS and provides a real-time sub-meter level augmentation service (SLAS) signal to code-phase single-frequency GPS L1 C/A users. The other unique QZSS signal is LEX (L-band Experiment), aka L6. It shares the 1278.75 MHz frequency with Galileo E6. LEX provides a regional open real-time centimeter-level accuracy augmentation service (CLAS) to multifrequency carrier-phase users for machinery control, precision farming, surveying, etc., at a high data rate, 2 kbps.

Glossary

A

Absolute Positioning (aka **Autonomous Positioning, Point Positioning, Point Solution, or Single Point Positioning**): A single receiver pseudorange positioning derived from observations on at least four GNSS satellites.

Accuracy: The closeness of a measured, computed, or estimated quantity to the quantity's standard or accepted true value. In an absolute sense, the true value is unknowable so accuracy must be estimated.

Acquisition: Detection and identification of signals received from a GNSS satellite accompanied by Doppler and code phase estimations of sufficient exactness to initialize the receiver's phase and delay tracking loops.

Active Marks (aka **Active Control**): A station occupied by equipment intended continuously collecting geodetic quality data.

Aeronautical Radio Navigation Services (ARNS): Designated in Article, Section 1.46 of the International Telecommunication Union (ITU) Radio Regulations as "A radionavigation service intended for the benefit and for the safe operation of aircraft." L5 users benefit from its place at 1176.45 MHz within the ARNS frequency band 1164–1215 MHz.

Allan Deviation (ADEV): The square root of the Allan variance.

Allan Variance (AVAR) (aka **Two-Sample Variance**): A well-known statistical technique described by David Allan in 1965 used to characterize the short-term stability of atomic clocks, crystal oscillators, etc.

Almanac: A brief and coarse summary of the orbital parameters that helps a GNSS receiver find the satellites. It can be acquired by a receiver's reading of the navigation message from a single GNSS satellite. In GPS, it is found in subframes 4 and 5 of the legacy Navigation message.

Alternative BOC (AltBOC): A signal multiplexing/mapping scheme with good noise and multipath suppression properties that was proposed in 2002/2003. It was first used in Galileo to combine two adjacent quadrature phase shift key signals, the complex upper (E5b) and lower (E5a) subcarriers, into one wideband signal with a common carrier frequency. A constant envelope is achieved on the transmit side by the addition of an intermodulation function (IM). It is similar to the binary offset carrier (BOC) modulation but differs from it in having better spectral isolation between the two upper and lower main lobes.

Ambiguity: The offset in a carrier-phase observation including both the initial phase and the integer ambiguity that remains constant from the moment a receiver locks onto the GNSS signal as long as that lock is maintained.

Ambiguity Fixed Solution: (*See* **Fixed Solution.**)

Ambiguity Float Solution: (*See* **Float Solution.**)

Analog: Representation of data by a continuous physical variable. For example, the modulated carrier wave used to convey information from a GNSS satellite to a GNSS receiver is an analog mechanism.

Analog-to-Digital Converter (ADC): A device on the front end of a GNSS receiver that converts the received analog signals to digital samples in preparation for subsequent signal processing tasks.

Anechoic Chamber: A term coined by Dr. Leo Beranek, who built the first such closed room within which the reflection of sound was minimized to facilitate acoustic experiments. Today, such rooms are also designed to block out and prevent reflection of electromagnetic energy such as radio signals. Tests of radars, antennas, etc., must be isolated from external interference to ensure valid results are conducted in such environments.

Antenna: An antenna is a resonant device that converts electrical energy to electromagnetic waves or vice versa. A GNSS receiver's antenna collects and often amplifies a satellite's signals. It converts the faint GNSS signal's electromagnetic waves into electric currents sensible to a GNSS receiver. Microstrip, aka patch antennas, are the most often used with GNSS receivers. Choke-ring antennas are intended to minimize multipath error.

Antenna Exchange: (*See* **ANTEX.**)

Antenna Gain Pattern: A nearly hemispherical shape, highest at zenith and lower near the horizon, is typical in GNSS antennas. It is a spatial representation of a relative measure of the antenna's capability in converting radio waves into electrical power.

Antenna Phase Center (APC) Receiver: The signals received by an idealized GNSS receiver's antenna would reach equal phase on the surface of an imagined perfect sphere. In that model, the antenna phase center would be the center of that sphere. However, real GNSS antennas are not ideal and the equal phase pattern is not a perfect sphere, it's a bit irregular. Further, it's actual phase center varies with the frequency, azimuth, and elevation of the received signals. In other words, as they change the GNSS antenna's phase center moves slightly in what is known as the Phase Center Variation (PCV). Finally, the coordinates derived from a GNSS observation are the coordinates of the receiver's antenna phase center which is unlikely to be exactly at the physical center of the antenna. The vector between the physical center and the antenna phase center is called the Phase Center Offset (PCO).

Antenna Reference Point (ARP) Receiver: An accessible point marked on a GNSS receiver antenna, usually where the vertical axis of the antenna intersects the bottom mount, ground plane, or choke ring. It is the point to which instrument heights can be measured and from which the vector between the antenna phase center and the antenna reference point for particular antennas can be acquired online, i.e., National Geodetic Survey (NGS) Antenna Calibration tables, International GNSS Service (IGS) ANTEX.

Antenna Splitter (*See also* **Zero Baseline Test**)**:** An antenna splitter is an attachment that divides a single GNSS antenna's signal in two. The divided signal goes to two GNSS receivers. This is the foundation of zero baseline test. The two receivers are using the same antenna so the baseline length should be measured as zero; since perfection remains elusive, it is usually a bit more.

ANTEX: The International GNSS Service's ANTenna EXchange format used to distribute GNSS receiver and satellite antenna calibration values often used in positioning and orbit determination software. The included phase center offsets (PCO) and phase center variations (PCV) are kept up to date as new antennas and satellites come into service. These data are useful in GNSS precise point positioning (PPP), satellite orbit and clock computations in modeling for high-accuracy carrier-phase observations.

Anti-Spoofing (AS): The implementation of an unpredictable modulation sequence, i.e., the encryption of the GPS P(Y) code, to render the structured interference that is spoofing less effective.

Aphelion: The point on Earth's, or any planet's, asteroid's, or comet's, orbit when it is farthest from the Sun.

Apogee: The most distant an orbiting object is from the Earth, i.e., the point on a GNSS satellite's orbit furthest from the Earth.

Apogee Kick Motor (AKM): A low thrust solid propellant system that is fired at the apogee of the satellite's elliptical transfer orbit to move it toward and eventually into its operational orbit. All GPS Block I and II, except Block IIF, utilized needed a boost from apogee kick motors to reach their medium Earth orbit (MEO). GPS III satellites, more massive than Block IIF satellites, will rely on liquid apogee engines to reach their operational orbit.

AS: Anti-spoofing.

Ascending Node: The point where an orbiting satellite crosses the Equatorial plane from south to north.

Astronomic Latitude: A measurement of latitude based on the angle between the Earth's equatorial plane and a plumb line (direction of gravity) so that it differs from the geodetic latitude by the meridional component of the deflection of the vertical.

Atmosphere: The layers of gases and aerosols that envelope the Earth. The propagation of GNSS signals is affected by their trip through the atmosphere.

Atomic Clock: A clock regulated by the electronic transition of an atom or molecule between two quantized energy levels. It is a continuously operating atomic frequency standard with an initialization time and a counter to supply both time and frequency.

Atomic Frequency Standard (AFS): An oscillator that produces a periodic signal from the resonance of an atom or molecule transitioning between two quantized energy levels. GNSS satellite's frequency standards are usually based on the electronic transition of cesium, hydrogen, or rubidium atoms.

Attitude: The angular position of a rigid body in space.

Attosecond: One-quintillionth (10^{-18}) of a second.

Attribute: Information about features of interest. Attributes such as date, size, material, color, and so forth, are frequently recorded during data collection for Geographic Information Systems (GIS).

Australian Positioning Service (AUSPOS): A free online GPS data processing facility provided by Geoscience Australia. It takes advantage of both the IGS Stations Network and the IGS product range. AUSPOS works with data collected anywhere on Earth.

Autocorrelation: The correlation of a signal with a copy of itself.

Azimuth: A horizontal angle from a reference direction. Typically measured clockwise from north or south, positive to the east.

B

Ballistic Camera: A camera used in the photogrammetric tracking of artificial Earth satellites often against the background of fixed stars.

Bandpass Filter: A device that a specified range of frequencies to pass and rejects all others.

Bandwidth: The range of frequencies required for a modulated carrier wave to convey its signal free of distortion or attenuation and for the receiver to efficiently process the signal.

Base Station (aka **Reference Station**): A GNSS receiver at a known location. The base station monitors satellite signals, their errors and is often intended to provide data with which to differentially correct GNSS measurements.

Baseband: The higher-frequency band to which an information signal is upshifted for transmission by a modulated carrier. To transfer information from the usually very low-frequency original signal onto a carrier, aka bandpass signal, the higher-frequency baseband becomes the envelope within which the modulated carrier is created. When it arrives at the receiver, the baseband processing block then down-converts it in order to provide the observables: code pseudoranges, carrier-phase measurements, navigation data, etc.

Baseband Signal: An information signal at its original frequency. It is aka a lowpass signal.

Baseline: A line between two stations from which simultaneous GNSS observations have been made. Often one of the two GNSS receivers is the base station at a known position and the other a rover whose position is to be determined relative to the base. A three-dimensional vector between this pair of stations is one expression of a baseline.

Beat Frequency: When two signals of different frequencies are combined, additional frequencies are created.

BeiDou: China's GNSS. In the early 2000s, China considered collaborating in the European Union led global GNSS project that became Galileo but ultimately decided to build their own. So now the fourth global GNSS system, joining those undertaken by the United States—GPS, Russia—GLONASS, and Europe—Galileo, is the Chinese BeiDou system, named after the Big Dipper. It provides positioning, navigation, and timing (PNT) and search and rescue (SAR) services worldwide. BeiDou's development has been through three steps. Each step increased the coverage of their satellite navigation system.

BeiDou SBAS (BDSBAS): China's BeiDou GEO satellites provide a satellite-based augmentation service (SBAS) called BDSBAS via B1C and B2a signals for users in and around China, i.e., between 70°E longitude to 145°E longitude and 5°N latitude to 55°N latitude. There are two service modes, single-frequency with dual constellations and dual frequency with multiple constellations.

Between-Epoch Difference: The difference in the phase of the signal on one frequency from one satellite as measured between two epochs observed by one receiver. For example, a GNSS satellite and a GNSS receiver are always in motion relative to one another, and therefore, the frequency of the signal broadcast by the satellite is not the same as the frequency received. The Doppler observable in GNSS is the measurement of the change of phase between two epochs.

Between-Receiver Single Difference: The difference in the code or carrier GNSS measurements or parameters between two receivers simultaneously observing the signal from the same satellite on the same frequency. For a pair of receivers simultaneously observing the same satellite, a between-receiver single difference pseudorange or carrier-phase observable can virtually eliminate errors attributable to the satellite's clock. When baselines are short, the between-receiver single difference can also greatly reduce errors attributable to orbit and atmospheric biases.

Between-Satellite Single Difference: The difference in the code or carrier GNSS measurements or parameters between signals from two satellites on the same frequency simultaneously observed by one receiver. For one receiver simultaneously observing two satellites, a between-satellite single difference pseudorange or carrier-phase observable can virtually eliminate errors attributable to the receiver's clock.

Bias: A systematic error or measurement offset. Biases may originate from physical limitations in satellites or other hardware; imperfect modeling of orbits, atmospheric conditions, clock errors, and so forth. Biases affect all GNSS measurements and hence the coordinates and baselines derived from them.

BIH: Bureau l'International de l'Heure.

Binary Offset Carrier (BOC) Modulation: A split-spectrum modulation. The signal's spectrum is divided into two parts and can be shaped to accommodate inter-system-compatibility. It is characteristic for such modulation to be written BOC (α, β) where α indicates the frequency of the square wave modulation of the carrier, aka the subcarrier frequency factor, and β describes the frequency of the pseudorandom noise modulation, aka the spreading code factor. In the case of the M-code the notation BOC (10, 5). An important characteristic of BOC modulation is that the code's greatest power density is at the edges, that is, at the nulls.

Binary Phase-Shift Keying (BPSK) (Binary Biphase Modulation): Modulation of a carrier wave by shifting its phases from 0° or 180° to indicate the binary values of the data being transported, data such as pseudorandom noise codes and navigation information.

Bit: A unit of information. In a binary system either a 1 or a 0.

Block: A classification of the GPS satellite's generations.

BPS: Bits per second.

Broadcast: A modulated electromagnetic wave transmitted across a large geographical area.

Broadcast Ephemeris (*See also* Ephemeris): A table of values broadcast by a GNSS satellites that includes locations and related data from which it is possible to derive a satellite's position and velocity.

Bureau International de l'Heure: When the development of radio made it possible to coordinate time measurement, the Bureau International de l'Heure, hosted by the Paris Observatory, was entrusted with unifying the measurements of Universal Time (UT). It began its operation in 1912, although its statutes were officially settled only in 1919.

Byte: A sequence of eight binary digits that represents a single character, that is, a letter, a number.

C

C/A Code (Coarse/Acquisition Code): A binary code known by the names, Civilian/Access code, Clear/Access code, Clear/Acquisition code, Coarse/Acquisition code, and various other combinations of similar words. It is known as the Standard Positioning Service (SPS). It is a standard spread-spectrum pseudorandom noise code modulated on the GPS L1 carrier using binary biphase modulations in a 1023-bit pattern called a Gold Code. Each C/A code is unique to the particular GPS satellite broadcasting it. It has a chipping rate 1.023 MHz, more than a million bits per second. The C/A code is a direct sequence code and a source of information for pseudorange measurements. The GPS C/A code is not carried on L2. C/A codes are defined for GPS and GNSS systems such as SBAS and QZSS.

C/No (aka **Carrier-to-Noise Density Ratio**): Carrier-to-noise ratio per unit bandwidth (1 Hz) in dBHz in which typical values are 37 to 45 dBHz, the higher the better.

Carrier: A periodic electromagnetic wave, usually sinusoidal, on which data can be modulated. Common methods of modulation are frequency modulation and amplitude modulation. GNSS signals are commonly located in the L-band and employ frequencies in a range of about 1100–1600 MHz. In GPS, the phase of the carrier is modulated. There are currently three GPS carrier waves. They are L1, L2, and L5, which are broadcast at 1575.42, 1227.60, and 1176.45 MHz, respectively.

Carrier Beat Phase (*See also* **Beat Frequency**; **Doppler Shift**; **Reconstructed Carrier-Phase**): The phase of a beat frequency created when the carrier frequency generated by a GNSS receiver combines with an incoming carrier from a GNSS satellite. The two carriers have slightly different frequencies as a consequence of the Doppler shift of the satellite carrier compared with the nominally constant receiver generated carrier.

Carrier-Phase Ambiguity (aka **Integer Ambiguity**): The integer ambiguity N represents the number of full phase cycles between the receiver and the satellite at the first instant of the receiver's lock-on. It can also be labeled the carrier-phase ambiguity, integer-cycle ambiguity, or the cycle count at lock-on. It does not change from the moment of the lock to the end of the observation unless that lock is lost.

Carrier-Phase Ranging: (1) The term is usually used to mean a GNSS measurement based on the carrier signal itself rather than measurements based on the codes modulated onto the carrier wave. (2) Carrier-phase may also be used to mean a part of a full carrier wavelength. In GPS, an L1 wavelength is about 19 cm, an L2 wavelength is about 24 cm, and an L5 wavelength is about 25 cm. A part of the full wavelength may be expressed in a phase angle from 0° to 360° carrier-phase measurements. One or more GPS receivers occupy a base station, a known position. They collect the signals from the same satellites at the same time as other GPS receivers that may be stationary or moving. The other receivers occupy unknown positions in the same geographic area. Occupying known positions, the base station receivers find corrective factors that can either be communicated in real time to the other receivers, as in RTK or may be applied in postprocessing, as in static positioning. Relative positioning is in contrast to absolute, or point positioning. In relative positioning, errors that are common to both receivers, such as satellite clock biases, ephemeris errors, propagation delays, and so forth, are mitigated.

Carrier Tracking Loop (*See also* **Code Tracking Loop**): A feedback loop that a GNSS receiver uses to generate and match the incoming carrier wave from a GNSS satellite.

Cesium Clock (aka **Cesium Frequency Standard**): An atomic clock that is regulated by the element cesium. Cesium atoms, specifically atoms of the isotope Cs-133, are exposed to microwaves and the atoms vibrate at one of their resonant frequencies. Time is measured by counting the corresponding cycles based on the hyperfine frequency of the cesium atom's ground state of 9192631770 Hz.

Chandler Period: Oscillation of the Earth's axis for a period that varies from 412 to 442 days that is named after Seth Chandler U.S. Astronomer (1846–1913).

Channel: An interface on a chip, i.e., an application-specific integrated circuit (ASIC), in a GNSS receiver that includes algorithms for signal acquisition, tracking, and demodulation of the navigation message from one GNSS signal, from one GNSS satellite.

Check Point: A reference point from an independent source of higher accuracy used in the estimation of the positional accuracy of a data set.

Chi-Square Test: A statistical method used to evaluate the quality of the fit between the observed values and the expected values.

Chipping Rate: The rate at which chips, binary 1s and 0s, are produced. In GPS, the P code chipping rate is about 10 million bits per second. The C/A code chipping rate is about 1 million bits per second.

Choke-Ring Antenna: An omnidirectional GNSS receiver antenna invented at the Jet Propulsion Laboratory that has good multipath rejection due to the concentric cylinders, usually a quarter wavelength deep, that surround the central antenna.

Circular Error Probable (CEP): A description of two-dimensional precision. The radius of a circle, with its center at the actual position, that is expected to be large enough to include half (50%) of the normal distribution of the scatter of points observed for the position.

Circular T: A monthly publication of Coordinated Universal Time (UTC) by the Time Department of the International Bureau of Weight and Measurements (Bureau International des Poids et Mesures, BIPM).

Civil Navigation Message (CNAV): Four modernized NAV messages available for user familiarization and equipment development. While the navigational data content they provide is like that in LNAV, the new civil navigation message format is pseudo-packetized, more flexible, robust, offers forward error correction (FEC) and can support 63 satellites, compared with 32 in the LNAV message.

Civilian Code: (*See* **C/A Code**.)

Civilian/Access Code: (*See* **C/A Code**.)

Clear/Access Code: (*See* **C/A Code**.)

Clear/Acquisition Code: (*See* **C/A Code**.)

Clock Offset (aka **Clock Bias**)**:** A difference between the same moment of time indicated by two different clocks. In GNSS, the difference between a moment of time determined by a receiver or satellite clock and the same moment determined by a reference, such as GPS Time or other system time maintained by a GNSS control segment.

CNAV: Civil NAVigation message.

Coarse/Acquisition (C/A): (*See* **C/A Code**.)

Code Division Multiple Access (CDMA): In this scheme, all the satellites in a GNSS constellation broadcast on the same frequency (or frequencies) and still ensure the receivers can uniquely pick out each satellite's signal from the others because while the satellites share the common carrier frequency (or frequencies), each has its own distinct pseudorandom noise (PRN) code.

Code Phase: The phase of the pseudorandom noise (PRN) code modulated on a GNSS signal and also the pseudorange measurement derived from the alignment of the received code with the locally generated code and the calculated time. Sometimes used to mean pseudorange measurements expressed in units of cycles.

Code States: With the formula code state = 1−2X code chips (0 and 1) are transformed into code states (+1 and −1). X is the code chip value. For example, a normal code state is +1, corresponds to a code chip value of 0. A mirror code state is −1. It corresponds to a code chip value of 1.

Code Tracking Loop: A feedback loop used by a GNSS receiver to generate pseudorandom noise codes to compare with the incoming pseudorandom noise codes from a GNSS satellite, such as C/A or P codes, from a GPS satellite.

Codeless: A technique in which the receiver extracts useful information from a signal without specific knowledge of the code. The approach has been used by civil receivers to obtain dual-frequency measurements with the use of the GPS P(Y) signal. It has been proposed monitoring of the Galileo Public Regulated Service (PRS) for signal quality.

Cold Start (aka **Factory Start**)**:** The name of the slow time to first fix (TTFF) that results when a GNSS receiver has been inactive for some time. In that case, it does not know its own position. It is without a good time estimate and a recent almanac. In this state, the receiver cannot quickly find the necessary satellites for a fix. It must search for satellites somewhat blindly. When the receiver does acquire a satellite it can download an almanac, a brief and coarse summary of the orbital parameters, which helps the GNSS receiver find other satellites. Then as the necessary number of satellites are acquired and tracked the receiver can download their ephemerides, clock, etc., data and eventually achieve a fix. A cold start can take as long as 750 seconds (12½ minutes) or so.

Combined Factor (aka **Grid Factor**)**:** Multiplying the elevation factor and the scale factor produces a single ratio that is usually known as the *combined factor* or the *grid factor*. Using this grid factor, the measured line is converted from a ground distance to a grid distance in one jump.

Compatibility: The ability of the various GNSS positioning, navigation, and timing (PNT) services to be used separately or together without interfering with each other.

Complete Instantaneous Phase Measurement (*See* **Fractional Instantaneous Phase Measurement**.)**:** A measurement including the integer number of cycles of carrier beat phase since the initial phase measurement.

Confidence Level: The probability that the true value is within a particular range of values expressed as a percentage.

Confidence Region: A region within which the true value is expected to fall, attended by a confidence level.

Conformal: The representation of the shape of small areas is substantially unchanged on a conformal map.

Constellation: (1) All GNSS satellites on orbit. (2) The group of satellites used to derive a position. (3) The satellites available to a GNSS receiver at a particular moment.

Conterminous United States (CONUS): The contiguous, or adjoining states, of the United States and Washington, DC excluding Hawaii and Alaska.

Conterminous: Sharing a common boundary, for example, the conterminous United States.

Continuously Operating Reference Stations (CORS): A network of GNSS receivers with antennas that constantly track satellites, whose data support the differential correction of three-dimensional positioning and/or sometimes supporting real-time positioning applications. Each station is on a very stable mount in a secure site.

Continuous-Tracking Receiver: (*See* **Multichannel Receiver.**)

Control Segment: The ground-based infrastructure and network of control, monitor, and uploading stations responsible for tracking GNSS satellites, monitoring their status and transmissions; analyzing data and uploading commands to the constellation. Typically, several monitoring stations are located across a wide area. They collect satellite information and relay it to the master control station, where it is analyzed and from which satellite ephemerides, clock corrections, atmospheric corrections, etc., are prepared and subsequently sent to upload stations for transmission back to the satellites.

Coordinate Epoch: Tectonic plate motion and/or crustal deformation continually change coordinates so, they always must always be qualified with the epoch at which they are valid—the coordinate epoch.

Coordinated Universal Time (aka **UTC, Zulu Time**): Universal Time Coordinated (UTC). Universal Time systems are international, based on atomic clocks around the world. Coordinated atomic clock time called International Atomic Time (Tempes Atomique International, TAI) was established in the 1970s and it is very stable. It is more stable than the actual rotation of the Earth. TAI would drift out of alignment with the planet if leap seconds were not introduced periodically as they are in UTC. UTC is, in fact, one of several Universal Time standards. There are standards more refined than UTC's tenth of a second. UTC is maintained by the U.S. Naval Observatory (USNO).

Correlation: When two GNSS signals match.

Correlation Channel: A channel in a GNSS receiver used to shift or compare the incoming signal with an internally generated signal. The code generated by the receiver is correlated with the incoming signal to find the correct delay. Once they are aligned, one way to keep them so is a delay lock loop. Correlator designs are sometimes optimized for acquisition of signal under foliage, accuracy, multipath mitigation, and so forth.

Correlator: A tool using a combination of an early and a late correlator inside a GNSS receiver to measure the relationship of the receiver's replica code and phase to the incoming signal from the satellite. The tracking loops use the early-minus-late difference of the correlator outputs to discriminate tracking errors and help them continuously align the signals.

Correlator Spacing: The separation in PRN units between the early and late correlators in a delay lock loop.

COSPAS-SARSAT: An international satellite-aided distress alert detection and routing system for search and rescue. Transponders on the satellites transfer the user's distress signals to the dedicated ground segment broadcast centers, which initiate rescue operations.

Costas Loop: A phase-lock loop (PLL) invented in 1956 by John P. Costas of General Electric. It recovers the carrier and demodulates binary phase shift keying (BPSK) modulation without being affected by data bit transitions.

Cross-Correlation: A measure of the similarity between two signals. Cross-correlation checks the inner product at all relative shifts between two GNSS signals and thereby revealing their similarity at every shift. By finding the index of the cross-correlation with the largest value the true shift can be estimated.

CTS: Conventional Terrestrial System.

Cutoff Angle: (*See* **Mask Angle**.)

Cycle Ambiguity (aka **Integer Ambiguity**): The number of full wavelengths, the integer number of wavelengths, between a particular receiver and satellite is initially unknown in a carrier-phase measurement. It is called the cycle or integer ambiguity. If a single-frequency receiver is tracking several satellites, there is a different ambiguity for each. If a multifrequency receiver is tracking several satellites, there is a different ambiguity for each of the satellite's broadcast frequencies, i.e., L1, L2, and L5. However, in every case, the ambiguity is a constant number as long as the tracking lock is not interrupted. However, should the signal be blocked, if a cycle slip occurs, there is a new ambiguity to be resolved. Once the initial integer ambiguity value is resolved in a fixed solution for each satellite-receiver pair, the integrated carrier-phase measurements can yield very precise positions.

Cycle Slip: A discontinuity of an integer number of cycles in the carrier-phase observable. A jump of a whole number of cycles and a temporary loss of lock in the carrier tracking loop of a GPS receiver. Usually, the result of a temporarily blocked GPS signal. A cycle slip causes the cycle ambiguity to change suddenly. The repair of a cycle slip includes the discovery of the number of missing cycles during an outage.

D

Data Channel: A signal that carries the navigation message usually transmitted together with a dataless pilot signal.

Data Logger (DL) (aka **Data Recorder and Data Collector**): A data entry computer, usually small, lightweight, and often handheld. A data logger stores information supplemental to the measurements of a GNSS receiver.

Data Message: (*See* **Navigation Message**.)

Data Transfer: Transporting data from one computer or software to another, often accompanied by a change in the format of the data.

Dataset: An organized collection of related data compiled specifically for computer processing.

Datum: (1) Parameters, constants, and conventions that form the foundation for the realization of a coordinate system. (2) A means of relating coordinates determined by any means to a well-defined reference frame. (3) Any reference point line or surface used as a basis for calculation or measurement of other quantities. (4) The singular of data.

Decibel (dB): Decibel, a tenth of a bel. The bel was named for Alexander Graham Bell. Decibel does not actually indicate the power of the antenna; it refers to a comparison. Most GNSS antennas have a gain of about 3 dB (decibels). This indicates that the GNSS antenna has about 50% of the capability of an *isotropic* antenna. An isotropic antenna is a hypothetical, lossless antenna that has equal capabilities in all directions.

Declination: Like the latitude of a point on the Earth, the declination of a celestial body is an angle measured at the center of the Earth from the plane of the equator, positive to the north and negative to the south to the subject.

Deflection of the Vertical: At a point on the Earth the angle between the direction of gravity and a line perpendicular to the reference ellipsoid.

Delay Lock Loop (DLL). A controller in a GNSS receiver that aligns the replica code with the satellite's signal after the carrier is removed.

Developable Surface: A surface used that can be smoothly flattened to a plane in a map projection without unpredictable deformation.

Differential GNSS (DGNSS) (aka **Relative GNSS**): (1) A method that improves GNSS pseudorange accuracy. A GNSS receiver at a base station, a known position, measures pseudoranges to the same satellites at the same time as other roving GNSS receivers. The roving receivers occupy unknown positions in the same geographic area. Occupying a known position, the base station receiver finds corrective factors that can either be communicated in real time to the roving receivers or may be applied in postprocessing. (2) The term Differential GNSS is sometimes used to describe relative positioning. In this context, it refers to more precise carrier-phase measurement to determine the relative positions of two receivers tracking the same GNSS signals in contrast to absolute, or point, positioning.

Digital: Involving or using numerical digits. Information in a binary state, either a one or a zero, a plus or a minus, is digital. Computers utilize the digital form almost universally.

Digital Elevation Model (DEM): A digital representation of the topography of bare Earth without buildings, trees, etc.

Dilution of Precision (DOP): In GNSS positioning, an indication of geometric strength of the configuration of satellites in a particular constellation at a particular moment and hence the expected reliability of the results. A high DOP anticipates poorer results than a low DOP. A low DOP indicates that the satellites are widely separated. Since it is based solely on the geometry of the satellites, DOP can be computed without actual measurements. There are various categories of DOP, depending on the components of the position fix that are of most interest. Standard varieties of Dilution of Precision include:
- PDOP: Position
- GDOP: Geometric
- RDOP: Relative
- HDOP: Horizontal
- VDOP: Vertical
- TDOP: Time

Dipole Antenna: A simple antenna made of two conductors with a feedline connecting them. It can have acceptable performance indoors but have high directivity and virtually no reception on the perpendicular. Dipole antennas should not be used for navigation.

Distance Root Mean Square (drms): A statistical measurement that can characterize the scatter in a set of randomly varying measurements on a plane. The drms is calculated from a set of data by finding the root-mean-square value of the radial errors from the mean position. In other words, the root-mean-square value of the linear distances between each measured position and the ostensibly true location. In GNSS positioning, 2 drms is more commonly used. Two drms does *not* mean two-dimensional (2D) rms. Two drms does mean twice the distance root mean square. In practical terms, a particular 2 drms value is the radius of a circle that is expected to contain from 95% to 98% of the positions a receiver collects in one occupation, depending on the nature of the particular error ellipse involved. Two drms is convenient to calculate. In a non-biased dataset, it can be predicted using covariance analysis by multiplying the HDOP by the standard deviation of the observed pseudoranges.

Dithering: Intentional introduction of digital noise. Dithering the satellite's clocks is the method the U.S. Department of Defense (DoD) uses to degrade the accuracy of the GPS Standard Positioning Service (SPS). This degradation was known as Selective Availability (SA). SA was switched off on May 2, 2000, by presidential order.

Diurnal Tide: A cycle of one high and low tide each lunar day.

DoD: Department of Defense.

DOP: (*See* **Dilution of Precision**.)

Doppler Effect: In GNSS, a systematic change in the apparent frequency of the received signal caused by the motion of the satellite and receiver relative to one other.

Doppler Orbitography and Radiopositioning Integrated by Satellite (DORIS): A French system in which the transmitters are on the ground and the receiver is on board the satellite. It is designed to determine satellite's orbit but is used for a wider range of applications such as radar station calibration, evaluation of satellite maneuvers, point positioning, and navigation.

Doppler-Aiding: A method of receiver signal processing that relies on the Doppler shift to smooth tracking.

DOT: Department of Transportation.

Double-Difference: A method of GNSS data processing. In this method, simultaneous measurements made by two GNSS receivers, either pseudorange or carrier-phase, are combined. Both satellite and receiver clock errors are virtually eliminated from the solution. Most high-precision GNSS positioning methods use double-difference processing in some form.

Dual Frequency: Receivers of GNSS measurements that utilize more than one carrier. Dual frequency implies that advantage is taken of pseudorange and/or carrier phase on the L-band frequencies. Dual frequency allows modeling of ionospheric bias and attendant improvement in long baseline measurements particularly.

Dynamic Heights: A height based on the geopotential surface through the benchmark. It is calculated by dividing the geopotential number of the benchmark by the normal gravity value (G) computed on the GRS 80 ellipsoid at 45° latitude (G = 980.6199 gal). They are most relevant near large bodies of water.

Dynamic Positioning: (*See* **Kinematic Positioning**.)

E

Earth-Centered Earth-Fixed (ECEF): A Cartesian system of coordinates or reference frame with three axes in which the origin of the three-dimensional system is the Earth's center of mass, the geocenter. The z-axis passes through the North Pole, that is, the International Reference Pole (IRP) as defined by the International Earth Rotation and Reference Systems Service (IERS). The x-axis passes through the intersection of zero longitude, the International Reference Meridian (IRM), and the equator. The y-axis extends through the geocenter along a line perpendicular from the x-axis. It completes the right-handed system with the x- and z-axes and all three rotate with the Earth.

Electronic Distance Measuring (EDM) Device: A surveying instrument that calculates distances by measuring the time elapsed between broadcast and reception of reflection of stable electromagnetic waves and phase shifts in them.

Effective Isotropic Radiated Power (EIRP): Also called Equivalent Isotropic Radiated Power. The power and gain of a real antenna compared with the power and gain of a hypothetical isotropic antenna or expressed in another way, the power an isotropic antenna would need to achieve a signal of the same level and in the same direction as the maximum radiation of a real antenna.

Elevation: The distance measured along the direction of gravity above a surface of constant potential aka the zero datum. Usually, the reference surface is the geoid. Mean Sea Level (MSL), once utilized as a reference surface approximating the geoid, is known to differ from the geoid up to a meter or more. The term height, sometimes considered synonymous with elevation, most often refers to the distance above a reference ellipsoid in geodesy.

Elevation Factor (aka **Ellipsoid Factor**): SPCS83 name given to the factor used to reduce a measured distance from the topographic surface to the ellipsoid.

Elevation Mask Angle: (*See* **Mask Angle**.)

Ellipsoid (aka **Ellipsoid of Revolution**) (aka **Spheroid**): A closed surface generated by rotating an ellipse about its minor axis whose planar sections are either ellipses or circles. Two quantities can fully define a biaxial ellipsoid of revolution, the semi-major axis, a, and the flattening, $f = (a - b)/a$, where b is the length of the semi-minor axis. For example, the reference ellipsoid for the North American Datum of 1983 is GRS80, the Geodetic Reference System of 1980.

Ellipsoid Factor (aka **Elevation Factor**): SPCS83 name given to the factor used to reduce a measured distance from the topographic surface to the ellipsoid.

Ellipsoidal Height (aka **Geodetic Height**): The distance from an ellipsoid of reference to a point on the Earth's surface, as measured along the perpendicular from the ellipsoid. It is symbolized by h. Ellipsoidal height is not the same as elevation above mean sea level nor orthometric height. Ellipsoidal height also differs from geoidal height.

Ephemeris: Generally, data from which the path of a celestial body and its position on that path at a moment of time can be calculated. Specifically, the orbital parameters of a GNSS satellite from which a receiver can calculate its position at a moment of time. Specifically, a table of values including locations and related data from which it is possible to derive a satellite's position and velocity. In GPS broadcast, ephemerides are compiled by the Master Control Station, uploaded to the satellites by the Control Segment, and transmitted to receivers in the navigation messages. It is designed to provide orbital elements quickly and is not as accurate as the precise ephemeris. Precise ephemerides are provided via postprocessed tables.

Epoch: (1) A time interval. In GNSS, the period of each observation in seconds. (2) A moment in time, similar to a date but expressed as a year and a decimal of that year, i.e., July 4, 2022, would be written as 2022.51.

Equipotential: A continuous surface on which every point has the same potential, i.e., water that is stable—at rest, and is perpendicular to the direction of gravity everywhere. There are an infinite number of equipotential surfaces that do not intersect each other but converge toward the poles. The geoid is an equipotential surface that varies around a reference ellipsoid. MSL is not an equipotential surface due to variation in temperature, salinity, currents, etc., effects on its topography.

Error Budget: A summary of the magnitudes and sources of statistical errors that can help in approximating the actual errors that will accrue when observations are made.

European Geostationary Navigation Overlay Service (EGNOS): A European satellite-based augmentation system (SBAS) which has been operational since 2009 and available for safety-of-life service since March of 2011. It has 39 monitoring stations in Europe, Africa, and North America; four master control centers and six navigation stations from which three geostationary satellites are controlled. It typically delivers 1.2 m differentially corrected horizontal accuracy.

F

FAA: Federal Aviation Administration.

Fast-Switching Channel: (*See* **Multiplexing Channel**.)

Federal Radionavigation Plan: A document mandated by National Defense Authorization Act for Fiscal Year 1998 that is published every other year by the Departments of Defense (DoD), Homeland Security (DHS), and Transportation (DOT). In an effort to reduce the functional overlap of federal initiatives in radio navigation, it summarizes plans for promotion, maintenance, and discontinuation of domestic and international systems. It is the official source of positioning, navigation, and timing (PNT) policy and planning for the U.S. federal government.

Femtosecond: One-millionth of a nanosecond (10^{-15}) of a second.

Fixed Solution: The solution where the ambiguities are resolved to the actual integer values. The fixed solution is superior to the float solution.

Float Solution: The initial ambiguity estimates yield floating-point numbers, not integers, so the result is called a float solution. The fixed solution is superior to the float solution.

FM: Frequency modulation.

FOC: Full operational capability.

Forward Error Correction (FEC): A technique to detect and correct errors in transmitted data without retransmission. A redundant error-correcting block code or convolution code accompanies the transmitted data frame. They allow the receiver to detect and correct some transmission errors. It can then remove the redundant bits and generate the actual frame.

Fractional Initial Cycle (Phase Measurement): A carrier beat phase measurement not including the integer cycle count. It is always between zero and one cycle.

Frame Reference Epoch (aka **Reference Epoch**)**:** When a reference frame is established, recognizing that there is constant movement, an epoch is chosen on which that movement is theoretically paused. It is sometimes known as the frame epoch. It is as if there is an instantaneous snapshot of the station's coordinates. They are caught in a sort of freeze frame. The epoch chosen for NAD83 (2011) was midnight on January 1 of 2010.

Frequency (*See also* **Wavelength**)**:** The number of cycles per unit of time, typically seconds, with frequency described in Hertz (Hz). For example, in GPS, the frequency of the unmodulated carrier waves are L1 at 1575.42 MHz, L2 at 1227.60 MHz, and L5 at 1176.45 MHz.

Frequency Band: Within the electromagnetic spectrum, a particular range of frequencies.

Frequency Division Multiple Access (FDMA): A strategy in which all the satellites in the constellation transmit the same codes, but each transmits them on its own unique assigned frequency.

Frequency Lock Loop (FLL): A controller that keeps the frequency of the signal coming into a GNSS receiver aligned with the carrier replica, the reference signal, generated by the GNSS receiver. It includes a frequency discriminator, a loop filter, and a numerically controlled oscillator (NCO). The controller sends a smoothed phase error estimate back to the NCO.

FRP: Federal Radionavigation Plan.

Full Operational Capability (FOC): When a GNSS can meet all operational and performance requirements.

Fundamental Clock Rate: In GPS 10.23 MHz symbolized F_o.

G

Gain: The gain, or gain pattern is the ratio of a GNSS antenna's power gain in a particular direction compared with a perfectly spherical theoretical isotropic antenna that, if such a thing could exist, would radiate equally in all directions.

Galileo: The European Union's GNSS with governance divided between the GSA, the European Commission (EC), and the European Space Agency (ESA). It is approaching full operational status, intended to have worldwide coverage and be compatible with the other GNSS constellations.

GDOP: Geometric Dilution of Precision.

General Theory of Relativity (*See also* **Special Theory of Relativity**):
A physical theory published by A. Einstein in 1916. In this theory, space and time are no longer viewed as separate but rather form a four-dimensional continuum called space-time that is curved in the neighborhood of massive objects. The theory of general relativity replaces the concept of absolute motion with that of relative motion between two systems or frames of reference and defines the changes that occur in length, mass, and time when a moving object or light passes through a gravitational field. General Relativity predicts that as gravity weakens the rate of clocks increase; they tick faster. On the other hand, Special Relativity predicts that moving clocks appear to tick more slowly than stationary clocks since the rate of a moving clock seems to decrease as its velocity increases. Therefore, for GPS satellites, General Relativity predicts that the atomic clocks on orbit in GPS satellites tick faster than the atomic clocks on Earth by about 45,900 ns/day. Special Relativity predicts that the velocity of atomic clocks moving at GPS orbital speeds tick slower by about 7200 ns/day than clocks on Earth. The rates of the clocks in GPS satellites are reset before launch to compensate for these predicted effects.

GEO: Geosynchronous Equatorial Orbit, aka a geostationary orbit is a circular at 35,785 km (22,236 miles) above Earth and on the equatorial plane. In this configuration, the satellite's orbital period is equal to Earth's rotation period and it appears to be stationary in the sky.

Geocenter: The center of mass of the total Earth including both the solid and fluid portions of the planet.

Geocentric Radius: The distance from the geocenter to a point on the reference ellipsoid surface at geodetic latitude.

Geodesy: The science concerned with the size and shape of the Earth. Geodesy also involves the determination of positions on the Earth's surface and the description of variations of the planet's gravity field.

Geodetic Coordinates: Latitude, longitude, and height based on a reference frames ellipsoid.

Geodetic Datum: A model defined by an ellipsoid and the relationship between the ellipsoid and the surface of the Earth, including a Cartesian coordinate system. In modern usage, eight constants are used to form the coordinate system used for geodetic control. Two constants are required to define the dimensions of the reference ellipsoid. Three constants are needed to specify the location of the origin of the coordinate system, and three more constants are needed to specify the orientation of the coordinate system. In the past, a geodetic datum was defined by five quantities: the latitude and longitude of an initial point, the azimuth of a line from this point, and the two constants to define the reference ellipsoid.

Geodetic Height (aka **Ellipsoidal Height**): The distance from an ellipsoid of reference to a point on the Earth's surface, as measured along the perpendicular from the ellipsoid. It is symbolized by h. Geodetic height is not the same as elevation above mean sea level nor orthometric height. Geodetic height also differs from geoidal height.

Geodetic Reference System 1980 (GRS80): A geodetic reference system including a global reference ellipsoid and a gravity field model that was adopted by the International Union of Geodesy and Geophysics in 1979.

Geodetic Survey: A survey that considers the size and shape of the Earth. A geodetic survey may be performed using terrestrial or satellite positioning techniques.

Geographic Information System (GIS): A system used to acquire, store, manipulate, analyze, and display spatial data in layers. A GIS can also be used to conduct analysis, display the results of queries and create feature maps.

Geoid: The equipotential surface of the Earth's gravity field which best fits mean sea level in the least squares sense. The geoid surface is everywhere perpendicular to gravity.

Geoid Height: The distance from the ellipsoid of reference to the geoid measured along a perpendicular to the ellipsoid of reference. It is symbolized by N.

Geoid Model: Representation as point values, contours, or gridded data of the geoid height across a region or the world.

Geomagnetic Latitude: A system of latitude from the geomagnetic equator rather than the geographic equator.

Geometric Range: The true distance between a receiver and a satellite.

Geometric Dilution of Precision (GDOP): (*See* **Dilution of Precision**.)

Geostationary Satellite (GEO): A satellite in orbit on the same plane as the equator at an altitude of 35,786 km that travels at about 3 km per second so that it appears to be over a fixed position on the Earth.

GigaHertz (GHz): A measure of frequency, specifically one billion cycles per second.

Global Navigation Satellite System (GNSS): The GPS system is one component of the worldwide effort known as the Global Navigation Satellite System (GNSS). Another component of GNSS is the GLONASS system of the Russian Federation, a third is the Galileo system administered by the EU and the fourth is the Chinese

BeiDou Satellite Navigation and Positioning System. It is likely that more constellations will be included in GNSS definition, such as the Japanese Quasi-Zenith Satellite System (QZSS), the Indian Regional Navigation Satellite System (IRNSS). All are positioning, navigation, and timing systems. The first four mentioned have global scope, QZSS, and NavIC are regional systems.

Globalnaya Navigationnaya Sputnikovaya Sistema, Global Orbiting Navigation Satellite System (GLONASS): The Russian Federation's GNSS did not reach full operational status before the collapse of the Soviet Union. Today, GLONASS is operational and has worldwide coverage.

GNSS Receiver: An apparatus that captures GNSS signals to derive measurements of time, position, and velocity.

GPS Aided GEO Augmented Navigation (GAGAN): An Indian satellite-based augmentation system (SBAS).

GPS Navigation Message: (*See* **Legacy Navigation Message**.)

GPS Operational Control Segment (OCS): The U.S. government tracking and uploading facilities distributed around the world. These facilities not only monitor the L-band signals from the GPS satellites and update their Navigation Messages but also track the satellite's health, maneuvers, and many other things, even battery recharging.

GPS Time (GPST): GPS Time is the time given by all GPS Monitoring Stations and satellite clocks. GPS Time is regulated by Coordinated Universal Time (UTC). GPS Time and Coordinated Universal Time were the same at midnight UT on January 6, 1980. Since then, GPS Time has not been adjusted as leap seconds were inserted into UTC approximately every 18 months. These leap seconds keep UTC approximately synchronized with the Earth's rotation. GPS Time has no leap seconds and is offset from UTC by an integer number of seconds; the number of seconds is in the Navigation Message and most GPS receivers apply the correction automatically. The exact difference is contained in two constants within the Navigation Message, A0 and A1, providing both the time difference and rate of system time relative to UTC. Disregarding the leap second offset, GPS Time is mandated to stay within one microsecond of UTC.

GPS Week: GPS weeks are expressed in integers counted consecutively from the first GPS week. GPS week 0, began at the start of GPS Time, 00 hour UTC on January 6, 1980, and ended on January 12, 1980. It was followed by GPS week 1, GPS week 2, and so on. However, at the end of GPS week 1023—August 15 through August 21, 1999—it was necessary to start the numbering again at 0 because the LNAV GPS week field has only 10 bits, so 1024 is the largest week count it could accommodate. When it reaches that limit, its reckoning of the GPS week starts again at 0. Such a rollover was required at 00 hours on August 22, 1999. A second rollover was required on the night of April 6 to 7, 2019, GPS Week 2047. The next such rollover will be on November 20, 2038. To alleviate this problem, the modernized messages L2-CNAV, CNAV-2, L5-CNAV, and MNAV have a 13-bit field for the GPS week count. This means that the GPS week, as counted by these messages, will not need to roll over until 157 years after 1980 in 2137.

GPSBM (GPS On Bench Marks): A benchmark in the United States that has been established by leveling and also been measured by GPS.

GPT2: A tropospheric model recommended by the International Earth Rotation and Reference Systems Service (IERS). It is global. It relies on a global 5° grid of numerical mean weather data from which it provides estimates of the pressure, temperature, water vapor pressure, etc., at a receiver site.

Grid Distance: In State Plane Coordinate work, the distance between the two stations on the grid, the mapping plane, typically contrasted with the distance between two stations represented on the topographic surface of the Earth.

Grid Factor (aka **Combined Factor**): Multiplying the elevation factor and the scale factor produces a single ratio that is usually known as the *combined factor* or the *grid* factor. Using this grid factor the measured line is converted from a ground distance to a grid distance in one jump.

Ground Antennas (GAs): The S-band antennas co-located at some of the GPS ground stations.

Ground Distance: In State Plane Coordinate work the distance between the two stations on the topographic surface of the Earth typically contrasted with the distance between two stations represented on the grid, the mapping plane.

Ground Plane: Usually a metal sheet, which is used with many antennas to reduce multipath interference by eliminating signals from low elevation angles. Generally, larger ground planes, multiple wavelengths in size, have a more stabilizing influence than smaller ground planes.

Ground Segment: The GPS Master Control Station, Alternate Master Control Station, 11 Command-and-Control antennas, and 16 monitoring sites.

Ground-Based Augmentation System (GBAS): Ground infrastructure that collect, calculate, and broadcast GNSS differential corrections via a very high frequency (VHF) data link known as the VHF data broadcast (VDB) to assist aircraft in approach and landing at airports.

Group Delay: The delay of the GNSS codes through the dispersive medium ionosphere.

GRS80: Geocentric Reference System 1980 is a geodetic reference system adopted by the XVII. General Assembly of the International Union of Geodesy and Geophysics (IUGG) in Canberra, Australia, in December of 1979. Its reference ellipsoid has these parameters:

$$\text{Major semiaxis } (a) = 6{,}378{,}137 \text{ m}$$
$$\text{Flattening } (1/f) = 298.257222101$$

H

Handover Word (HOW): Information used to transfer tracking from the C/A-code to the P code. This time synchronization information is in the second word of each subframe of the GPS Legacy Navigation Message.

Hatanaka Compression: A method developed by Yuri Hatanaka that is used to compress Receiver Independent Exchange (RINEX) files.

Height, Ellipsoidal: (*See* **Ellipsoidal Height**.)

Height, Orthometric: (*See* **Orthometric Height**.)

Height of Instrument (HI): The vertical height of a surveying instrument such as a GNSS receiver antenna over the station above which it is centered.

Helmert Transformation: Named after Professor Dr. Friedrich Robert Helmert, the transformation changes three-dimensional coordinates from one reference frame typically using 7 parameters: 3 translations, 3 rotations, and scale.

Hertz (Hz): A unit used in the measure of frequency named after Heinrich Hertz. Specifically, a wavelength with a duration of 1 second, known as 1 cycle per second, has a frequency of 1 Hertz (Hz) in the International System of Units (SI).

Heuristic Approach: Trial and error.

High Accuracy Reference Networks (HARN): The network of control that results from cooperative ventures between NGS and the states and often includes other organizations as well. With heavy reliance on GPS observations, these networks were intended to provide accurate, vehicle-accessible, regularly spaced control points with good overhead visibility. To ensure coherence, when the GPS measurements were complete, they were submitted to NGS for inclusion in a statewide readjustment of the existing NGRS covered by the state. Coordinate shifts of 0.3 to 1.0 m from NAD83 values were typical in these readjustments.

Horizontal Distance at Mean Elevation: In State Plane Coordinate work the distance between two stations corrected to an averaged horizontal plane.

Horizontal Time-Dependent Positioning (HTDP): An NGS online utility with an underlying crustal dynamics model and given the necessary information can provide estimates of the movement of a mark through time. It also supports 14-parameter Helmert transformations from one NAD 83, ITRF, and WGS84 reference frame to another and from one realization to another.

Hot Start (aka **Standby**)**:** The name of the nearly immediate time to first fix (TTFF), sometimes known as time to subsequent fix (TTSF) in this case, achieved by a GNSS receiver possessed of prior information about not only its own position and time but also the broadcast ephemerides and clock data of the GNSS satellites. This sort of start typically happens when access to GNSS signals that were temporarily blocked by obstructions are once again available.

Hydrogen Maser: An atomic clock. A device that uses microwave amplification by stimulated emission of radiation is called a maser. Actually, the microwave designation is not entirely accurate since masers have been developed to operate at many wavelengths. In any case, a maser is an oscillator whose frequency is derived from atomic resonance. One of the most useful types of masers is based on transitions in hydrogen, which occur at 1421 MHz. The hydrogen maser provides a very sharp, constant oscillating signal and thus serves as a time standard for an atomic clock. The active hydrogen maser provides the best known frequency stability for a frequency generator commercially available. At a 1-hour averaging time, the active maser exceeds the stability of the best known cesium oscillators by a factor of at least 100 and the hydrogen maser is also extremely environmentally rugged.

I

IERS: International Earth Rotation and Reference Systems Service.

International Foot: As of January 1, 2023, the U.S. Survey Foot is no more. From then on 1 foot will be exactly equal to 0.3048 m, in other words, it will be the International Foot. This decision was taken under the authority of the National Institute of Standards and Technology (NIST) and the National Geodetic Survey (NGS), National Ocean Service (NOS), and National Oceanic and Atmospheric Administration (NOAA).

IGS: International GNSS Service—a collaboration between more than 350 organizations in more than 118 countries. It includes agencies, universities, and research institutions and operates a worldwide GNSS tracking network of over 500 Continuously Operating Reference Stations (CORS).

Inclined Geosynchronous Orbit (IGSO): A geosynchronous satellite can be inclined with respect to the equator, whereas a satellite in a geostationary orbit (GEO) is on the same plane as the equator. IGSO satellites are used in conjunction with satellites in geostationary satellites by regional GNSS.

Independent Baselines (aka **Nontrivial Baselines**): These are baselines observed using GNSS Relative Positioning techniques. When more than two receivers are observing at the same time, both independent (nontrivial) and trivial baselines are generated. For example, where r is the number of receivers, every complete static session yields $r - 1$ independent (nontrivial) baselines. If four receivers are used simultaneously, six baselines are created. However, three of those lines will fully define the position of each occupied station in relation to the others. Therefore, the user can consider three $(r - 1)$ of the six lines independent (nontrivial), but once the decision is made, only those three baselines are included in the network.

In-Fill: An RTK survey style used when there is a risk of radio link interruption. When the correction signals from the base station fail to reach the rover, the collected raw data is stored in the memory of the receiver, perhaps at a different rate than the real-time data, for postprocessing after the work is completed.

Inmarsat: International Maritime Satellite.

Integer Ambiguity (aka **Integer-Cycle Ambiguity**): The number of wavelengths (cycles) between the satellite and the receiver when a carrier-phase range is being measured.

Integrity: A quality measure of GNSS performance including a system to provide a warning when the system should not be used for navigation because of some inadequacy.

Interferometry: The measurement of the interference between the phases of signals that originate at a common source but travel different paths to the receivers. The combination of such signals, collected by two separate receivers, invariably finds them out of step since one has traveled a longer distance than the other. Analysis of the signal's phase difference can yield very accurate ranges, and interferometry has become an indispensable measurement technique in several scientific fields.

Intermediate Frequency (IF): The lower frequency to which the received GNSS signal is down converted to facilitate the subsequent processing.

International Association of Geodesy (IAG): A part of the International Union of Geodesy and Geophysics that promotes international cooperation and knowledge. It studies geodetic problems and global change, establishes reference systems, monitors the gravity field, rotation, and deformation of the Earth's surface including ocean and ice.

International Atomic Time (TAI): TAI is the basis of Coordinated Universal Time (UTC). It is a weighted average of the data from approximately 400 atomic clocks and is computed monthly by the International Bureau of Weights and Measurements (BIPM).

International Bureau of Weight and Measurements (BIPM): An organization established in 1875 to unify measurement systems, preserve international standards and national standards.

International Earth Rotation and Reference Systems Service (IERS): IERS was created in 1988 by the International Union of Geodesy and Geophysics (IUGG) and the International Astronomical Union (IAU). It is an interdisciplinary service organization that includes astronomy, geodesy, and geophysics. IERS maintains the International Celestial Reference System and the International Terrestrial Reference System.

International GNSS Service (IGS): The IGS is a federation of over 350 self-funding agencies, universities, and research institutions in more than 118 countries that operates a worldwide GNSS tracking network of over 500 Continuously Operating Reference Stations (CORS). IGS produces post-mission ephemerides, tracking station coordinates, Earth orientation parameters, satellite clock corrections, and tropospheric and ionospheric models that support the realization of the International Terrestrial Reference Frame. Originally. It was established in 1994 as an initiative of the International Association of Geodesy that functions as a component of the Global Geodetic Observing System (GGOS) and is a member of the World Data System (WDS).

International Great Lakes Datum (IGLD): IGLD is expressed as dynamic heights. It is based on an adopted elevation at Point Rimouski/Father's Point and is a reference system for the Great Lakes-St. Lawrence River Basin.

International Reference Meridian (IRM) (aka IERS Reference Meridian): The prime meridian (0° longitude) maintained by the International Earth Rotation and Reference Systems Service (IERS) 5.31 arcseconds, 102.5 m east of the Greenwich Meridian.

International Reference Pole (IRP): The IERS Reference Pole (IRP) corresponds with the BIH Conventional Terrestrial Pole (CTP) (epoch 1984.0), uncertainty of 0.005″.

International Terrestrial Reference Frame (ITRF): A geocentric global realization of the International Terrestrial Reference System (ITRS) created by the IERS and made available for observation via points attached to the Earth's surface with defined three-dimensional coordinates and velocities at a moment in time, an epoch and renewed every few years.

International Terrestrial Reference System (ITRS): As specified by the IUGG resolution No. 2 adopted in Vienna, 1991, a system of instructions, principles, and methods necessary to the definition of the origin, scale, orientation, and time evolution of a geocentric CTRS based on the coordinates and velocities derived from the observation of four space geodetic systems; VLBI, LLR, GPS, SLR, and DORIS and realized by the International Terrestrial Reference Frame (ITRF). Its length unit is the meter. Its axes are consistent with the Bureau International de l'Heuer (BIH) System at 1984.0 within +/–3 milliarcseconds.

International Union of Geodesy and Geophysics: An organization dedicated to geophysical and geodetic study of, and dissemination of knowledge about, the Earth, space, and their dynamics. It does research, conducts observations, coordinates activities, and works with other scientific bodies.

Interoperability: The idea that properly equipped receivers will be able to obtain useful signals from all available satellites in all the constellations and have their solutions improved rather than impeded by the various configurations of the different satellite broadcasts.

Ionosphere: A layer of atmosphere extending from about 50 to 1000 km above the Earth's surface in which gas molecules are ionized by electromagnetic and particle radiation from the sun. The apparent speed, polarization, and direction of GNSS signals are affected by the density of free electrons in this nonhomogeneous and dispersive band of atmosphere which effect varies with season, location, and time.

Ionospheric Delay (aka **Ionospheric Refraction**): The difference in the propagation time for a signal passing through the ionosphere compared with the propagation time for the same signal passing through a vacuum. The magnitude of the ionospheric delay changes with the time of day, the latitude, the season, solar activity, and the observing direction. For example, it is usually least at the zenith and increases as a satellite from which the signal emanates gets closer to the horizon. Unlike the troposphere effect, the ionospheric delay is frequency dependent. There are two categories of ionospheric delay, phase, and group. Group delay affects the GNSS codes, the modulations on the carriers; phase advance affects the carriers themselves. Group delay and phase advance have the same magnitude but opposite signs. Since the ionospheric delay is frequency dependent, it can be mitigated by multifrequency receivers. Even so, over long baselines, the residual ionospheric delay can remain a substantial bias for high precision work. Single-frequency GNSS receivers cannot significantly mitigate the error and must depend on the ionospheric correction available in the Navigation Message to remove even 50% of the effect.

Ionospheric Pierce Point (IPP): The point on a line between a GNSS satellite and receiver at which that line intersects an imagined thin layer where all the ionospheric electrons are assumed to be concentrated. That layer is presumed to be 350 km above the surface of the WGS84 ellipsoid.

Isostatic Rebound: This effect is literally the Earth's crust rising slowly, rebounding, from the removal of the weight and subsurface fluids caused by the retreat of the glaciers from the last ice age.

Isotropic Antenna: An isotropic antenna is a hypothetical, lossless antenna that has equal capabilities in all directions.

Issue of Data Clock (IODC): When the clock parameters are changed in subframe 1 of the GPS LNAV, the issue number of the satellite clock correction changes. It thereby provides a means of detecting any change in the clock correction parameters. The drift of each satellite's clock varies. The broadcast clock correction cannot be updated with enough frequency to adequately represent the drift, so the IODC provides a representation of the reliability of the broadcast clock correction.

Issue of Data Ephemeris (IODE): The issue number of the satellite ephemeris parameter data set in subframes 2 and 3 of the GPS LNAV provides a means of ensuring that the ephemerides in these subframes are consistent with each other. It also provides a way of detecting changes in them. The IODEs in subframes 2 and 3 are also compared with Issue of Data Clock (IODC). When the IODE in subframes 2 and 3 do not match with the IODC in subframe 1 that indicates a data set cutover has occurred.

Iteration: Repeating a process with the intention of successively improving the result.

ITRF: International Terrestrial Reference System.

J

Joint Program Office (JPO): The U.S. Department of Defense office that managed GPS system from 1973 to 2006.

JPL: Jet Propulsion Laboratory (NASA).

Julian Date (aka **Julian Day Number**): Most practitioners of GNSS use the term Julian date to mean the day of the year (DOY) counted consecutively from January 1 of that year. With this method, January 1 is day 1 and December 31 is day 365, excepting in leap years. Because Julian date represents days in consecutive integers.it is most useful in calculating the difference between days. However, the more correct meaning is the number of days since noon on January 1, 4713 BC.

K

Kalman Filter: In GNSS, a numerical data combiner used in determining an instantaneous position estimate from multiple statistical measurements on a time-varying signal in the presence of noise. The Kalman filter is a set of mathematical equations that provides an estimation technique based on least squares. This recursive solution to the discrete data linear filtering problem was proposed by R.E. Kalman in 1960. Since then, the Kalman filter has been applied to radio navigation in general and GNSS in particular, among other methods of measurement.

Keplerian Elements: The six Keplerian orbital elements are semi-major axis, eccentricity, inclination, longitude of the ascending node, argument of perigee, and true anomaly.

KiloHertz (kHz): A measure of frequency, specifically 1000 Hertz, 1000 cycles per second.

Kinematic Positioning (aka **Stop & Go Positioning**) (*See also* **Real-Time Kinematic Positioning**): (1) A version of relative positioning in which one receiver is a stationary reference and at least one other roving receiver coordinates unknown positions with short occupation times while both track the same satellites and maintain constant lock. If lock is lost reinitialization is necessary to fix the integer ambiguity. (2) GNSS applications in which receivers on vehicles are in continuous motion.

Klobuchar Model: A straightforward empirical model of eight parameters developed in the 1970s to characterize and mitigate ionospheric delay for single-frequency receivers. The model is based on the assumption that the ionospheric electron content is concentrated in a thin layer at 350 km above the WGS84 ellipsoid. The corrections derived are broadcast within navigation messages on a daily basis. It was first used in the Global Positioning System and continues to be used by other GNSS with some modifications.

L

LAMBDA (Least-Squares AMBiguity Decorrelation Adjustment): It is a two-step process. Since it is difficult to find the integer least squares ambiguities when done in the search space supplied by the double-difference, the first step is to decorrelate them in a process also known as reduction. This produces new transformed ambiguities that cover a smaller spectrum than the original ambiguities. That way, the space within which the search for the solution is done can be scaled down. With the number of candidate coordinates reduced, the estimation of the actual integer ambiguities is more easily solved. In the second step, the search itself is thereby made faster and more efficient. The circumstances and redundancy of the observations themselves also affect the process. In other words, ambiguity resolution is most efficient when the baselines are short, there are lots of satellites, multiple frequencies, uninterrupted tracking, the dilution of precision is low, there is limited multipath, and the observation sessions are long.

Lambert Conformal Conic Projection: A map projection that typically has two parallels, standard lines of exact scale, which intersect straight meridians at right angles. Scale varies along the meridians, which radiate out from a point outside the limits of map.

Laser Retro-Reflector Array (LRA): Small optical instruments that provide a target that reflects laser signals from ground Satellite Laser Ranging stations back to their source to provide tracking data for precise orbit determination.

Latency: The time required for corrections to be prepared and transmitted by a base receiver and applied by the rover.

Latitude: An angular coordinate, the angle measured from an equatorial plane to a line. The angle measured northward from the equator is positive, southward is negative. The geodetic latitude of a point is the angle between the equatorial plane of the ellipsoid and a normal to the ellipsoid through the point. At that point, the astronomic latitude differs from the geodetic latitude by the meridional component of the deflection of the vertical.

L-Band: The frequency band from 1 to 2 GHz (1000 to 2000 MHz) as defined by the Institute for Electrical & Electronics Engineers. It was defined in the old (pre-1970) U.S. military classification from 390 to 1550 MHz.

Leap Second: A second that is added (a positive leap second) or omitted (a negative leap second) from Coordinated Universal Time (UTC).

Least-Squares: A method of estimating a quantity by minimizing the sum of the squares of the residuals.

Left Hand Circular Polarized (LHCP): A signal rotates counter-clockwise when observed in the direction of propagation.

Legacy Accuracy Improvement Initiative (L-AII): A National Geospatial-Intelligence Agency (NGA)-led project completed in 2008 that increased the data available to the GPS Master Control Station (MCS). It added ten NGA GPS monitoring stations which had been built to help the agency in its definition of the Earth reference frame. The additional stations augmented the existing 6, more than doubling the number to 16, and tripling the amount of data on GPS satellite orbits which increased the accuracy of the broadcast ephemeris by 10 to 15%. L-AII also produced improvement in the MCS Kalman filtering and its ability to monitor the performance of the GPS space vehicles, benefitting all users of GPS.

Legacy Navigation Message (LNAV): Each one of the 25 frames of the GPS legacy navigation message is comprised 5 subframes. Subframes 1, 2, and 3 tell the receiver information about the particular satellite that sent it. The first subframe contains the satellite clock correction. Subframes 2 and 3 communicate the transmitting satellite's ephemeris, so the receiver can find it. The information in these three subframes is repeated with every frame, but subframes 4 and 5 are different. They each contain 25 pages that concern all of the satellites in the constellation. Each of the 5 subframes consists of ten 30-bit words, 24 bits of data, and a 6-bit parity block. The first two words in every subframe are the telemetry (TLM) word and the handover word (HOW). HOW is known as the handover word because it assists a receiver locked onto the C/A code to be handed over to the appropriate portion of the week-long P(Y) code and begin tracking it, that is, if it is authorized to do so. At 50 bps, it takes 750 seconds (12½ minutes) for the receiver to get everything in all 37,500 bits of the LNAV.

Lines-of-Exact-Scale: Lines along which the distance on the mapping plane, the state plane grid, are the same as the distance on the reference ellipsoid.

Line-of-Sight (LOS): Typically refers to the straight line between a GNSS satellite and the antenna of a GNSS receiver. The LOS signal is also referred to as the direct-path signal.

Liquid Apogee Engine (LAE): A low thrust liquid propellant system that is fired at the apogee of satellite's elliptical transfer orbit to move it toward and eventually into its operational orbit. GPS III satellites, more massive than Block IIF satellites, will rely on liquid apogee engines to reach their operational orbit.

LNAV: Legacy NAVigation message.

Local Accuracy (aka **Relative Accuracy**): Local accuracy represents the uncertainty of a position relative to other positions nearby. In other words, local accuracy would be useful in knowing the accuracy of adjacent points in relation to each other. Local Accuracy is aka relative accuracy. In other words, local horizontal and vertical accuracies represent the averaged uncertainty in points relative to adjacent points to which they are directly connected. Within a well-defined geographical area, local accuracy may be the most immediate concern.

Local Coordinate System: Cartesian coordinates within a set of x, y, and z axes defined for a survey of a limited extent with the third axis along the local vertical. Its relationship with a global coordinate system may not be defined.

Local Geodetic Horizon System (LHGS): Coordinates expressed on a plane with a point origin, north along the meridian, east perpendicular to the meridian, and up normal to the reference ellipsoid at the point origin.

Longitude: An angular coordinate. The dihedral angle from the plane of reference, $0°$ meridian, to a plane through a point of concern, and both planes perpendicular to the plane of the equator. In 1884, the Greenwich Meridian was designated the initial meridian for longitudes. At that point, the astronomic longitude differs from the geodetic longitude by the amount of the component in the prime vertical of the local deflection of the vertical divided by the cosine of the latitude.

Loop Closure: A procedure by which some aspects of the internal consistency of a GNSS network is discovered. A series of baseline vector components from more than one GNSS session, forming a loop or closed figure, is added together. The closure error is the ratio of the length of the line representing the combined errors of all the vector's components to the length of the perimeter of the figure.

Loran-C (LOng RAnge Navigation System-C): A positioning system developed in the United States during World War II. It used low-frequency radio waves, which could bounce off the ionosphere at night providing over the horizon capability. It had long range (1400 miles) but low accuracy (10's of miles). One station is referred to as the master and the others as slaves. A group consisting of a master and up to four slaves is called a chain. The lines of position are hyperbolas. A new improved system was developed and named Loran-C. The original LORAN system became known as Loran-A. As the Loran-C system expanded, the Loran-A system declined. Loran C was designed to provide radio navigation service for U.S. coastal waters and was later expanded to include complete coverage of the continental U.S. and Alaska. Twenty-four U.S. Loran-C stations worked in partnership with Canadian and Russian stations to provide coverage in Canadian waters and in the Bering Sea.

Low Distortion Projection (LDP): A map projection grid created to minimize the difference between distances calculated by grid coordinates inverses (grid distances) and the equivalent horizontal distances measured on the ground.

Low-Noise Amplifier (LNA): An electronic amplifier that amplifies low strength signals without degrading their signal-to-noise ratio.

Lunar Laser Ranging (LLR): Laser ranging from ground stations to corner cube reflector arrays on the Moon that were left there during missions. There are four available arrays. Three of them set during the Apollo missions and one during the Soviet Lunokhod 2 mission. These techniques can achieve positions of centimeter precision when information is gathered from several stations. However, one drawback is that the observations must be spread over long periods, up to a month, and they, of course, depend on two-way measurement.

M

Map Projection: A systematic representation of all or part of the surface of the Earth on a plane, usually including meridians and parallels. Today, the work is most often in a mathematical procedure. A map projection cannot be done without distortion.

Mask Angle: An elevation angle below which satellites are not tracked. The technique is used to mitigate atmospheric, multipath, and attenuation errors. A usual mask angle is 15°.

Master Control Station (MCS): For GPS, it is a facility manned by the 2nd Space Operations Squadron at Schriever Air Force Base (formerly Falcon AFB) in Colorado Springs, Colorado. The Alternate GPS Master Control Station is at Vandenberg Air Force Base. For all GNSS, the Master Control Station is responsible for command and control of the constellation. It is in a network of worldwide tracking and upload stations that comprise a Control Segment.

Mean Sea Level (MSL): The arithmetic mean of hourly tide gauge readings of the ocean's surface taken over a long period of time, 19 years.

Medium Earth Orbit (MEO): An orbit above low Earth orbit at 2000 km and below geostationary orbit at 35,786 km.

MegaHertz (MHz): A measure of frequency, specifically 1,000,000 Hertz, one million cycles per second.

Michibiki Satellite Augmentation System (MSAS): Japan's SBAS, which has been operational since 2007.

Microsecond (μsec): One-millionth of a second.

Microstrip Antenna: A low-profile, low-cost antenna. A thin conductive strip on a ground plane with a dielectric substrate material in-between, usually square or rectangular for efficient analysis and fabrication.

Military NAVigation message (MNAV): A modernized packetized military navigation message created as the companion for the M-code signals.

Millisecond (ms or msec): One-thousandth of a second.

Minimally Constrained (*See* Network Adjustment.): A least squares adjustment of all observations in a network with only the constraints necessary to achieve a meaningful solution. For example, the adjustment of a GNSS network with the coordinates of only one station fixed in position and height.

Minitrack: A tracking system described by John T. Mengel, one of the inventors of the system, as having within the satellite "an oscillator of minimum size and weight ... to illuminate pairs of antennas at a ground station which measures the angular positions of the satellites using phase-comparison techniques." In an article, Mengel and Paul Herget compared the antennas of Vanguard's Minitrack stations to human ears. "An individual locates the source of sound by virtue of the phase differences in the sound waves, arriving at different times at his two ears. Similarly the listening units of the minitrack system are pairs of receiving antennas, set a measured distance apart, which indicate the direction of the signal by phase differences in the radio waves. ..."

MNAV: Military NAVigation Message.

Modem (Modulator/Demodulator): A device that converts digital signals to analog signals and analog signals to digital signals. Computers sharing data usually require a modem at each end of a phone line to perform the conversion.

Modulation: Variation of one or more aspects, phase, frequency, amplitude, pulse, etc., of a carrier wave by an information signal and thereby impressing that information onto the carrier for transmission.

Multichannel Receiver (aka Parallel Receiver): A receiver with many independent channels. Each channel can be dedicated to tracking one satellite continuously.

Multipath (aka Multipath Error): When the GNSS signal reaches the receiver by two or more different paths including the direct path signal, reflected and diffracted paths, aka non-line-of-sight or NLOS signals. Positioning errors result. Multipath is mitigated with various preventative antenna designs and filtering algorithms.

Multiple Signal Messages (MSM): Seven different types were added in RTCM-3.2 in a standardized framework, one set for each constellation. They range from MSM1 for simple pseudorange delivery to MSM7 for delivering pseudorange, carrier-phase ranges, range rate, and more.

Multiplexing Channel (aka **Fast-Switching, Fast-Sequencing,** and **Fast-Multiplexing**)**:** A channel of a GPS receiver that tracks through a series of a satellite's signals, from one signal to the next in a rapid sequence.

Multiplexing Receiver: A GPS receiver that tracks a satellite's signals sequentially and differs from a multichannel receiver in which individual channels are dedicated to each satellite signal.

N

NAD83 (North American Datum, 1983): The horizontal control datum for positioning in Canada, the United States, Mexico, and Central America based on a geocentric origin and the Geodetic Reference System 1980 (GRS80) ellipsoid. The values for GRS 80 adopted by the International Union of Geodesy and Geophysics in 1979 are $a = 6378137$ m and the reciprocal of flattening $= 1/f = 298.257222101$. It was designed to be compatible with the Bureau International de l'Heuer (BIH) Terrestrial System BTS84. The origin was defined by satellite laser ranging (SLR), orientation by astronomic observations, and scale by both SLR and very long baseline interferometry (VLBI). NAD83 was actually realized through Doppler observations using internationally accepted transformations from the Doppler reference frame to BTS84 and adjustment of some 250,000 points. VLBI stations were also included to provide an accurate connection to other reference frames. While NAD83 is similar to the World Geodetic System of 1984 (WGS 84), it is not the same as WGS 84. Defined and maintained by the U.S. Department of Defense, WGS84 is a global geodetic datum used by the GPS Control Segment.

NAD83 (2011) Epoch 2010.00: The latest realization of NAD83.

Nadir: Down.

Nanosecond (ns or nsec): One-billionth (10^{-9}) of a second.

NANU: Notice Advisory to NAVSTAR Users.

National Geodetic Survey (NGS): A U.S. federal agency within the National Oceanic and Atmospheric Administration (NOAA), Department of Commerce, that governs and maintains the National Spatial Reference System (NSRS). Its authority is contained in U.S. Code, Title 33, USC 883a.

National Geodetic Vertical Datum of 1929 (NGVD 29): MSL fixed at 26 tide gauges was adjusted and the resulting datum was not mean sea level, the geoid, or an equipotential surface, so the Sea Level Datum of 1929 renamed NGVD29 on May 10, 1973.

National Geospatial-Intelligence Agency (NGA): NGA was the National Imagery and Mapping Agency (NIMA) from 1996 to 2003. It is a U.S. Department of Defense agency that produces geospatial intelligence in support of national security.

National Marine Electronics Association (NMEA): A U.S. standards association, formed in 1957, that has defined several data message formats widely used to communicate GNSS derived information and measurements through many interfaces.

National Spatial Reference System (NSRS): A consistent coordinate system maintained by the National Geodetic Survey. It defines latitude, longitude, height, scale, gravity, and orientation throughout the United States and provides the foundation for all national geospatial products used in mapping, charting, science, and engineering applications. It includes the network of Continuously Operating Reference Stations (CORS), a network of permanently marked points, a consistent, accurate, and up-to-date national shoreline and a set of accurate models describing dynamic, geophysical processes that affect spatial measurements.

NavIC: India's GNSS, formerly IRNSS, is now known by the operational name Navigation with Indian Constellation, NavIC, a Sanskrit word for sailor or navigator. It was authorized by the Indian government in 2006 and developed by the Indian Space Research Organization (ISRO). The system provides positioning, navigation, and timing (PNT) service over India and the region of approximately 1500 km around it, specifically the area from 30°S latitude to 50°N latitude and from 30°E longitude to 130°E longitude.

Navigation Data Units (NDU): NDU support the creation of new navigation messages with improved broadcast ephemeris and clock corrections on Block IIF GPS satellites.

Navigation Message (NAV) (aka Data Message): Ground control uploads data necessary for pseudorange positioning to each GNSS satellite which transmits this data to user's receivers in a navigation message. The message includes an ionospheric model, particularly important for single-frequency receivers and it includes the satellite's ephemeris, which provides the satellite's location. The message contains time and clock corrections to allow the receiver to compute satellite clock offsets and time conversions. It has information on the satellite's health. It also includes data called almanacs. The almanacs provide a receiver with enough snippets of ephemeris information of all the satellites in the constellation to approximate their positions within a few kilometers.

NAVSAT: A European radio navigation system under development.

NAVSTAR NAVigation Satellite Timing and Ranging: The GPS satellite system.

Network Accuracy: Network accuracy is the uncertainty of a position relative to a datum or reference frame. Network accuracy is less about the accuracy of the positions at each end of the line relative to each other than both relative to the whole datum or reference frame. Network accuracy is aka absolute accuracy. In other words, network accuracy represents the accuracy of a point with respect to the reference system. Those tasked with constructing a control network that embraces a wide geographical scope will most often need to know the position's relationship to the realization of the reference frame on which they are working. Typically network horizontal and vertical accuracies require that a point's accuracy be specified with respect to an appropriate national geodetic system. In the United States, as a practical matter, this most often means that the work is tied to at least one of the more of the Continuously Operating Reference Stations (CORS), which represent the most accessible realization of the National Spatial Reference System in the nation.

Network Adjustment: A least squares solution in which baseline vectors are treated as observations. It may be minimally constrained. It may be constrained by more than one known coordinates, as is usual in a GNSS survey to densify previously established control or a geodetic framework.

Non-Developable Surface: A surface that cannot be flattened to a plane in a map projection without introducing irregular distortion from shrinking, creasing, or stretching.

North American Datum 1983: A horizontal and geometric control datum that covers not only the United States but also Central America, Canada, Greenland, and Mexico that was completed on July 31, 1986. Since then, NAD83 has gone through several realizations due to the constant improvement in geodesy. It is currently NAD83(2011) epoch 2010.00.

North American-Pacific Geopotential Datum of 2022 (NAPGD2022): A geopotential datum that provides consistent geodetic data related to the gravity field including orthometric heights, geoid undulations, gravity anomalies, deflections of the vertical.

North American Vertical Datum of 1988 (NAVD88): Established in 1991, it is a minimally constrained adjustment of U.S., Canadian, and Mexican leveling observations holding fixed the height of the primary tidal benchmark of the new International Great Lakes Datum of 1985 at Father Point/Rimouski, Quebec, Canada. NAVD88 and IGLD85 are now the same. Between NAVD88 orthometric heights and those referred to the National Geodetic Vertical Datum of 1929 (NGVD29), there are differences ranging from −40 to +150 cm within the lower 48 states. The differences range from +94 to +240 cm in Alaska. GPS derived orthometric heights estimated using the precise geoid models now available are compatible with NAVD88. NAVD88 includes 81,500 km of new leveling data never before adjusted to NGVD29. The principal impetus for NAVD88 was minimizing the recompilation of national mapping products. The NAVD88 datum does not correspond exactly to the theoretical level surface defined by the GRS80 definitions.

Numerically Controlled Oscillator NCO: A programmable linear frequency generator through which the phase or frequency of a waveform is controlled.

O

Observable: The signals whose measurement contributes to solving the range or distance between the satellite and the receiver. The word is used to draw a distinction between the thing being measured, "the observable," and the measurement, "the observation."

Observation Session (aka **Observation**)**:** Continuous and simultaneous collection of GPS data by two or more receivers.

OCXO: Oven-controlled crystal oscillator.

Omega: A radio navigation system that can provide global coverage with only eight ground-based transmitting stations.

Online Positioning User Service (OPUS): An online service made available through the National Geodetic Survey that provides free access to high-accuracy National Spatial Reference System (NSRS) coordinates. OPUS uses the same software which computes coordinates for the nation's geodetic control marks and the National Oceanic and Atmospheric Administration (NOAA) Continuously Operating Reference Stations (CORS) Network (NCN).

On-the-Fly (OTF): A method of resolving the carrier-phase ambiguity very quickly. The method requires dual-frequency GNSS receivers capable of making both carrier-phase and precise pseudorange measurements. The receiver is not required to remain stationary, making the technique useful for initializing in carrier-phase kinematic GPS.

Operational Control Segment (OCS): (*See* **Ground Segment**.)

Orthometric Height: The vertical distance from the geoid to the surface of the Earth. GPS heights are ellipsoidal. Ellipsoidal heights are the vertical distance from an ellipsoid of reference to the Earth's surface. Ellipsoidal heights differ from leveled, orthometric heights. And conversion from ellipsoidal to orthometric heights requires the vertical distance from the ellipsoid of reference to the geoid, the geoidal height. The vertical distance from the ellipsoid of reference to the geoid around the world varies from +75 to –100 m; in the coterminous United States, it varies from –8 to –53 m. The geoidal heights are negative because the geoid is beneath the ellipsoid. In other words, the ellipsoid is overhead. The relationship between these three heights is:

$$h = H + N$$

where h is the ellipsoid height, H is the orthometric height, and N is the geoidal height.

Orthomorphic: The representation of the shape of small areas is substantially unchanged on an orthomorphic map.

Oscillator (*See also* **Quartz Crystal Controlled Oscillator**): An oscillator is an electronic or mechanical device that produces an alternating output of a stable frequency. They are used in both EDM and GNSS measurements and are clocks in the most general sense.

Outage: GNSS positioning service is unavailable. Possible reasons for an outage include enough satellites are not visible, the dilution of precision value is too large, or the signal-to-noise ratio value or C/No is too small.

P

P Code (aka **Protected Code**): A binary code known by the names P Code, Precise Code, and Protected Code. It is a standard spread spectrum GPS pseudorandom noise code. It is modulated on the L1 and the L2 carrier using binary biphase modulations. Each weeklong segment of the P Code is unique to a single GPS satellite and is repeated each week. It has a chipping rate 10.23 MHz, more than 10 million bits per second. The P Code is sometimes replaced with the more secure Y-Code in a process known as anti-spoofing.

Packetize: To break data into small bundles, known as packets, according to a specific protocol. Each packet is then transmitted separately.

Parallel Receiver: (*See* **Multichannel Receiver**.)

Passive Marks (aka **Passive Control**): A survey control point, often a metal disk set in concrete or a stainless steel rod, whose position was likely originally derived from conventional surveying as contrasted with an active control station that constantly tracks GNSS satellites that is part of a Continuously Operating Reference Station (CORS) Network.

PDOP-Position Dilution of Precision: (*See* **Dilution of Precision**.)

Perigee: The point at which the moon or a satellite is nearest to the center of the Earth.

Perihelion: The point on Earth's, or any planet's, asteroid's, or comet's, orbit when it is closest to the Sun.

Perturbation: Deviation in the path of an object in orbit. Deviation being departure of the actual orbit from the predicted Keplerian orbit. Perturbing forces of Earth's orbiting satellites are caused by atmospheric drag, radiation pressure from the sun, the gravity of the moon and the sun, the geomagnetic field, and the noncentral aspect of Earth's gravity.

Phase Center Offset (PCO): The vector that defines the position of the "mean antenna phase center" with respect to this ARP.

Phase Center Variation (PCV): A function that models the location of antenna phase center (APC) relative to the "mean antenna phase center" depending on the observation direction.

Phase Delay (aka **Phase Advancement**): The effect in the ionosphere that causes the carrier phases to be advanced while the pseudoranges are delayed by the group delay.

Phase Lock: When two modulated carrier waves reach the same phase angle at exactly the same time, they are said to be in phase, coherent, or phase locked.

Phase Lock Loop (PLL): A controller that keeps the signal coming into a GNSS receiver aligned with the carrier replica, the reference signal, generated by the GNSS receiver. It includes a phase discriminator, a loop, and a numerically controlled oscillator (NCO). The controller sends a smoothed phase error estimate back to the NCO.

Phase Observable: (*See* **Carrier Phase**.)

Phase Shift: When two waves reach the same phase angle at different times, they are out of phase or phase shifted.

Phase Smoothed Pseudorange: A pseudorange measurement with its random errors reduced by combination with carrier-phase information.

Picosecond: One-trillionth (10^{-12}) of a second, one-millionth of a microsecond.

Pilot Signal: A dataless signal modulated with a ranging code that supports longer integration times.

PLL: Phase Lock Loop.

Point Positioning: (*See* **Absolute Positioning.**)

Polar Motion: The motion of the rotation pole, with respect to the crust and mantle of the Earth.

Posigrade Orbit: A satellite is in a posigrade orbit when it rotates in the same direction as the rotation of the Earth.

Position: A description, frequently by coordinates, of the location and orientation of a point or object.

Postprocessed GNSS: A method of deriving positions from GNSS observations in which base and roving receivers do not communicate in real time as they do in Real-Time Kinematic (RTK) GNSS. Each receiver records the satellite observations independently. Their collections are combined later. The method can be applied to pseudorange or carrier-phase measurements.

Potential Energy: Energy stored in an object due to its position, for example, the potential energy an object derives from the force of gravity is related to its height.

PPP: Precise Point Positioning.

Precession: The slow circling motion of the rotation axis of a spinning rigid body around another line intersecting it due to a torque (i.e., gravitation).

Precise Code: (*See* **P Code.**)

Precise Ephemeris: (*See* **Ephemeris.**)

Precise Point Positioning (PPP): A carrier-phase positioning technique that requires precise satellite orbit and clock error information from sources like the International GNSS Service to determine the GNSS rover position. It can be implemented through postprocessing or in real time.

Precise Positioning Service (PPS): GPS positioning for the military at a higher level of absolute positioning accuracy than is available to C/A code receivers, which relies on Standard Positioning Service (SPS). PPS is based on the dual-frequency P(Y) code.

Precision: Agreement among measurements of the same quantity; widely scattered results are less precise than those that are closely grouped. The higher the precision, the smaller the random errors in a series of measurements. The precision of a GNSS survey depends on the network design, surveying methods, processing procedures, and equipment.

PRN: (*See* **Pseudorandom Noise.**)

PRN Number: A unique GPS satellite ID based on the signals it transmits.

Protected Code: (*See* **P Code.**)

Pseudolite (aka **Pseudo Satellite**): A ground-based differential station which simulates the signal of a GNSS satellite with a typical maximum range of 50 km. Pseudolites can enhance the accuracy and extend the coverage of the GNSS constellation. Pseudolite signals are designed to minimize their interference with the GNSS signal.

Pseudo-Packetized: A format with 12-second 300-bit message packets.

Pseudorandom Noise (aka **PRN**): A sequence of digital 1s and 0s that appear to be randomly distributed but can be reproduced exactly is used in the codes GNSS systems. The binary signals have noise-like properties and are modulated on the GNSS carrier waves. Each GPS satellite has unique C/A and P codes. A satellite may be identified according to its PRN number. Thirty-two GPS satellite pseudorandom noise codes are currently defined.

Pseudorange: In GNSS a time biased distance measurement. It is based on code transmitted by a GNSS satellite, collected by a GNSS receiver, and then correlated with a replica of the same code generated in the receiver. However, there is no account for errors in synchronization between the satellite's clock and the receiver's clock in a pseudorange. The precision of the measurement is a function of the resolution of the code.

P(Y) Code: The GPS Precision (P) code, aka the Precise Positioning Service (PPS), is encrypted into the Y code and is known as the P(Y) code. It is a spread-spectrum pseudorandom noise code modulated onto L1 and L2. The basic code generation produces a set of 37 mutually exclusive P code sequences 7 days long at its chipping rate of 10.23 Mbps. It is designed for authorized users.

Q

Quadrature (Q): When a signal has two components, one in-phase (I) and one quadrature-phase (Q) the quadrature component is broadcast 90° phase shifted from the in-phase component.

Quadrifilar Helix: An antenna that has two orthogonal bifilar helical loops on a common axis.

Quartz Crystal Controlled Oscillator: GNSS receivers rely on a quartz crystal oscillator to provide a stable reference so that other frequencies of the system can be compared with or generated from this reference. The fundamental component is the quartz crystal resonator. It utilizes the piezoelectric effect. When an electrical signal is applied, the quartz resonates at a frequency unique to its shape, size, and cut. The first study of the use of quartz crystal resonators to control the frequency of vacuum tube oscillators was made by Walter G. Cady in 1921. Important contributions were made by G.W. Pierce, who showed that plates of quartz cut in a certain way could be made to vibrate so as to control frequencies proportional to their thickness.

Quasi-Zenith Satellite System (QZSS). The regional Japanese position, navigation, and timing system. QZSS is designed to augment the accuracy and reliability of GPS throughout the Asia-Oceania region, particularly in Japan's mountains and urban canyons. Therefore, one of the QZSS satellites is always nearly directly overhead in Japan and dwells there for eight hours, the origin of the name quasi-zenith name. Japan Aerospace Exploration Agency (JAXA) plans to increase the four-satellite constellation seven satellites in the future.

R

R95: A representation of positional accuracy. The radius of a theoretical circle centered at the true position that would enclose 95% of the other positions.

Radio-Determination Satellite Service (RDSS): A satellite-based location service using whose radio frequencies are regulated by the International Telecommunication Union (ITU). RDSS usually has a paired uplink. Systems can offer Radio Determination Satellite Service (RDSS) or Radio Navigation Satellite Service (RNSS), ITU Radio Regulations define RNSS as a subset of RDSS.

Radio Frequency (RF): The rate electromagnetic radio waves in the communications and broadcasting band (3 kHz to 300 GHz) oscillate and, more generally, a term that simply means radio.

Radio Navigation: The determination of position, direction, and distance using the properties of transmitted radio waves.

Range (aka **Geometric Range**)**:** A distance between two points, particularly the distance between a GPS receiver and satellite.

Range Rate: The rate at which the range between a GNSS receiver and satellite changes. Usually measured by tracking the variation in the Doppler shift.

RDOP: Relative (normalized to 60 seconds).

Real-Time Kinematic (RTK): A method of determining relative positions between known control and unknown positions using carrier-phase measurements. A base station at the known position transmits corrections to the roving receiver or receivers. The procedure offers high accuracy immediately, in real time. Atmospheric biases are minimized over short RTK baselines.

Real-Time Network (RTN aka **NRTK):** A real-time positioning strategy that incorporates control from a network of GNSS receivers. The system benefits from the capability to build a correction model over an area and thereby minimize some biases.

Receiver Channel: (*See* **Channel**.)

Reconstructed Carrier Phase: (*See* **Carrier-Phase Ranging**.)

Reference Ellipsoid: An ellipsoid of revolution generated when an ellipse is revolved around the minor axis is an approximation the Earth's shape. A reference ellipsoid is used in the creation of a reference frame which supports geodetic calculations.

Reference Epoch: (*See* **Frame Reference Epoch**.)

Reference Epoch Coordinates (RECs): U.S. National Geodetic Survey (NGS) estimated coordinates for a reference epoch defined by NGS. These epochs will be defined every five or ten years.

Reference Frame: A reference frame is the realization of a reference system through the establishment of accessible control from which surveys can begin and maps made.

Reference Signal: The GNSS receiver generated local replicas of the code and carrier.

Reference System: A reference system is an unequivocal specification of conventions, algorithms, and numerical constants. A reference system remains unattached to the real world, that is, done in the realization of a reference frame.

Relative Positioning: A GNSS surveying method that improves the precision of GNSS measurements. One or more GNSS receivers occupy a base station, a known position. They collect the signals from the same satellites at the same time as other GNSS receivers that may be stationary or moving. The other receivers occupy unknown positions in the same geographic area. Occupying known positions, the base station receivers find corrective factors that can either be communicated in real-time to the other receivers, as in RTK, or may be applied in postprocessing, as in static positioning. Relative positioning is in contrast to absolute, or point positioning. In relative positioning, errors that are common to both receivers, such as satellite clock biases, ephemeris errors, propagation delays, and so forth, are mitigated.

Right-Hand Circular Polarized (RHCP): A signal rotates clockwise when observed in the direction of propagation.

RINEX (Receiver Independent Exchange Format): An internationally accepted ASCII-based format for GNSS observation, navigation and meteorological data that allows interchangeable use of data from dissimilar receiver models and postprocessing software developed by the Astronomical Institute of the University of Berne in 1989. More than 60 receivers from 4 different manufacturers were used in the GPS survey EUREF 89. RINEX was developed for the exchange of the GPS data collected in that project.

Root Mean Square (RMS): The square root of the mean of squared errors for a sample.

Rover (aka **Mobile Receiver**). A GNSS receiver that is in motion relative to a stationary base station during a session.

RTCM (Radio Technical Commission for Maritime Services): In DGPS, the abbreviation RTCM has come to mean the correction messages transmitted by some reference stations using a protocol developed by the Radio Technical Commission for Maritime Services Special Committee 104. These corrections can be collected and decoded by DGPS receivers designed to accept the signal. The corrections allow the receiver to generate corrected coordinates in real time. There are several sources of RTCM broadcasts.

Rubidium Clock (aka **Rubidium Frequency Standard**): An environmentally tolerant and very accurate atomic clock whose working element is gaseous rubidium. The resonant transition frequency of the Rb-87 atom (6,834,682,614 Hz) is used as a reference. Rubidium frequency standards are small, light, and have low power consumption.

S

Satellite: The origin of GNSS signals. A GNSS measurement is the distance between the satellite's antenna phase center at the moment the signal is emitted and the receiver's antenna phase center at the moment that signal is received. As the phase centers do not coincide with the mechanical antenna reference points (ARP), high-precision GNSS applications require phase center corrections (PCCs) to relate the GNSS measurements to the ARP of the satellite and receiver antenna. These typically include PCOs and PCVs.

Satellite-Based Augmentation System (SBAS): Differential GNSS services that rely on a network of satellite monitoring stations to track constellations, collect data, and upload to satellites that, in turn, broadcast navigation messages to users. Those in service include the U.S. Wide Area Augmentation System (WAAS), available since 2003, Japan's Michibiki Satellite Augmentation System, MSAS, which was operational in 2007; Europe's European Geostationary Navigation Overlay System, EGNOS, available in 2011 and India's GPS-aided GEO-augmented navigation (GAGAN) certified in 2013. Many more augmentation systems are in development; South Korea's Korea Augmentation Satellite System (KASS); Africa's ASECNA SBAS for Africa and Indian Ocean, A-SBAS; Russia's System for Differential Corrections and Monitoring, SDCM; China's BeiDou SBAS, BDSBAS and Australia's and New Zealand's Southern Positioning Augmentation Network (SPAN).

Satellite Clocks: Two rubidium (Rb) and two cesium (Cs) atomic clocks are aboard Block II/IIA satellites. Three rubidium clocks are on Block IIR satellites. Hydrogen Maser time standards may be used in future satellites.

Satellite Constellation: In GPS, four satellites in each of six orbital planes. In GLONASS, eight satellites in each of three orbital planes.

Satellite Laser Ranging (SLR): Direct unambiguous range measurement of the round-trip time-of-flight from ground stations to the retro-reflectors on orbiting satellites using very short pulsed lasers. SLR measurements make a significant contribution to the International Terrestrial Reference Frame (ITRF).

S-Band: Defined by IEEE as a frequency band in the range of 2 to 4 GHz.

S-Code: (*See* **C/A-Code**.)

Sea-Level Factor: SPCS27 name given to the factor used to reduce a measured distance from the topographic surface to the ellipsoid. It is now known as the ellipsoid factor.

Secant Projection: A map projection in which the conic or cylindrical developable surface cuts through the reference ellipsoid in contrast to a tangent projection where the developable surface only touches the reference ellipsoid.

Second: Base unit of time in the International System of Units. The duration of 9,192,631,770 periods of radiation corresponding to the transition between the two hyperfine levels of the ground state of the cesium-133 atom undisturbed by external fields.

Selective Availability (SA): Intentional manipulation of the GPS broadcast ephemerides and the dithering of satellite clock stability by the U.S. government to deny access to the full GPS accuracy for Standard Positioning Service (SPS) users. Selective availability began in March of 1990, removed during the Persian Gulf War from August 10, 1990, to July 1, 1991, re-enabled in November 1991 and removed by presidential order on May 2, 2000. GPS Block III satellites will not have SA capability.

Semi-Codeless Tracking: Like codeless tracking in which a receiver does not need the knowledge of the signal code to extract useful measurements but in semi-codeless tracking, at least partial knowledge of a signal code is required.

Sidereal Day: The time it takes the Earth to rotate once about its axis relative to the stars instead of the sun. The distant stars appear in the same positions in the sky at the end of a sidereal day that they had at the beginning. At a meridian, a sidereal day is the period between two successive upper transits of the vernal equinox.

Sidereal Time: The time measured relative to the stars instead of the sun.

Sideshot: A surveying measurement that establishes the position of a station that is either not part of the main body of the survey, i.e., traverse, or not a basis from which the survey will be extended.

Signal-in-Space Range Error (SISRE): SISRE includes the errors in pseudorange measurements considered to be attributable to the space and control segment such as signal imperfections, satellite ephemeris errors, clock errors, etc. It does not include errors contributed by atmospheric delays, multipath, or errors specific to the user's receiver and environment.

Signal Squaring: There was a method that did not use the codes carried by the satellite's signal. It was codeless tracking that relied on signal squaring. It was first used in the earliest civilian GPS receivers such as the Macrometer Model V-1000. The method supplanted proposals for a TRANSIT-like Doppler solution. It made no use of pseudoranging and relied exclusively on the carrier-phase observable. Like other methods, it also depended on the creation of an intermediate or beat frequency. But with signal squaring, the beat frequency was created by multiplying the incoming carrier by itself. The result had double the frequency and half the wavelength of the original. It was squared.

Single Difference: (*See* **Between-Receiver Difference**) (*See also* **Between-Satellite Single Difference**).

Single Point GNSS: The most familiar and ubiquitous application of the technology. It is the solution used by cell phones; GNSS enabled cameras, car navigation systems, and many more devices. Single point positioning is aka autonomous positioning, point positioning, the point solution, the navigation solution, or absolute positioning. Typically, the results of this solution are in real time or near real time. It is characterized by a single receiver measuring pseudoranges to a minimum of four satellites simultaneously. In this solution, the receiver must rely on the information it receives from the satellite's navigation messages to learn the positions of the satellites, the satellite clock offset, the ionospheric correction, etc.

Site Calibration (aka **Localization**): Typically, a *site calibration*, aka *localization*, is performed to prepare such a GNSS project to be done using plane coordinates. A site calibration establishes the relationship between geographical coordinates—latitude, longitude, and ellipsoidal height—with plane coordinates—northing, easting, and orthometric heights across the area. In the final analysis, the relationship is expressed in three dimensions: translation, rotation, and scale. Because of the inevitable distortion that a site calibration must model, one of the prerequisites for such localization is the enclosure of the area by the control stations that will be utilized during the work.

Small Circles: Parallels of latitude other than the equator. The planes they describe do not include the center of the Earth as do great circles.

Space Segment: The portion of the GPS system in space, including the satellites and their signals.

Space Vehicle (SV): A, manned or unmanned, man-made craft capable of traveling in outer space. In GNSS, specifically, satellites that generate signals used for positioning and timing by GNSS receivers.

Spatial Data (aka **Geospatial Data**): Information that identifies the geographic location and characteristics of natural or constructed features and boundaries of the Earth. This information may be derived from, among other things, remote sensing, mapping, and surveying technologies.

SPCS27: The State Plane Coordinate System based on NAD27 and its reference ellipsoid Clarke 1866.

SPCS83: The State Plane Coordinate System based on NAD83 and its reference ellipsoid GRS80.

Special Theory of Relativity (*See also* **General Theory of Relativity**): A theory developed by Albert Einstein, predicting, among other things, the changes that occur in length, mass, and time at speeds approaching the speed of light. General relativity predicts that as gravity weakens the rate of clocks increase and they tick faster. On the other hand, special relativity predicts that moving clocks appear to tick more slowly than stationary clocks since the rate of a moving clock seems to decrease as its velocity increases. Therefore, for GPS satellites, general relativity predicts that the atomic clocks in orbit on GPS satellites tick faster than the atomic clocks on Earth by about 45,900 ns/day. Special relativity predicts that the velocity of atomic clocks moving at GPS orbital speeds ticks slower by about 7200 ns/day than clocks on Earth. The rates of the clocks in GPS satellites are reset before launch to compensate for these predicted effects.

Spherical Error Probable (SEP): A description of three-dimensional precision. The radius of a sphere with its center at the purported actual position that is expected to be large enough to include half (50%) of the normal distribution of the scatter of points observed for the position.

Spoofing: The generation and transmission of structured interference masquerading as an authentic GNSS signal. Military GNSS signals were considered the only ones under threat in the past, but civil GNSS signals are also affected today. Low-cost software-defined radio hardware and open civilian GNSS protocols make deliberate spoofing easier than in the past. It is now possible for a programmer with inexpensive off-the-shelf equipment to create and broadcast sufficiently realistic GNSS signals to degrade a receiver's position, velocity, and timing solutions or even produce entirely incorrect results. Spoofing can also be accidental such as a receiver misconstruing a wayward signal from a GNSS simulator or repeater as an authentic GNSS satellite transmission.

Spread Spectrum Signal: A signal spread over a frequency band wider than needed to carry its information. A spread spectrum signal is used to reduce noise, increase security and prevent jamming, mitigate multipath, and allow unambiguous satellite tracking.

SPS: (*See* **Standard Positioning Service.**)

Sputnik 1: The first artificial satellite to orbit the Earth was launched on October 4, 1957, by the Soviet Union and remained in orbit until January 4, 1958.

Standard Deviation (aka **1-sigma, 1σ**): An indication of the dispersion of random errors in a series of measurements of the same quantity. The more tightly grouped the measurements around their average (mean), the smaller the standard deviation. In an unbiased dataset, approximately 68% of the individual measurements will be within the range expressed by the standard deviation.

Standard Positioning Service (SPS): Civilian absolute positioning accuracy using pseudorange measurements from a single-frequency C/A code receiver. One of two services provided by GPS that is intended for civilian use and is based upon the C/A-code signal on one frequency, GPS L1 (1575.42 MHz).

State Plane Coordinate System (SPCS): A plane Cartesian coordinate system customized for each U.S. state. Each SPCS includes multiple zones whose boundaries generally follow county boundaries (except in Alaska). The central meridian is in the approximate center of each zone and the origin is a sufficient distance south and west from the zone's central meridian to ensure that all coordinates are positive. The Transverse Mercator projection is assigned to states that are longer north–south the Lambert Conformal projection is assigned to those that are longer east–west.

Static Positioning: A relative, differential surveying method in which at least two stationary GNSS receivers collect signals simultaneously from the same constellation of satellites during long observation sessions. Generally, static GNSS measurements are postprocessed and the relative position of the two units can be accurately determined. Static positioning is in contrast to kinematic and Real-Time Kinematic positioning, where one or more receivers track satellites while in motion. Static positioning is also in contrast to absolute, or point, positioning, which has no Earthbound relative positioning component.

Station Visibility Diagram: A diagram for recording obstructions in preparation for a GNSS observation. The concentric circles are meant to indicate $10°$ increments along the upper half of the celestial sphere, from the observer's horizon at $0°$ on the perimeter to the observer's zenith at $90°$ in the center. The hemisphere is cut by the observer's meridian, shown as a line from $0°$ in the north to $180°$ in the south. The prime vertical is signified as the line from $90°$ in the east to $270°$ in the west. The other numbers and solid lines radiating from the center, every $30°$ around the perimeter of the figure, are azimuths from the north and are augmented by dashed lines every $10°$.

Stop-and-Go Positioning: (*See* **Kinematic Positioning**.)

Survey Epoch Coordinates (SECs): U.S. National Geodetic Survey (NGS) computed coordinates for one survey epoch.

SV: Space vehicle.

SV Time: Time per an individual satellite's clock.

System for Differential Corrections and Monitoring (SDCM): The Russian SBAS, with 19 reference stations in the Russian Federation, provides correction to both GPS and GLONASS. It typically delivers 1.5 m differentially corrected horizontal accuracy.

T

TAI: International Atomic Time.

Telemetry: Automatic recording and transmission of data from GNSS satellites to the Ground Control Segment for monitoring and analysis.

Time Dilation (aka **Relativistic Time Dilation**) (*See* **Special Theory of Relativity**.) (*See* **General Theory of Relativity**.): Systematic variation in time's rate on an orbiting GNSS satellite relative to time's rate on Earth. The variation is predicted by the special theory of relativity and the general theory of relativity as presented by A. Einstein.

Time to First Fix (TTFF): The time that elapses from the moment a GNSS receiver is switched on to its delivery of a two- or three-dimensional position, called a fix, and sets its clock to the correct time. To accomplish this the receiver must search for and find enough satellites (four satellites for a three-dimensional fix), track them, decipher the pertinent portions of the navigation message, including the clock correction and calculate a navigation solution. The TTFF is the longest at a cold start, less at a warm start, and least at a hot start. A receiver that has a current almanac, a current ephemeris, time, and position in its memory can have a hot start.

Time-of-Week (TOW): The number of 1.5-second periods elapsed since the beginning of the GPS week.

TLM: Telemetry.

Total Electron Content (TEC): A measurement of the number of electrons in a cross-sectional area of one square meter along the path through an ionized atmosphere between a transmitter like a GNSS satellite and a receiver.

Total Station: A surveying instrument that is the integration of an electronic theodolite, an electronic distance measuring (EDM) device, a microprocessor, an electronic data collector, and an onboard storage system.

TOW: Time-of-week.

Transit Navigation System (aka **Navy Navigational Satellite System**): Led by ARPA (aka DARPA in 1972) at the same institution and Dr. Richard Kershner of Johns Hopkins the finding that the Doppler shift of Sputnik 1's signal provided enough information to determine the exact moment of its closest approach to the Earth was inspiring. It led to the creation of the Navy Navigational Satellite System (NNSS or NAVSAT) and the subsequent launch of six satellites specifically designed to be used for navigation of military aircraft and ships. This same system, eventually known as TRANSIT, was classified in 1964, declassified in 1967, and was widely used in civilian hydrographic and geodetic surveying for many years until it was switched off on December 31, 1996.

Transverse Mercator Projection (aka **Gauss-Krüger Projection**): An adaptation of the Mercator projection in which the axis of the cylinder is perpendicular to the polar axis. It is a conformal projection. Unlike the Lambert Conic projection, the Transverse Mercator represents meridians of longitude as curves rather than straight lines on the developed grid.

Trilateration: A surveying method in which the length of the sides of triangles are measured and angles are computed to determine the positions of points on the Earth's surface.

Triple-Difference (aka **Receiver-Satellite-Time Triple Difference and Between-Epochs Difference**): The combination of two double differences. Each of the double differences involves two satellites and two receivers. The triple difference is derived between two epochs. In other words, a triple difference involves two satellites, two receivers, and time. Triple differences ease the detection of cycle slips.

Trivial Baseline: Baselines observed using GNSS Relative Positioning techniques. When more than two receivers are observing at the same time, both independent (nontrivial) and trivial baselines are generated. For example, where r is the number of receivers, every complete static session yields $r-1$ independent (nontrivial) baselines, and the remaining baselines are trivial. For example, if four receivers are used simultaneously, six baselines are created. Three are trivial and three are independent. The three independent baselines will fully define the position of each occupied station in relation to the others. The three trivial baselines may be processed, but the observational data used for them has already produced independent baselines. Therefore, only the independent baseline results should be used for network adjustment or quality control.

Troposphere: The lowest of the Earth's atmospheric layers when combined with the tropopause and the stratosphere it is from the surface to about 9 km over the poles and about 16 km over the equator.

Tropospheric Effect (aka **Tropospheric Delay**)**:** The troposphere comprises approximately 9 km of the atmosphere over the poles and 16 km over the equator. The tropospheric effect is non-dispersive for frequencies under 30 GHz. Therefore, it affects both L1 and L2 equally. Refraction in the troposphere has a dry component and a wet component. The dry component is related to the atmospheric pressure and contributes about 90% of the effect. It is more easily modeled than the wet component. The GNSS signal that travels the shortest path through the atmosphere will be the least affected by it. Therefore, the tropospheric delay is least at the zenith and most near the horizon. GNSS receivers at the ends of short baselines collect signals that pass through substantially the same atmosphere and the tropospheric delay may not be troublesome. However, the atmosphere may be very different at the ends of long baselines.

U

U.S. Survey Foot: As of January 1, 2023, the U.S. Survey Foot is no more. From then on 1 foot will be exactly equal to 0.3048 m, in other words, it will be the International Foot. This decision was taken under the authority of the National Institute of Standards and Technology (NIST) and the National Geodetic Survey (NGS), National Ocean Service (NOS), and National Oceanic and Atmospheric Administration (NOAA).

United Kingdom Offshore Operators Association (UKOOA): Member companies of the UKOOA are licensed by the government to search for and produce oil and gas in U.K. waters. It publishes, among other things *The Use of Differential GPS in Offshore Surveying.* These guidelines cover installation and operation, quality measures, minimum training standards, receiver outputs, and data exchange format.

University NAVSTAR Consortium (UNAVCO): A university research and education consortium.

UNB3m: A tropospheric model developed at the University of New Brunswick that can be used to present an example of the relative weight of the two tropospheric delay components. According to the UNB3 model, the dry component of the delay at the zenith of a receiver at the equator is about 2.3 m, whereas the wet component is about 0.27 m.

United States Geological Survey (USGS): A scientific bureau within the United States Department of the Interior that produces national topographic map series, among many other functions.

Universal Polar Stereographic (UPS) Projection: The polar regions are covered by two polar stereographic secant map projections, one on the north, whose central point is at the North Pole, one on the south, whose central point is at the South Pole.

Universal Time Coordinated (UTC): (*See* **Coordinated Universal Time**.)

Universal Transverse Mercator (UTM): The Transverse Mercator projection represents geodetic latitude and longitude positions, as grid coordinates, northing, and easting, on a cylindrical surface that can be developed into a flat surface. Universal Transverse Mercator is a particular type of Transverse Mercator projection. It was adopted by the U.S. Army for large-scale military maps and is shown on all 7.5-minute quadrangle maps and 15-minute quadrangle maps prepared by the U.S. Geological Survey. The Earth is divided into 60 zones between 84°N latitude and 80°S latitude, most of which are 6° of longitude wide. Each of these UTM zones has a unique central meridian.

USCG: U.S. Coast Guard.

User Equipment Error (UEE): Refers to errors included in the UERE that are attributable to atmospheric delays, multipath; errors specific to the user's receiver and environment in pseudorange measurements.

User Equivalent Range Error (UERE) (aka **User Range Error**): Refers to errors in pseudorange measurements from two categories. The first category is the signal-in-space range error (SISRE), which includes the errors attributable to the space and control segment, such as signal imperfections, satellite ephemeris errors, clock errors, etc. It does not include errors contributed by atmospheric delays, multipath; errors specific to the user's receiver and environment. Those are in the second category, the User Equipment Error (UEE). The total UERE can then be written.

User Interface: The software and hardware that activate displays and controls that are the means of communication between a GNSS receiver and the receiver's operator.

User Range Accuracy (URA): A statistical indicator of the worst-case ranging accuracy available from a specific satellite's signal within that satellite's footprint. It is an original integrity monitoring parameter in the GPS broadcast navigation message that communicates an estimate of the standard deviation (1-sigma) value of the User Range Error (URE). For example, when the 1-sigma URE is likely to be less than 2.4 m, the URA index value is 0. When the 1-sigma URE will probably be greater than 2.4 m but less than 3.4 m, the URA index value is 1. If the 1-sigma URE is expected to be between 3.4 m and 4.85 m, the URA index value is 2. A URA index of 15 means unbounded and is a warning to the user that reliance on the satellite is at their own risk.

User Segment: That component of the GNSS system that includes the user equipment, applications, and operational procedures. The part of the whole GNSS system that includes the receivers of GNSS signals.

User-Equivalent Range Error (UERE): The contribution in range units of individual uncorrelated biases to the range measurement error.

UTC: (*See* **Coordinated Universal Time**.)

UTC (SU): Coordinated Universal Time (former Soviet Union, now Russia.

V

Very Large-Scale Integration (VLSI): Hundreds of thousands of transistors being embedded on a single silicon semiconductor microchip.

Very Long Baseline Interferometry (VLBI): By measuring the arrival time of the front of microwaves emitted by a distant quasar at two widely separated Earth-based antennas, the relative position of the antennas is determined. Because the time difference measurements are precise to a few picoseconds, the relative positions of the antennas are accurate within a few millimeters and the quasar positions to fractions of a milliarcsecond. The International Earth Rotation and Reference Systems Service (IERS) calculates the polar motion coordinates as part of the Earth orientation parameters (EOP), which are predicted on the basis of VLBI measurements and satellite ranging. VLBI is used in Universal Time UT1 determination, the realizations of the International Terrestrial Reference Frame, and the International Celestial Reference System.

Voltage-Controlled Quartz Crystal Oscillator (*See also* **Quartz Crystal Controlled Oscillator**): A quartz crystal oscillator with a voltage-controlled frequency.

W

WAAS: Wide Area Augmentation System.

Warm Start (aka **Normal Start**): The name of the more rapid time to first fix (TTFF) that can be achieved by a GNSS receiver possessed of prior information about its own position, within approximately 100 km; velocity, within about 25 m/s; a fairly accurate time estimate and a recent almanac. With these data, the receiver can quickly find the necessary satellites and get their ephemeris and clock information which they repeat every 30 seconds. Therefore, a warm start can reduce the TTFF to 30 seconds or less.

Wavelength (*See also* **Frequency**): Along a sine wave, the distance between adjacent points of equal phase. The distance required for one complete cycle.

Waypoint: A two-dimensional coordinate to be reached by GNSS navigation.

WGS: World Geodetic System.

Wheels: In GNSS satellites, attitude and rotational rates are controlled with momentum wheels. Two rotations are necessary during the satellite's orbit to ensure the solar panels are pointing at the Sun. The angular momentum builds up and it accumulates in the wheels. It must be unloaded with magnetorquers during station-keeping to avoid orbital perturbations. Magnetorquers are magnetic coils that introduce the needed torque through interaction with the Earth's magnetic field.

Wide Area Augmentation System (WAAS): A U.S. Federal Aviation Authority (FAA) Satellite-Based Augmentation System (SBAS) that promotes GPS accuracy, availability, and integrity. It provides a satellite signal for users to support en route and precision approach aircraft navigation. Similar systems are Europe's European Geostationary Navigation Overlay System and Japan's MT-SAT.

World Geodetic System 1984 (WGS84): A World Geodetic Earth-Fixed Terrestrial Reference system defined by the U.S. Department of Defense. Though it was not in its original realization, today, the origin of the WGS 84 reference frame is the center of mass of the Earth, the geocenter. The GPS Control Segment has worked in WGS84 since January 1987; and therefore, single point GPS positions are said to be in this datum.

X

XO: A crystal controlled clock oscillator.

Y

Y-Code: When anti-spoofing is on the P code is encrypted into the Y code and transmitted on L1 and L2. Anti-spoofing was first activated on all Block II satellites on January 31, 1994.

Z

Z-Count Word: The GPS satellite clock time at the leading edge of the next data subframe of the transmitted GPS message (usually expressed as an integer number of 1.5 second periods).

Zero Baseline Test: Two or more receivers are connected to one antenna with a signal or antenna splitter. An observation is done with the divided signal from the single antenna reaching both receivers simultaneously. Since the receivers are sharing the same antenna, satellite clock biases, ephemeris errors, atmospheric biases, and multipath are all canceled. In the absence of multipath, the only remaining errors are attributable to random noise and receiver biases.

Zulu Time: (*See* **Coordinated Universal Time**.)

References

Brunner, F.K., and W.M. Welsch. 1993. Effect of the Troposphere on GPS Measurements. *GPS World* 4(1): 42–51, Advanstar Communications.

Federal Geodetic Data Committee (FGDC). 1998. Geospatial Positioning Accuracy Standards, Part 2: Standards for Geodetic Networks. National Ocean and Atmospheric Administration (NOAA). Available at https://www.fgdc.gov/standards/projects/accuracy/part2/chapter2 (accessed Nov. 3, 2022).

Fotopoulous, G. 2000. *Parameterization of DGPS Carrier Phase Errors over a Regional Network of Reference Stations.* MS thesis, University of Calgary. Department of Geomatic Engineering (UCGE Reports, Number 20142), Calgary, AB, Canada.

Lazar, S. 2002. Modernization and the Move to GPS III *Crosslink* 3(2): 42–46. Available at https://www.scribd.com/document/367599076/Crosslk-Mag-02 (accessed Nov. 3, 2022).

Wells, D., Ed. 1986. *Guide to GPS Positioning.* Canadian GPS Associates, Fredericton, NB, Canada.

Index

Note: Locators in *italics* represent figures and **bold** indicate tables in the text.